Kenig / Ophir

Processing of Biodegradable Polymers

Samuel Kenig
Amos Ophir

Processing of Biodegradable Polymers

HANSER

Print-ISBN: 978-1-56990-873-0
E-Book-ISBN: 978-1-56990-874-7

Bibliographic information of the German National Library:
The German National Library lists this publication in the German National Bibliography; detailed bibliographic data are available on the Internet at *http://dnb.d-nb.de*.

www.hanserpublications.com
www.hanser-fachbuch.de
Editor: Dr. Mark Smith
Production Management: le-tex publishing services GmbH, Leipzig
Cover concept: Marc Müller-Bremer, *www.rebranding.de*, Munich
Cover design: Max Kostopoulos
Cover picture: © Eliya Yam Nisim, Shenkar Graphic Studio
Typesetting: Eberl & Koesel Studio, Kempten
Printed and bound by: CPI Books GmbH, Leck
Printed in Germany

Progress in Polymer Processing (PPP) Series

Satinath Bhattacharya, *Series Editor*
RMIT University, Australia

Editorial Advisory Board

Ica **Manas-Zloczower**
Case Western Reserve University
U.S.A.

Masami **Okamoto**
Toyota Technological Institute
Japan

Chul **Park**
University of Toronto
Canada

Luiz A. **Pessan**
Universidade Federal de São Carlos
Brazil

Changyu **Shen**
Dalian University of Technology
China

Mark **Smith**
Carl Hanser Verlag GmbH & Co. KG
Germany

Giuseppe **Titomanlio**
Università degli Studi di Salerno
Italy

Lih-Sheng (Tom) **Turng**
University of Wisconsin
U.S.A.

John **Vlachopoulos**
McMaster University
Canada

List of Contributors

Abdellah **Ajji**	École Polytechnique de Montréal, Canada
Paloma **Cabesa Segura**	University of Mons, Belgium
Anna **Dotan**	Shenkar College, Israel
M. Naushad **Emmambux**	University of Pretoria, South Africa
Fabiana **Foglia**	University of Salerno, Italy
Adel Ramezani **Kakroodi**	University of Toronto, Canada
Yasamin **Kazemi**	University of Toronto, Canada
Sam **Kenig**	Shenkar College, Israel
Baptiste **Leroy**	University of Mons, Belgium
Shih-Jung **Liu**	Chang Gung University, Taiwan
Giada **Lo Re**	Chalmers University of Technology, Sweden
Chiara **Magnani**	University of Mons, Belgium
Mondli Abednicko **Masanabo**	Council for Scientific and Industrial Research, South Africa
Manjusri **Misra**	University of Guelph, Canada
Amar K. **Mohanty**	University of Guelph, Canada
Mpho Phillip **Motloung**	Council for Scientific and Industrial Research, South Africa
Ramani **Narayan**	Michigan State University, USA
Debarshi **Nath**	University of Guelph, Canada
Amos **Ophir**	Shenkar College, Israel
Akhilesh K. **Pal**	University of Guelph, Canada
Roberto **Pantani**	University of Salerno, Italy
Chul B. **Park**	University of Toronto, Canada

Ehsan **Pesaranhajiabbas**	University of Guelph, Canada
Bandaru **Ramarao**	State University of New York, USA
Jean-Marie **Raquez**	University of Mons, Belgium
Suprakas Sinha **Ray**	Council for Scientific and Industrial Research, South Africa; University of Pretoria, South Africa; University of Johannesburg, South Africa
Ramin Yousefzadeh **Tabasi**	Sabert Corporation, USA
Valentina **Volpe**	University of Salerno, Italy
Ruddy **Wattiez**	University of Mons, Belgium
Fengwei **Xie**	Newcastle University, United Kingdom

Preface

With the emergence of commercial biobased and biodegradable polymers (BDPs) and their blends, new potential routes have been opened to replace conventional fossil-based polymers, with the aim of reducing the carbon footprint of polymeric materials and the overall ecological burden of plastics.

This book covers the basic properties of BDPs according to their classifications: polyester-based, polysaccharides and proteins, the rheology of BDPs and their blends, and their numerous applications, with an emphasis on processing and reactive processing. However, BDPs are sensitive to the processing conditions due to their composition, which is tuned to bio-degradation. Hence, special attention has been directed to minimize the in-process degradation and enhance their final processed properties. To remedy some of the BDP processing shortcomings, special additives, fillers, and blends have been incorporated and developed with minimal effect on the BDPs' bio-degradation rate. All of these aspects of BDP processing are considered in this book, including their characteristics in extrusion, reactive extrusion, injection molding, compression molding, casting, electrospinning, and 3D printing, as well as the processing of recycled BDPs.

The processing of biodegradable polymers is an important stage bridging the technology of the BDP's composition, its final properties, and derived applications due to the structuring during processing of the polymer.

The book is directed toward scientists, engineers, and practitioners that deal with BDPs. It offers a comprehensive approach that includes not only a review of the science and technology of existing BDP compositions but also the appropriate processing methods and the derived applications.

Contents

List of Contributors VII

Preface IX

1 Biodegradable Polymers' Composition, Properties, Processing, and Applications 1

Anna Dotan, Amos Ophir, and Samuel Kenig

1.1 Introduction to Biodegradable Polymers and Biodegradability 2

1.1.1 Defining Biodegradable Polymers and Biodegradability 2

1.1.2 Fundamental Mechanisms of Degradation 2

1.1.3 Composting 3

1.2 Biodegradable Polymers' Composition, Processing, and Properties 3

1.2.1 Poly(lactic acid) (PLA) 4

1.2.1.1 Composition and Properties 4

1.2.1.2 Biodegradation Mechanisms 6

1.2.1.3 Processability 6

1.2.2 Polybutylene Succinate (PBS) 7

1.2.3 Modified PET 8

1.2.4 Polyhydroxyalkanoate (PHA) 9

1.2.4.1 Composition and Properties 9

1.2.4.2 PHA Stabilization 10

1.2.4.3 Branched PHA 11

1.2.4.4 Properties of PHB 11

1.2.4.5 Biodegradability 12

1.2.4.6 Processability 12

1.2.5 PHBH Polyester 13
1.2.6 Water-Soluble, Biodegradable Polymers 13
1.2.6.1 Starch 13
1.2.6.2 Polyvinyl Alcohol (PVOH) 14
1.2.6.3 Cellulose Ether 15
1.2.6.4 Polyethylene Glycol (PEG) 15
1.2.6.5 Polyacrylamide 16
1.2.6.6 Agarose 16
1.2.6.7 Polyethylene Oxide 17
1.2.6.8 Chitosan 17
1.2.6.9 Sodium Polyacrylate 18
1.3 Biodegradable Polymers – Summary of Properties 18
1.4 Biodegradable Polymer Blends, Copolymers, and Additives 20
1.4.1 PLA Blends, Copolymers, and Additives 20
1.4.1.1 PLA Additives 22
1.4.1.2 PLA Fillers 23
1.4.1.3 Biodegradable PLA Nanocomposites 24
1.4.2 PHA Blends, Copolymers, and Additives 25
1.4.2.1 PHA Blends 25
1.4.2.2 PHA Additives 26
1.4.3 PTMAT Blends 27
1.5 Conclusions 27

2 Polyhydroxyalkanoates (PHAs): Advancements in Blends, Biocomposites, and Their Commercialization 31

Debarshi Nath, Ehsan Pesaranhajiabbas, Akhilesh K. Pal, Bandaru Ramarao, Manjusri Misra, and Amar K. Mohanty

2.1 Introduction 31
2.2 PHA-Based Blends 35
2.2.1 Non-Compatibilized Blends of PHAs and Biodegradable Polymers 38
2.2.2 Compatibilized PHA-Based Blends 41
2.3 PHA-Based Biocomposites and Nanocomposites 47
2.3.1 Filler-Based Biocomposites and Nanocomposites 47
2.3.2 Lignocellulosic Filler- and Fiber-Reinforced Biocomposites 52

2.4 Biodegradation of PHA-Based Materials ... 58
2.4.1 Biodegradation Studies on PHA-Based Blends and Composites ... 60
2.4.1.1 Biodegradation Studies on PHAs and Their Blends ... 61
2.4.1.2 Biodegradation Studies on PHA-Based Biocomposites ... 62
2.5 Commercial Applications of PHAs ... 63
2.6 Conclusion ... 65

3 Polyhydroxyalkanoates as Biodegradable Biopolymers for a Circular Economy: Characteristics, Challenges, and Recent Advances ... 73

Chiara Magnani, Paloma Cabesa Segura, Baptiste Leroy, Giada Lo Re, Ruddy Wattiez, and Jean-Marie Raquez[1]

3.1 Introduction to PHAs ... 74
3.1.1 Production ... 75
3.1.2 Structural Aspects ... 77
3.1.3 Applications ... 80
3.1.4 Drawbacks ... 84
3.2 Reducing Production Costs of PHAs ... 86
3.2.1 Cheaper Carbon Sources ... 87
3.2.2 Photosynthetic Microorganisms ... 89
3.2.3 Mixed Cultures ... 90
3.2.4 Recombinant Bacteria ... 91
3.2.5 Valorization of By-Products ... 92
3.2.6 Improved Extraction Techniques ... 92
3.3 Improving Production Reproducibility of PHAs ... 94
3.4 Improving Performance of PHAs ... 94
3.4.1 Controlling Crystallization ... 95
3.4.2 Improving Mechanical Properties ... 96
3.4.3 Improving Thermal Stability and Processability ... 98
3.5 Conclusions ... 100

4 **Manufacture of Biodegradable Polymers Using Reactive Extrusion Processing** 113

Neha Mulchandani and Ramani Narayan

4.1 Introduction 113
4.2 Biodegradable Polymers for Managed End-of-Life 115
4.3 Manufacturing Biodegradable Polymers by Reactive Extrusion 116
4.4 Ring-Opening Polymerization by Reactive Extrusion 119
4.4.1 Ring-Opening Polymerization of ε-Caprolactone 119
4.4.2 Ring-Opening Polymerization of Lactide 121
4.5 Chemical Modification by Grafting 122
4.5.1 Poly(ε-caprolactone)-Grafted Chitosan 123
4.5.2 Poly(lactic acid)-Grafted Maleic Anhydride 123
4.5.3 Functionalization of Poly(lactic acid) 124
4.5.4 Poly(lactic acid)-Grafted Vinyltrimethoxysilane 126
4.5.5 Chemical Modification of Starch 128
4.6 Stereocomplexation in Poly(lactic acid) 131
4.7 Copolymerization by Reactive Extrusion 134
4.8 Preparation of Composites by Reactive Extrusion 135
4.9 Conclusions and Outlook 137

5 **Thermal Processing of Natural-Based Biodegradable Polymers** 143

Mondli Abednicko Masanabo, Mpho Phillip Motloung, M. Naushad Emmambux, and Suprakas Sinha Ray

5.1 Introduction 143
5.2 Thermal Processability of Natural Polymers 144
5.2.1 Starch 144
5.2.2 Cellulose 145
5.2.3 Proteins 146
5.3 Processing Techniques of Natural Polymers 147
5.3.1 Extrusion and Batch Processing 147
5.3.1.1 Starch 147
5.3.1.2 Cellulose 149
5.3.1.3 Proteins 149

5.3.1.4 Extrusion of Natural Polymer Blends ... 151
5.3.2 Extrusion Film-Blowing ... 152
5.3.2.1 Starch ... 152
5.3.2.2 Cellulose ... 153
5.3.2.3 Proteins ... 153
5.3.3 Injection Molding ... 156
5.3.3.1 Starch ... 156
5.3.3.2 Cellulose ... 157
5.3.3.3 Proteins ... 158
5.4 Discussion ... 159
5.5 Conclusions ... 160

6 Processing of Polysaccharides ... 165

Fengwei Xie

6.1 Introduction ... 165
6.2 Fundamentals of Polysaccharides ... 168
6.2.1 Starch ... 168
6.2.2 Cellulose ... 171
6.2.3 Chitin and Chitosan ... 173
6.2.4 Alginate ... 175
6.2.5 Pectin ... 176
6.3 Strategies to Formulate Polysaccharide-Based Materials ... 177
6.3.1 Dissolution and Plasticization ... 177
6.3.2 Polysaccharide-Based Blends ... 177
6.3.3 Polysaccharide-Based Nanocomposites ... 179
6.4 Processing of Polysaccharide-Based Materials ... 180
6.4.1 Wet Processing ... 180
6.4.2 Thermomechanical Processing ... 181
6.4.3 3D Printing ... 182
6.4.4 Electrospinning ... 184
6.5 Concluding Remarks ... 187
6.6 Abbreviations Used in This Chapter ... 188

7 Foam Injection Molding of Poly(lactic acid) Based on Physical Blowing Agents ... 203

Valentina Volpe, Fabiana Foglia, and Roberto Pantani

7.1 Introduction ... 203
7.2 A General Description of Injection Molding ... 204
7.3 Foam Injection Molding ... 205
7.3.1 Structural Foam ... 205
7.3.2 Surface Appearance ... 206
7.4 Processing of Structural Foams ... 207
7.4.1 Low-Pressure Process ... 207
7.4.2 Low-Pressure with Co-Injection ... 208
7.4.3 Gas Counterpressure ... 208
7.4.4 High-Pressure Process ... 209
7.4.5 Mold Opening ... 209
7.5 Foaming Agents ... 210
7.5.1 Chemical Blowing Agents ... 210
7.5.2 Physical Blowing Agents ... 211
7.5.3 Other Foaming Methods ... 211
7.6 Machine Requirements ... 212
7.6.1 Control of Back Pressure ... 212
7.6.2 Screw Modification ... 213
7.6.3 Shut-Off Nozzle ... 213
7.6.4 Special Modification of the Injection System ... 213
7.7 Process Variables ... 214
7.8 Foam Injection Molding of PLA ... 215
7.8.1 Solubility of Physical Blowing Agents in PLA ... 215
7.8.2 Mold Temperature ... 217
7.8.3 Injection Flow Rate ... 219
7.8.4 Back Pressure ... 221
7.8.5 Foam Injection Molding with Mold Opening: Effect of Opening Rate ... 225
7.9 Foam Injection Molding of PLA: Additives and Compounds ... 227
7.9.1 Chemical Modification of PLA ... 227

7.9.2 Foam Injection Molding of PLA with Talc 227
7.9.3 Foam Injection Molding of PLA with Natural Fibers 230
7.9.4 Foam Injection Molding of PLA-Based Nanocomposites 230
7.9.5 PLA Blends and Compounds 231
7.10 Conclusions and Perspectives 232

8 Microfibrillation as a Facile Processing Technique to Produce Improved Biobased Thermoplastic Blends 239

Adel Ramezani Kakroodi, Yasamin Kazemi, and Chul B. Park

8.1 Introduction 239
8.2 Microstructure of Microfibrillar Composites 241
8.3 Effects of Microfibrillation on Rheological Properties of Bioplastics 243
8.4 Effects of Microfibrillation on Crystallization Behavior of Bioplastics .. 246
8.5 Effects of Microfibrillation on Mechanical Properties of Bioplastics 249
8.6 Effects of Microfibrillation on Foaming Ability of Bioplastics 251
8.7 Conclusions 252

9 Electrospinning of Biodegradable Polymers 255

Shih-Jung Liu

9.1 Introduction 255
9.2 Variants of the Electrospinning Process 257
9.2.1 Coaxial Electrospinning 257
9.2.2 Other Variants of Electrospinning 258
9.2.3 Needleless Electrospinning 260
9.2.4 Electrospraying 260
9.3 Effects of Processing Parameters on Electrospinning 261
9.4 Biodegradable Polymers for Electrospinning 263
9.4.1 Natural Polymers 263
9.4.1.1 DNA 263
9.4.1.2 Silk Fibroin 264
9.4.1.3 Fibrinogens 264
9.4.1.4 Dextran 265
9.4.1.5 Chitin and Chitosan 266
9.4.1.6 Collagen 267

9.4.1.7 Gelatin ... 268
9.4.1.8 Hyaluronic Acid ... 268
9.4.1.9 Cellulose ... 269
9.4.2 Synthetic Polymers ... 269
9.4.2.1 Poly(lactic acid) (PLA) ... 270
9.4.2.2 Poly(lactic-*co*-glycolic acid) (PLGA) ... 271
9.4.2.3 Polycaprolactone (PCL) ... 273
9.5 Summary ... 274

10 Heat Sealability of Biodegradable Polymers for Packaging Applications and Their Degradation ... 281

Ramin Yousefzadeh Tabasi and Abdellah Ajji

10.1 Introduction ... 281
10.2 Results and Discussion ... 283
10.3 Conclusions ... 287

Index ... 289

1 Biodegradable Polymers' Composition, Properties, Processing, and Applications

Anna Dotan, Amos Ophir, and Samuel Kenig

This chapter aims to elucidate the relationship between composition, properties, processing, and applications of biodegradable polymers, which have attracted a growing interest in the commercial as well as the academic sector, with the goal of reducing and even eliminating the dependence on fossil fuels. Renewable, sustainable-type polymers trigger polymer biodegradability and all of these attributes can make a difference in the environment. Biodegradable polymers degrade by biological activity, particularly by enzymatic action leading to significant changes in the material's composition. In essence, biodegradable plastics should degrade, in a defined period, to simple molecules such as carbon dioxide and water. Polyesters are generally prone to biodegradation due to their potentially hydrolyzable ester bonds. Among the biodegradable polymers, the polyesters PLA (poly(lactic acid)), PHA (polyhydroxyalkanoate), and starch copolymers appear promising due to their natural and renewable sources. PLA is a linear aliphatic polyester produced by synthetic polycondensation of naturally produced lactic acid (corn). PHAs are aliphatic polyesters produced by a microbial process in a sugar-based medium, where the PHAs act as carbon and energy storage material in the bacteria. Such polymers are synthesized by microbes, with the polymer accumulating in the microbes' cells during growth.

This chapter reviews the chemical types of biodegradable polymers, their properties, blends, additives, and processability. Furthermore, the chapter indicates some important positive and feasible key application aspects related to the various end products, required properties, and their unique processing attributes.

1.1 Introduction to Biodegradable Polymers and Biodegradability

1.1.1 Defining Biodegradable Polymers and Biodegradability

The American Society for Testing and Materials (currently known as ASTM International), the European Committee for Standardization (CEN), the International Organization for Standardization (ISO), the German Institute for Standardization (DIN), and the Organic Reclamation and Composting Association (ORCA) are all actively involved in trying to develop definitions and testing methods for both biodegradability and compostability. Biodegradable polymers constitute only one subcategory of degradable polymers. Degradable polymers are broadly defined as polymers designed to undergo a physical or chemical change under specific conditions, resulting in a loss of some properties. Biodegradable polymers are designed to degrade by the action of living organisms such as bacteria, fungi, and algae. ASTM, ISO, CEN, and others have drafted a definition of compostability based on a respirometry test that measures carbon dioxide generated from biodegradation in a "compost-like" environment [1].

The ASTM definition, updated in 2000 by the ASTM Standard D883 and in 2023 by ASTM 6400-23, describes a *biodegradable plastic* as "a degradable plastic in which the degradation results from the action of naturally occurring microorganisms such as bacteria, fungi, and algae."

In essence, biodegradable plastics should degrade, in a defined period, into carbon dioxide and water. Biodegradation rates are dependent on the thickness and geometry of the tested articles. Rapid breakdown rates are often quoted for thin films. Thick articles such as plates, food trays, and cutlery can take up to a year to degrade biologically.

1.1.2 Fundamental Mechanisms of Degradation

Microbes catalyze the degradation of organic molecules and are greatly affected by the type, location, and number of functional groups. The ease of degradation in relation to the functional group follows the sequence:

$$OH > COOH > NH_2 > SH > CH_3 > Cl$$

Compounds with OH substitutions are most easily degraded, while those with halogen substitutions (Br, Cl, I) are the most difficult to degrade, and compounds with more substitutions are more difficult to degrade. Generally, polymers containing double bonds or ether, ester, and peptide bonds in the backbone, such as natural rubber, polyethers, polyesters, and polyamides, are biodegradable. Since only the polymer structure and not the source of the monomeric building blocks is responsible for biodegradability, the presence of hydrolytic linkages along the polymer chain susceptible to enzymatic cleavage is essential. The interaction of the active site of the enzyme with the polymer must also be possible, in addition to the degree of crystallinity and the molecular weight being favorable. Biodegradability is generally favored by hydrophilic and hydrolytic backbone and end groups, lower crystallinity, porosity, and thinness.

1.1.3 Composting

Composting is currently the most acceptable disposal method for biodegradable polymers. It requires collection systems and composting facilities. Compostable plastics are a subgroup of biodegradable plastics [2].

The ASTM definition, updated in 2000 by the ASTM Standard D883 and in 2023 by ASTM 6400-23, describes a *compostable plastic* as "a plastic that undergoes degradation by biological processes during composting to yield CO_2, water, inorganic compounds, and biomass at a rate consistent with other known compostable materials and leaves no visible, distinguishable or toxic residue."

■ 1.2 Biodegradable Polymers' Composition, Processing, and Properties

Aliphatic polyesters play a dominant role in biodegradable plastics due to their hydrolyzable ester bonds. Biodegradable polyesters, which have been developed commercially, are [1]: PLA – poly(lactic acid) or polylactide, PBS – polybutylene succinate, PTMAT – polytetramethylene adipate/terephthalate, PHA – polyhydroxyalkanoate, PHB – poly(3-hydroxybutyrate), PHH – polyhydroxyhexanoate, PHV – poly(3-hydroxyvalerate), PCL – polycaprolactone, PBSA – polybutylene succinate adipate, AAC – aliphatic-aromatic copolyesters, and PBAT – polybutylene adipate/terephthalate.

Generally, polyesters degrade eventually, with hydrolysis (degradation induced by water) being the dominant mechanism. Synthetic aliphatic polyesters are synthesized from diols and dicarboxylic acids via condensation polymerization. These polyesters can be blended with starch-based polymers for cost-competitive biodegradable plastics applications. The rates of degradation of PBSA, PHB/PHV, and PCL are similar, with the rate of PBS and PLA being respectively slower [1].

1.2.1 Poly(lactic acid) (PLA)

1.2.1.1 Composition and Properties

Poly(lactic acid) (PLA) is a linear aliphatic polyester made by polycondensation of naturally produced lactic acid or by catalyzed ring opening of the lactide moiety. Lactic acid is prepared (via starch fermentation) as a coproduct of wet corn milling. The ester linkages in PLA are prone to hydrolysis and enzymatic chain scission. PLA is often blended with starch to enhance biodegradation and decrease costs. However, the brittleness of the starch/PLA blend is a major shortcoming in many applications. Several low-molecular-weight plasticizers such as glycerol, sorbitol, and triethyl citrate are used to remedy this limitation. PLA does not have food contact approval due to its bacteria-based manufacturing method [1].

Lactide is a cyclic dimer of lactic acid possessing two optical isomers, D and L. L-lactide is the natural isomer and DL-lactide is a synthetic blend of D-lactide and L-lactide. The L-lactide (LPLA) homopolymer is a semicrystalline polymer. These materials exhibit high tensile strength and low elongation and consequently have a high modulus, making them suitable for load-bearing applications like orthopedic fixation and sutures. Poly(DL-lactide) (DLPLA) is amorphous as a result of a random distribution of D and L isomeric forms of lactic acid. The amorphous material has lower tensile strength, higher elongation, and enhanced degradation time, making it more suitable for drug delivery. Poly(L-lactide) has a crystallinity of 37%, with a melting point of 175–178 °C and a glass transition temperature (T_g) of 60–65 °C. The degradation time of LPLA is significantly longer than that of DLPLA, requiring more than 2 years to degrade completely. Copolymers of L-lactide and DL-lactide disrupt the crystallinity of L-lactide and accelerate the degradation process. A unique advantage of PLA is that the basic raw ingredient comes from corn, a very abundant crop and a renewable resource. Cornstarch is first converted to dextrose, which is fermented to lactic acid. Wheat, maize, sugar beet, and agricultural wastes have also been used as the starting feedstock. The composition of lactic acid is depicted in Figure 1.1.

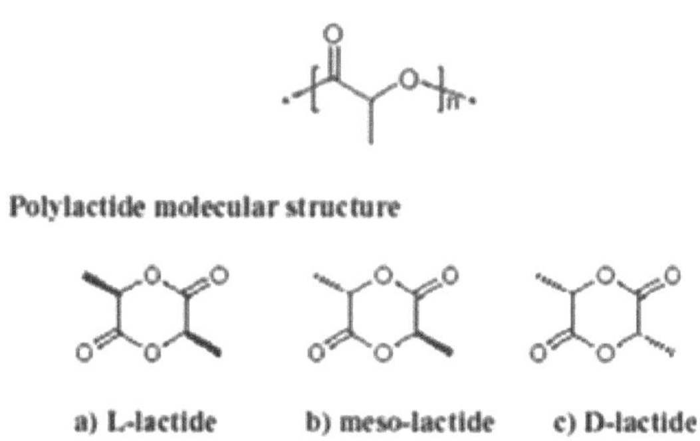

Figure 1.1 Different compositions of lactic acid

In many respects, PLA is analogous to polyethylene terephthalate. Both can be oriented, strain crystallized, and heat set to produce tough, solid, and transparent fibers or films. They both can reach levels of crystallinity of approximately 37%. Both polymers can be copolymerized to control the ultimate level of crystallinity achieved by annealing or orienting and heat setting. The significant difference is that polyethylene terephthalate has a glass transition temperature of 70–80 °C and a melting point of 255 °C. The control of the crystallinity of PLA is essential not only to maximize the service temperature of the polymer but also in regard to its degradation time. Crystalline poly(L-lactide) is more resistant to hydrolytic degradation than the amorphous random L – D copolymer form. Suture materials requiring long holding times are typically higher in L content than those intended for rapid absorption. Molecular weight is also a significant factor in the durability of the polymer. The composition of the copolymer ratios also results in different mechanical properties when processed by different methods. PLA can be processed much like polyolefins and can be extruded. It has low melt strength, but processors can work around this shortcoming. However, PLA crystallization rates are slow, resulting in longer injection molding cycles. This is shown clearly in Table 1.1 [3].

The poor thermal and hydrolytic stability of PLA-based materials, which limits their potential for durable applications, can be overcome using stereo-complex crystallization of enantiomeric poly(L-lactide) (PLLA) and poly(D-lactide) (PDLA). Stereo-complex crystallization provides a significant enhancement in thermal stability and biocompatibility through strong intermolecular interactions between L-lactyl and D-lactyl units. Stereo-complex crystallization is obtained using enantiomeric pairs of poly(L-lactide) (PLLA) and poly(D-lactide) (PDLA) from both melt and solution states. A blend of PLLA and PDLA enables hydrogen bonding interactions, leading to a helical chains arrangement between L-lactyl and D-lactyl units with opposite chiral conformation for PLA stereo-complexation (SC). SC increases the melting temperature by 50 °C compared to neat PLLA or PDLA [4].

Table 1.1 Mechanical Properties of PLA [3]

(L-)/(D,L) Copolymer Ratio	Process Condition[a]	Physical Condition[b]	Tensile Strength, MPa	Tensile Modulus, GPa	Elongation at Break, %
100/0	IM	C	64.8	4.00	-
90/10	IM	A	53.4	1.03	4.6
90/10	IM	C	58.6	1.29	5.1
90/10	EX, BO	SC	80.9	3.41	41.2
90/10	EX, BO, HS	SC	70.1	2.76	20.7
95/5	EX, BO	SC	68.6	1.88	56.7
95/5	EX, BO, HS	SC	60.7	1.63	63.8
80/20	IM	A	51.7	2.10	5.7

Table 1.1 Mechanical Properties of PLA [3] *(continuation)*

(L-)/(D,L) Copolymer Ratio	Process Condition[a]	Physical Condition[b]	Tensile Strength, MPa	Tensile Modulus, GPa	Elongation at Break, %
80/20	EX, BO	SC	84.1	2.94	18.2
80/20	EX, BO, HS	SC	80.1	2.66	32.3
Polystyrene	IM	A	48.3	3.10	2.0

a. IM - injection molded, EX - extruded, BO - biaxially oriented (3X), HS - heat set at 110C
b. A - amorphous, C - crystalline, SC - strain crystallized

1.2.1.2 Biodegradation Mechanisms

PLA is biodegradable by composting at temperatures above 60 °C. The first two weeks of degradation are by hydrolysis of water-soluble compounds and lactic acid. The rapid metabolization of these products into CO_2, water, and biomass by various microorganisms occurs following hydrolysis. An important feature in using polylactide polymers is the ability of the polymers to maintain their base properties until discarded and then degrade as rapidly as possible [1].

It has been reported that PLAs are biodegradable by combining hydrolysis and microbial metabolism. Few studies have dealt with the degradation of PLAs in natural soil. In general, crystalline LPLA is more resistant to hydrolytic degradation than the amorphous DL form. The biodegradation rate of copolymers depends on the molar ratio of lactide/glycolide, the molecular weight of the polymers, the degree of crystallinity, and the T_g of the polymers. The possible approaches to controlling the rate and occurrence of degradation include adding an activator or a water absorber, coating the part with either a hydrophobic or a hydrophilic material, copolymerizing the PLA, and controlling crystallinity. The hydrolytic degradation of PLA-SC is slower than that of neat PLLA or PDLA, due to the high degree of crystallinity.

The hydrolytic degradation of polyesters depends on their chemical composition, molecular weight and molecular weight distribution, pH, ionic strength, temperature, crystallinity, glass transition temperature, dimensions, hydrolysis mechanism (autocatalytic, noncatalytic, enzymatic), and processing history (injection molding, extrusion, sterilization) [5].

1.2.1.3 Processability

Lactide polymers are subjected to degradation during melt processing due to hydrolysis, which results in lactide formation and decreases the molecular weight of the polymer. Furthermore, as processing temperatures are increased above about 230 °C, lactide polymer degradation is substantially accelerated. During melt processing operations, the linear polylactide exhibits necking under tension. This leads to problems with maintaining consistency in film thickness. Linear polymers, such as PLA, exhibit hydrodynamic instability or draw resonance at high draw ratios. This can cause a periodic variation in a film width and gauge. The low melt strength of PLA can be enhanced by an increase in molecular weight average, branching, and increased bridging by free radical reaction and a polydispersity of at least about 2.5.

Polylactide copolymerized with epoxidized linseed oil has generated a polymer with reduced necking using low power consumption. Under similar extrusion temperatures, the control of polylactide could not be run because the power consumption exceeded the maximum levels (> 15 A). The results show that polylactide copolymerized with epoxidized linseed oil has the benefit of processing at lower temperatures and generates a polymer with increased melt strength, less necking, and a film of lower thickness [2].

1.2.2 Polybutylene Succinate (PBS)

Polybutylene succinate (PBS) is a synthetic aliphatic biodegradable polyester that has similar properties to those of PET. PBS is generally blended with other biodegradable materials, such as thermoplastic starch (TPS) and adipate copolymers (to form PBSA), to make its use economical [1]. PBS is a semicrystalline polymer with a T_m of 65 °C.

Degradation Mechanisms and Properties

PBS begins to biodegrade via a hydrolysis mechanism. Hydrolysis occurs at the ester linkages, which lowers the polymer's molecular weight, allowing for further degradation by microorganisms. Data based on SK Chemicals (Korea) PBS indicates a degradation rate of one month for 50% degradation, for a 40-micron-thick film in garden soil [47].

Polyester amides containing four-carbon diacid units tend to degrade through a mechanism implying cyclization to imides, probably by an intramolecular amidolysis reaction. Several researchers suggest that incorporating small amounts of ester groups into polyamide has enhanced degradability without losing most of their physical properties.

The hydrolytic degradability of the PBS series increases as a consequence of increasing the amide composition. Since amide groups are polar, their incorporation into the main ester chain increases hydrophilicity. The enhanced hydrophilicity increases the amount of water sorption during hydrolysis. Consequently, as a result of this increase in water sorption in the amorphous region, the hydrolytic degradation rate seems to be accelerated with increasing amide content. When the amount of the amide group is over 10%, the crystallinity dramatically decreases below 20% [6].

1.2.3 Modified PET

Modified PET (polyethylene terephthalate) contains comonomers, such as ether, amide, or aliphatic monomers, which provide "weak" linkages susceptible to biodegradation through hydrolysis. Typical modified PETs include PBAT (polybutylene adipate/terephthalate) and PTMAT (polytetramethylene adipate/terephthalate). PTMAT is an aliphatic-aromatic copolyester designed to perform in service and then completely degrade in a composting environment. It is a random copolymer derived from aliphatic acid, terephthalic acid (aromatic), and butanediol, resulting in polytetramethylene adipate-coterephthalate – PTMAT. This biodegradable polymer is commercially available. For instance, two commercial PTMAT polymers are Eastman's Eastar™ Bio and BASF's ecoflex®, composed of aromatic-aliphatic copolyesters based on butanediol, adipic acid, and terephthalic acid. Eastman's version is a highly linear polymer; BASF's polymers contain long-chain branching.

Eastman's PTMATs have a density of 1.21 g/cm^3, a melting point of 108 °C, and exhibit good clarity, adhesion, and elongation (up to 800%). They have high humidity and oil resistance and process much like LDPE. Eastar™ Bio is used in lawn-and-garden bags, agricultural films, netting, and paper coatings.

ecoflex® polymer processes efficiently and has a melting point of 110–115 °C and other properties that resemble those of LDPE. The film version demonstrates high elongation and dart impact and transparency with good welding and printability. Special additives can change the feel of ecoflex® from soft to stiff. ecoflex® exhibits high toughness and good cling properties for the replacement of vinyl polymers in vegetable, fruit, and meat wraps, having a 50% lower MVTR (moisture vapor transmission rate) than other biodegradables.

DuPont offers a modified PET (Biomax® series) encompassing three aliphatic monomers. Biomax® has a 1.35 g/cm^3 density and a 195 °C melting point, resulting in higher service temperature and faster processing rates than those of other biodegradable polymers. Furthermore, it offers high stiffness and 40% to 500% elongation. Biomax® has targeted disposable fast-food packaging, yard-waste bags, diaper backing, agricultural film, flowerpots, and bottles.

Degradation Mechanisms

Modified PET is hydro-biodegradable, with biodegradation following initial hydrolysis. The mechanism combines hydrolysis of the ester linkages and enzymatic attack on the ether and amide bonds. It is possible to control the degradation rates with modified PET by varying the comonomers used. Eastar™ Bio copolyester fully degrades to CO_2, H_2O, and biomass in a time frame comparable to that of paper, within 12 weeks in an active microbial environment such as a commercial composting site. In six months, the polymer will biodegrade completely [7].

PTMAT is designed to perform during its service life and then fully degrade in a composting environment. A 5-mil PTMAT film typically becomes invisible within 12 weeks of composting. Fibers have a greater surface area than films of the same weight and their biodegradation rate likely exceeds that of films. In Europe, the PTMAT copolyester is certified to be compostable. It is known that the degradation decreases as the amount of aromatic content increases [8]. Thus, in the range of high-aromatic-content polymers, they remain undegraded [9]. PTMAT ecoflex® contains approximately 9% w/w of aromatic moieties. It was also demonstrated that during the depolymerization of the polymer chains, either the aromatic sequences built water-soluble aromatic oligomers (e.g., if acid terminated) or the polymer chains are cleaved into smaller parts by the microbial enzymes. However, the short soluble oligomeric intermediates are hydrolyzed within three weeks into monomers, which are easily metabolized when a mixed microbial population from compost is present [9]. BASF commercialized a blend of PBAT and PLA under the trade name of ecovio®, as a partially biobased, compostable material.

1.2.4 Polyhydroxyalkanoate (PHA)

1.2.4.1 Composition and Properties

Polyhydroxyalkanoates (PHAs) are aliphatic polyesters produced via a microbial process using a sugar-based medium, serving as carbon and energy storage material for the bacteria. The two main polymers of the PHA family are poly(3-hydroxybutyrate) (PHB) and poly(3-hydroxyvalerate) (PHV). PHBs were the first biodegradable polyesters to be commercialized. Aliphatic polyesters, such as PHAs, are copolymers of hydroxybutyric acid and hydroxyvaleric acid and have been proven readily biodegradable. The most common commercial PHA is composed of a copolymer PHB/PHV, a plasticizer/softener (e.g., triacetin), and inorganic additives such as titanium dioxide and calcium carbonate. Production costs are a significant factor in the competition between PHAs and petroleum-based plastics. Lower-cost nutrient materials include corn liquor, molasses, and even activated sludge. The PHB homopolymer is a stiff and somewhat brittle polymer of high crystallinity. PHB copolymers are preferred for general purpose, as the degradation rate of PHB homopolymers is high at its normal melt processing temperature. PHB and its copolymers with PHV are melt-processable semicrystalline thermoplastics made by biofermentation from carbohydrates [1].

Depending on the carbon numbers in the monomeric constituents, PHAs can be classified as short-chain-length PHAs (SCL PHAs, C3–C5), which consist of 3–5 carbon monomers, or medium-chain-length PHAs (MCL PHAs, C6–C14), which consist of 6–14 carbon monomers in the 3-hydroxyalkanoate units. For example, PHB, PHV, and their copolymer poly(3-hydroxybutyrate-*co*-3-hydroxyvalerate)

(PHBV) are typical examples of SCL PHAs, whereas poly(3-hydroxyoctanoate) (PHO) and poly(3-hydroxynonanoate) (PHN), which are primarily formed as copolymers with 3-hydroxyhexanoate (HHx), 3-hydroxyheptanoate (HH), and/or 3-hydroxydecanoate (HD), are typical examples of MCL PHAs. More than 150 different PHA monomers have been identified, making them the largest group of natural polyesters [10]. This creates the possibility of producing biodegradable polymers with a wide range of properties. PHB is the best characterized PHA. It has the lowest molecular weight and is most common in nature. Its molecular weight can be up to two million (i.e., 20,000 monomers per polymer molecule). The monomer units of PHA are all in D configuration owing to the stereospecificity of the biosynthetic enzymes. Side chains comprise hydrogen or hydrocarbon up to C13 in length, and the repeating unit can range from 1 to 3 or more. These two parameters affect the hydrophobicity, T_g, T_m, and level of crystallinity. There are a variety of PHA homopolymers, copolymers, and terpolymers, obtained using high-productivity microbial biofactories. Copolymers vary in the type and proportion of monomers and are typically random in sequence. PHAs are available in molecular weights ranging from around 1000 to over two million [11, 12].

1.2.4.2 PHA Stabilization

The use of biodegradable aliphatic polyesters is limited due to their thermal degradation and low crystallization rate during melt processing and low toughness following solidification to room temperature. To remedy these shortcomings, the effects of hyperbranched polyesters (HBPs) and plasticizers were studied in the case of biodegradable PHA [13]. Results have shown that a synergistic effect takes place when a liquid HBP has been combined with a plasticizer based on citrate oil (CO), having a combined content of only 5%. In the optimized case, the molecular weight increased by 20% compared to the processed neat PHA. This is attributed to the reduction in viscous heating of the composite phase-separated melt. Furthermore, the crystallization rate constant increases 200-fold, while the total crystallinity is raised from 20% for the neat PHA to 40% for the composite PHA, and the crystallization temperature increases from 65 °C to 90 °C, respectively. This has been attributed to heterogeneous nucleation of the HBP and CO phase. As for the mechanical properties, the elongation at break was enhanced from a low level of 12% for the neat PHA to the level of 26% in the case of the composite PHA. This increase in toughness was accompanied with a moderate decrease of the modulus from 740 MPa to 570 MPa. The formation of a separated phase containing HBP and CO and its effect on the plasticization, nucleation, and mechanical properties was studied and verified by scanning electron microscopy (SEM) [14].

1.2.4.3 Branched PHA

Branched PHA (PBHB) copolymers (Nodax™) make them unique. Carbon side chains of C6 to C24 in length are grafted from the C4 backbone, with a comonomer content ranging from 2% to 20%. As in the case of the conventional LDPE copolymer, PBHB's long-chain branching allows a considerable range for tailoring crystallinity, melting point, stiffness, and toughness. Nodax™ C4/C6 is said to be fit for injection molding and extrusion of sheet or film. The polymer exhibits mechanical properties similar to those of a polyolefin and surface properties much like those of a PET – including high receptivity to printing inks and dyes. Adhesion to LDPE and PP is good enough to avoid tie layers in multilayer structures. Remarkably, Nodax™'s oxygen barrier approaches that of EVOH [15].

1.2.4.4 Properties of PHB

The comparison of properties between PHB and PP is summarized in Table 1.2. As is evident, PHB is water-insoluble and relatively resistant to hydrolytic degradation. This differentiates PHB from most other currently available biodegradable plastics, which are either water-soluble or moisture-sensitive. Furthermore, PHB has a melting point of 175 °C, a glass transition temperature of 15 °C, and a tensile strength of 40 MPa, which is close to that of polypropylene.

PHB exhibits good oxygen barrier properties and ultraviolet resistance and is nontoxic and biocompatible. However, it has poor resistance to acids and bases and is soluble in chloroform and other chlorinated hydrocarbons.

Table 1.2 PHB Properties Compared to those of PP

Property	Polypropylene (PP)	PHB
Melting point T_m [°C]	171-186	171-182
Glass transition temperature T_g [°C]	−5 to −15	5-15
Crystallinity [%]	65-70	65-80
Density [$g \cdot cm^{-3}$]	0.905-0.94	1.23-1.25
Molecular weight (× 10^5)	2.2-7	1-8
Molecular weight distribution	5-12	2.2-3
Flexural modulus [GPa]	1.7	3.5-4
Tensile strength [MPa]	39	40
Elongation at break [%]	400	6-8
UV resistance	poor	good
Solvent resistance	good	poor
Oxygen permeability [$cm^3 \cdot m^{-2} \cdot atm^{-1} \cdot d^{-1}$]	1700	45
Biodegradability	-	good
Other	due to low density, floats in aquatic system	due to high density, sinks in aquatic system

1.2.4.5 Biodegradability

PHAs are biodegradable via composting. The optimum conditions for the commercially available PHA (Biopol™) are a 10-week composting period at 60 °C and 55% moisture to achieve 100% degradation. These aliphatic polymers are suited for short-duration applications and high-degradation-rate requirements. It was found [16] that bacterial PHB/PHV (92/8 w/w) degraded nearly to completion within 20 days in anaerobic conditions, while synthetic aliphatic polyesters such as PLA, PBS, and PBSA did not degrade in 100 days. Cellophane, which was used as a control polymer, possessed a similar degradation rate as the PHB/PHV blend. Under landfill conditions, PHB/PHV degraded within 6 months [2].

PHAs offer a unique combination of biodegradability and hydrolytic stability under normal service conditions. Since carbon dioxide and water are the raw materials used to make PHAs, the PHA life cycle is a closed-loop process. For the first time, there is now a solution to the accumulation of plastics in the marine environment, which makes up the more significant part of the Earth's surface and is a source of much of the world's food [17].

The anaerobic decomposition is significant, for example, concerning biodegradation in landfills (where anoxic conditions exist) and anaerobic waste treatment processes. The natural polyhydroxyalkanoates PHB and PHBV were previously reported to be biodegradable under anaerobic conditions.

Thus, it can be stated that for PHB and PHBV, as well as PCL, a definite anaerobic attack was shown within the incubation period of 7 or 14 weeks. PHB and PHBV films, for example, were already wholly disintegrated at the sampling time in wastewater sludge and the anaerobic river sediment after 14 weeks of incubation. Increasing the degradation temperature may have a positive effect on biodegradability.

1.2.4.6 Processability

PHB can be processed in a narrow window, as it is susceptible to thermal degradation above its melting temperature, resulting in decreased molecular weight and poorer mechanical properties. As a result, PHB has been processed in solution [2]. This problem might be overcome by melting point reduction below the processing temperature or cross-linking. The melting point of PHB was reduced by copolymerizing 3-hydroxybutyrate with either 3-hydroxyvalerate or 4-hydroxybutyrate. Another approach is blending with various additives or cross-linkers, which can repolymerize fragmented polymer chains. Accordingly, some attempts to improve the melt stability of PHB have been made with additives, including various organic and inorganic compounds. A recent study found that polyhydroxyalkanoate with pendant epoxide groups has a higher thermal stability due to cross-linking reactions of pendant epoxide groups with fragmented carboxyl chain ends [1].

Blending 10% to 20% PLA into PHB resulted in upgraded processability, and PHV contents of 5–20% give a good range of properties broadly similar to those of the polyolefins (polyethylene and polypropylene). They have lower melt temperatures than the homopolymer, improving melt processing, and their other properties are close to those of PHB.

1.2.5 PHBH Polyester

Polyhydroxybutyrate-*co*-polyhydroxyhexanoates (PHBHs) resins are the newest types of naturally produced biodegradable polyesters. The PHBH resin is derived from fermentation of carbon sources such as sucrose, fatty acids, or molasses. These are aliphatic-aliphatic copolyesters, distinct from aliphatic-aromatic copolyesters. Besides being completely biodegradable, they exhibit barrier properties similar to those of ethylene vinyl alcohol. The commercial polymer Nodax™ was cast-extruded or blown to manufacture films. Potential applications are mono/multilayer film and nonwoven paper packaging at costs comparable to those of traditional materials such as EVOH.

PHBH resins biodegrade under aerobic and anaerobic conditions and are digestible in hot water under alkaline conditions [1].

1.2.6 Water-Soluble, Biodegradable Polymers

1.2.6.1 Starch

Starch is a food source produced and stored in most plants. It is a linear polymer (polysaccharide) made up of repeating glucose groups linked by glucoside linkages in the 1-4 carbon positions. The length of the starch chains varies with the plant source and generally has between 500 and 2000 glucose units. Amylose and amylopectin are the two major molecules of starch. The alpha linkage of amylose starch makes it flexible and degradable. Starch-based polymers are based on crops of corn (maize), wheat, or potatoes having a starch content exceeding 60% before significant material degradation occurs. Starch is available in abundant quantities at a very low cost.

Starch itself is very brittle and has poor mechanical properties for real applications. However, its elongation can improve significantly by blending it with plasticizers, such as water, amino acids, or glycerol, but the tensile strength is negatively impacted. Often, starch-based polymers are blended with aliphatic polyesters and polyvinyl alcohols to achieve the desired performance for various applications. Biodegradation of starch-based polymers results from an enzymatic action at the glucoside linkages between the sugar groups leading to a degradation in the chain length and the splitting off of sugar units readily utilized in biochemical degradation. At starch contents lower than 60%, the starch particles act as weak links in the polymer matrix and are sites for a biological attack. This results in the polymer matrix disintegrating into small fragments but not biodegrading.

There are several variants of biodegradable starch-based polymers, including thermoplastic starch (TPS) products, starch/aliphatic polyester blends, starch/PBS/PBSA polyester blends, and starch/PVOH blends.

Biodegradable TPSs have a starch (amylose) content greater than 70% and incorporate specific plasticizing additives that can produce thermoplastic materials with suitable properties and inherent biodegradability properties. Notably, such TPS compounds can be processed on existing plastics processing equipment. High-starch-content plastics are highly hydrophilic and readily disintegrate in contact with water. This can be overcome through blending and reacting with hydroxyl groups, which readily undergo acetylation, esterification, and etherification.

Polyvinyl alcohol (PVOH) is blended with starch to produce readily biodegradable plastics. An Australian company (Propak) produces loose-fill peanuts from a starch/PVOH blend (a grade of Mater-Bi™) that exhibits water-solubility in approximately 3 minutes. These expanded products have a closed-cell pore structure and densities ranging from 0.01 g/cm^3 to 0.1 g/cm^3.

PVOH is readily water-soluble, and the starch/PVOH blends are degraded via hydrolysis and biodegradation of the sugar molecules [7].

1.2.6.2 Polyvinyl Alcohol (PVOH)

PVOH is a readily biodegradable, not compostable, water-soluble polymer. Traditional extrusion technologies cannot process conventional PVOH since it decomposes close to its melting point of 230 °C. Even partially hydrolyzed grades, which melt at 180–190 °C, undergo some decomposition during melt processing.

Until recently, PVOH films were prepared by casting from an aqueous solution. Recently, however, new grades of PVOH have been commercialized, incorporating plasticizers to allow extrusion.

The degradation of PVOH is influenced by its crystallinity and molecular weight. Testing has shown that PVOH does not biodegrade but simply dissolves in water. Literature from PVOH manufacturers indicates that PVOH can be biodegradable by activated sludge treatment. Blends of PVOH with starch, glycerol, and chitosan exhibit degradation of the second constituents, with the PVOH almost intact. Some evidence of oxidation and possibly biodegradation of the amorphous phase of PVOH has been noticed [11, 17]. The poor mineralization of PVOH in the soil is attributed to the absence of PVOH-degrading microorganisms in the soil. However, PVOH has been shown to degrade quickly in an aqueous medium enriched with

PVOH-degrading microbes [18]. Interactions with solid substrates have been claimed responsible for the limited biodegradability of microbial polysaccharides and proteins. Their presence seems restricted to specific environments, such as PVOH-polluted textile or paper mill effluents.

1.2.6.3 Cellulose Ether

Cellulose ethers are high-molecular-weight compounds produced by replacing the hydrogen atoms of hydroxyl groups in the anhydroglucose units of cellulose with alkyl or substituted alkyl groups.

Cellulose ethers are water-soluble. The FDA (U.S. Food and Drug Administration) approved methylcellulose and hydroxypropyl methylcellulose as acceptable food additives. The maximum concentration is limited only by the solution viscosity. Cellulose ethers form clear, tough, and flexible films. Cellulose ethers are used as dispersing agents in agricultural fields, as suspending and dispersing aids for pesticide and fertilizer powders. They provide high wet tack and adhesion to waxy plant surfaces. Chemically inert and nonionic, cellulose ether is compatible with many active ingredients [19].

Despite their prolonged biodegradation rate, cellulose ether products should not pose any waste-/soil-recycling hazard. They are similar in behavior to wheat flour or sawdust. Because cellulose ether products and their aqueous solutions present no significant ecological problems, they can be disposed of by industrial incineration or in an approved landfill.

1.2.6.4 Polyethylene Glycol (PEG)

Polyethylene glycol is a water-soluble, waxy solid used in the cosmetic industry. As the molecular weight of PEG is raised, the viscosity and the freezing point increase as well.

Polyethylene glycol is obtained by condensation of ethylene oxide and water, having a general formula of $H(OCH_2CH_2)_nOH$, where n designates the number of oxyethylene groups, typically from 4 to about 180. The low-molecular-weight materials from $n = 2$ to $n = 4$ are diethylene glycol, triethylene glycol, and tetraethylene glycol, respectively. The materials with a low molecular weight up to 700 are colorless, odorless, viscous liquids with a freezing point of −10 °C (diethylene glycol). In comparison, polymerized compounds with a molecular weight higher than 1000 are waxlike solids with a melting point up to 670 °C for $n = 180$. PEG is also soluble in many organic solvents, including aromatic hydrocarbons (not aliphatic). It is nontoxic, odorless, neutral, lubricative, nonvolatile, and nonirritating, and is used in a variety of pharmaceuticals and medications as a solvent, dispersing agent, ointment and suppository base, vehicle, and tablet excipient.

The biodegradability of PEG polymers depends on the molecular weight. At low molecular weights, the polymers degrade almost completely. They will not degrade at high molecular weights (10,000 and higher). As the hydrophobicity of the polymer increases, the biodegradation is reduced. Microbial degradation in PEG polymers occurs from the chain end by an oxidative shortening of the polyether chain while leaving the alkyl portions of the molecules substantially intact. Under these conditions, only the PEG chain was attacked by a stepwise fermentative degradation, leaving behind fatty acid intermediates. The biodegradation of PEG-based polymers is affected by the terminal free hydroxyl group in the hydrophilic portion [20].

1.2.6.5 Polyacrylamide

Polyacrylamide (PAM) is a synthetic water-soluble polymer made from acrylamide monomers. PAM has infinite solubility in water but dissolves very slowly. It is used to purify water by flocculation of suspended organic materials, in irrigation water to improve soil texture, and in pesticide formulations to limit spray drift. Acrylamide is a nerve toxin in humans and affects male reproduction, causing congenital disabilities and cancer in animals. Polyacrylamide is designated a nontoxic additive, which may explain why there has been little concern over its accumulation in food crops. A limit of 500 ppm acrylamide in polyacrylamide preparations has been arbitrarily determined to be acceptable for use in agriculture or water treatment [9].

The biodegradation rates in soil often do not conform to the predictions of kinetic models. The biodegradation rate and the level of the initial compound degraded are lower in the presence of aggregates than in their absence.

1.2.6.6 Agarose

Agarose is a polysaccharide generally extracted from seaweed. Polysaccharide-based polymers include agarose, alginate, carrageenan, dextran, and chitosan. These polymers are highly water-soluble due to their large number of hydroxy groups, and the solutions tend to be low-melting-point aqueous gels. Sheets of agarose gels are readily prepared by casting a warm solution into a mold and are frequently used in molecular biology to separate large molecules by electrophoresis. A wide range of different agaroses of varying molecular weights and properties are commercially available. Agarose is used extensively in the food industry as a stabilizing ingredient in ice cream, instant whips, and dessert jelly.

Agarose is a nontoxic, fully biodegradable polymer. As a polysaccharide, the biodegradation process is done by hydrolytic enzymes.

1.2.6.7 Polyethylene Oxide

Polyethylene oxide is a nonionic, high-molecular-weight, water-soluble polymer with a molecular weight ranging from 100,000 to about 8,000,000. It exhibits many properties typical of other classes of water-soluble polymers: lubricity, binding, water retention, thickening, and film formation. The Federal Environmental Protection Agency identifies polyethylene oxide as an inert ingredient for unrestricted use. Considerable interest has been shown in food, drugs, and cosmetic products applications. Such uses fall within the scope of the Federal Food, Drug, and Cosmetic Act. The FDA has recognized and approved the use of polyethylene oxide polymers for specific food-packaging uses and as a direct additive to beer [12] and nonionic water surfactants.

Polyethylene oxide is a biodegradable polymer, and its biodegradation mechanisms are very similar to the PEG biodegradation mechanisms in aerobic and anaerobic ways.

1.2.6.8 Chitosan

Chitosan is a natural product synthesized from chitin, a polysaccharide found in the exoskeleton of shellfish like shrimp or crabs. Chitosan is chemically similar to cellulose. Like plant fibers, it possesses many of the properties of fiber; however, unlike plant fibers, it can bind fat significantly, acting like a "fat sponge" in digestion. Chitosan is an artificial molecule that is often used to dye textiles. It contains an amine group, or a group without carbons bonded to oxygen, whereas chitin contains an amide group. Chitosan has a history of about three decades of use in processes like detoxifying water and absorbing greases, oils, heavy metals, and other potentially toxic substances. Water-soluble chitosan is widely used in cosmetics, food, fruit-water keeping and preservation, flocculation of wastewater, relaxation agents of drugs, nontoxic adhesives, textile and dyeing, and paper assistant preparation, among others [15].

Chitosan is biodegradable. The molecules can be readily broken by a chitosan-hydrolyzing enzyme, chitosanase [17]. The range of the most appropriate temperatures for chitosan biodegradation is estimated as 30–36 °C. The shortest time of biodegradation and induction time were observed in chitosan with low molecular weight.

1.2.6.9 Sodium Polyacrylate

Sodium polyacrylate (PAC) is a water-soluble, high-molecular-weight compound, and its aqueous solution shows a very high viscosity [21]. In the dry state, the chains of the polymer are coiled and lined with carboxyl groups. When hydrated, the carboxyl groups dissociate into negatively charged carboxylate ions, repelling one another along the polymer chain and widening the polymer coils for more water to penetrate and contact more carboxyl groups. The ultimate swelling forms a gel-like material. Ordinary table salt is added to reliquefy the gelled polymer for disposal purposes. The added sodium chloride in the solution dramatically decreases the ability of polyacrylate to absorb and retain water. Once the gel has been liquefied, it can be safely poured down the drain. The practical application for this superabsorber was realized in the early 1980s for disposable diapers. Later, fuel filters and feminine hygiene products were commercialized. Sodium polyacrylate is considered nontoxic. However, inhalation of airborne particles of the powder or contact with the eyes can have severe adverse reactions [21]. In general, polyacrylate has low acute toxicity after oral and dermal administration.

Polyacrylates biodegrade slowly. No biodegradation of high-molecular-weight polyacrylates under anaerobic conditions has been reported. No experimental data describing the bioaccumulation potential of polyacrylates is available.

1.3 Biodegradable Polymers – Summary of Properties

Table 1.3 summarizes the physical and mechanical properties of the major biodegradable polymers compared to the conventional polystyrene (PS) and low-density polyethylene (LDPE). Figure 1.2 compares the composting times of the different biodegradable polymers.

The composting time reported in Figure 1.2 is an approximate time required for an acceptable level of disintegration of the material. The disintegration time is thus approximated as the time when the starting material could not be identified in the final compost, composed of a size smaller than 10 mm. The composting time does not reflect the complete biodegradation time of the polymers. However, the process could be completed during composting [22].

Table 1.3 Biodegradable Polymers – Physical and Mechanical Properties [22, 23]

Property	PHB (Biopol)	PHB-V (Bio-pol)	PCL (Tone-787)	PLA (Ecopia)	PAS (Bionolle 1000)	PEA (BAK 1095	Ecoflex	Eastar Bio	Mater-Bi (NF01U)	PS	LDPE
Melting point (°C)	177	135	60	177–180	114	125	110–115	108	110	-	100–110
Tensile strength (MPa)	40	25	4	45	60	25	36	22	25	25	8–10
Elongation at break (%)	6	25	800–1000	3	800	400	820	700	600	1–2.5	15–600
Tensile modulus (GPa)	4	1	0.386	2.8	0.5	0.18	0.08	100	0.12	2.8–3.5	0.20
Density (g/cm^3)	1.25	1.25	1.145	1.21	1.2	1.07	1.25	1.22	1.30	1.04	0.92

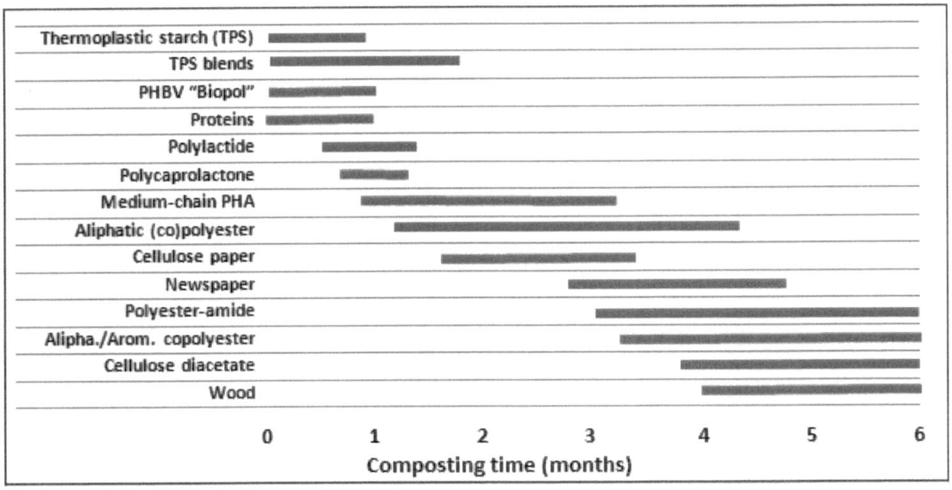

Figure 1.2 Composting times of biodegradable polymers

1.4 Biodegradable Polymer Blends, Copolymers, and Additives

1.4.1 PLA Blends, Copolymers, and Additives

Unmodified PLA exhibits brittleness and low heat distortion temperatures, especially below 450 °F. These shortcomings can be improved by copolymerization with caprolactone. The slow crystallization rates can be accelerated through nucleation or post-annealing of molded or extruded products. Modifiers can raise PLA's impact and tear strengths above normal high-impact polystyrene (HIPS) levels. At impact strengths equal to those of HIPS, PLAs reportedly provide higher tensile strength and modulus. High-molecular-weight PLAs that show a flex-crack resistance higher than that of polystyrene can be thermoformed. PLA's rheology, processing temperatures, and melt viscosity are similar to those of polystyrene, and it extrudes and thermoforms similarly to HIPS. Resin modification can also remedy neat PLA's low melt strength and high neck-in during extrusion coating. PLA has several attractive features for extrusion coating: higher clarity and gloss than polyethylene, high stiffness, low coating temperature, good heat sealability, low coefficient of friction, and the ability to be printed without pretreatment (due to PLA's polar nature). Unmodified PLAs tend to degrade rather slowly. Biodegradable shrink films have been prepared also by the melt mixing of PLA with ethylene vinyl alcohol (EVOH) copolymers. Foams also can be manufactured from PLA [24].

Biodegradable polymers using lactic acid copolymerized with biodegradable aliphatic polyester, which are termed CPLAs, have been developed (Dainippon Ink and Chemicals, Japan). The aliphatic polyester is made from dicarboxylic acid, glycol, and other materials. These copolymers with a molecular weight from 150,000 to 250,000 can be made soft or hard by changing the copolymer content. CPLAs are said to degrade into low-molecular-weight polymers by hydrolysis after three to six months of exposure in soil, sludge, or seawater and to decompose in six to twelve months by microbial action. CPLAs break down in about two weeks if composted with food waste. The materials can be transparent, flexible, or rigid, depending on the amount of polyester. A soft, flexible material contains large amounts of polyester, while a hard, rigid material has only a small percentage of polyester [24].

Investigating blend compositions of poly(lactic acid)/polyvinyl acetate (PVA) led to the conclusion that they are miscible for all compositions (single T_g) [25]. There is a synergistic effect in PLA/PVA blend compositions containing up to 30% PVA. The ones with 5% to 30% PVA demonstrate an increase in mechanical properties as a result of chemical interactions between the PLA/PVA blend components. However, the addition of PVA to PLA reduces the degradation rate. This may be due to the reduction in the surface tension, compared with pure PLA, leading to the reduction of the enzyme's ability to bind to the surface and hence detrimentally affecting the biodegradation rate [25].

PLA/PEO (poly(ethyleneoxide)) and PLA/PEG blends were also investigated [25], focusing on miscibility, mechanical properties, and biodegradation of melt-extruded samples. Their miscibility results were not conclusively attributed to the high crystallinity of both PEO and PEG and the fact that these two polymers significantly enhanced the crystallization of PLA. However, the results suggest that the blends are miscible. Blends of PLA and PEG with up to 30% PEG content are amorphous and miscible. Up to 50% PEG, a reduction in tensile strength is obtained. Above 50% PEG, the high crystallinity of PEG dominates the morphology and results in an increased modulus with almost zero elongation at break. Interestingly, the tensile strength and elongation at break of PLA/PEO blown films were shown to increase when the PEO content was at 20%. Blown films could not be produced from PLA/PEG blends. The degradation rates were measured as a function of weight loss, considering the dissolution of PEO, PEG, and the enzymatic degradation of PLA. It was found that the rate of enzymatic degradation of PLA was enhanced by the presence of PEO in the blend but decreased by the presence of PEG.

Films produced with 50% PLA and 50% PBSU-AD (polybutylene succinate adipate random copolymer) presented competitive properties similar to those in PE films. The processability, film toughness, and flexibility were improved with the addition of PBSU-AD. Higher ratios of PLA resulted in process instabilities. The elongation and toughness of the film are increased as the percentage of PLA is reduced and the PBSU-AD content increases. The blend can include a compatibilizer, including one or more polyesters, polyethers, or polyvinyl alcohols [26–28]. The blends exhibit an elongation at break of 50%, 100%, 200%, 300%, 400%, and up to 500% after 70 days of aging. In addition, the blends have a toughness of at least 20 MJ/m^3, 40 MJ/m^3, 60 MJ/m^3, and up to 120 MJ/m^3 or more. These plastic products can be used for food packaging, compost bags, and other disposable items. Soil testing in an artificial soil environment was performed on 0.3-mm-thick films of the blends using the respirometry method, where a standard soil mix (1:1:0.1 potting soil/sand/dehydrated cow manure by weight) was prepared at 30 ± 2 °C. Carbon dioxide production, expressed as a fraction of the theoretical carbon content of the test materials, was measured as a function of time. After 45 days, they degraded almost 100%, while poly(lactic acid) degraded only about 14% by loss in weight. The degradation rate was relatively fast for blends with 70% and 50% PBSU. After 45 days, the blends degraded about 77% to 25%, respectively, by loss in weight [28].

The thermal and mechanical properties of biodegradable blends of poly(L-lactic acid) (PLLA) and lignin were investigated. Both the results of DSC (differential scanning calorimetry) and FTIR (Fourier transform infrared) spectral analyses indicate the existence of intermolecular interaction between PLLA and lignin. A study of the blends' mechanical properties indicates that the material's maximum strength and elongation at break decrease and the Young's modulus remains almost constant when the lignin content reaches 20%. Thermogravimetry indicates that lignin accelerates the thermal degradation of PLLA when the content of lignin is more than 20% [29].

Biodegradable films were cast from homogenous solutions of PLLA and cellulose acetate with and without tetraisopropyl titanate (TPT), which was used as a compatibilizer. The addition of TPT was shown to produce smaller crystallites of PLLA, reduce the thermal decomposition temperature, and assist in the hydrolysis of the blend. The dynamic mechanical properties indicated that the addition of TPT and the reduction in crystallite size of PLLA did not affect the Young's modulus [30].

1.4.1.1 PLA Additives

The incorporation of additives has been used to overcome the inherent brittleness of PLA. The elongation at break of PLA is typically 3–5% due to physical aging, which occurs during storage at room temperature [31]. Blending PLA with suitable plasticizers has been explored, obtaining compatible blends to prevent phase separation and migration. It was shown that the presence of plasticizers in the PLA matrix contributed to a higher degree of crystallization and a depressed T_g. The more efficient the plasticizer, the more significant the decrease in T_g and the faster the aging/cold-crystallization process. Low-molecular-weight plasticizers, such as tributyl citrate (TBC) and diethyl bishydroxymethyl malonate (DBM), drastically decreased the T_g of PLA and the extent increased with an increasing amount of plasticizer. The ability of PLA to incorporate the plasticizer decreases with the increase in crystallinity, leading to the migration of the plasticizer. Increasing the molecular weight of the plasticizers by synthesizing oligoesters and oligoesteramides resulted in blends that displayed a T_g decrease slightly lower than with the low-molecular-weight plasticizers [32].

Polypropylene glycol (PPG) and epoxy-functionalized PPG (PPG-E) were evaluated as plasticizers for PLA. PPG, with a molecular weight of 4000, is biodegradable through enzymatic degradation. It is flexible and amorphous, with a glass transition temperature of approximately −90 °C. The mechanical properties of PLA/PEG blends deteriorate quickly due to physical aging. Blends of PLA/PEG show, at first, an increase in elongation up to 200%. After one month, the elongation decreased to 80%; after five months, the elongation was less than 10%. This is due to PLA's physical aging and the phase separation between PLA and PEG. Compared with PEG or PEO, the advantages of PPG are its amorphous structure and water-solubility. Furthermore, epoxy-capped PPG exhibited chemical interaction between PPG and PLA. Hence, blends containing 10–20 wt % PPG and 10–40 wt % epoxy-functionalized PPG were prepared by extrusion. Both plasticizers were effective in enhancing the toughness of PLA. The blends' glass transition, crystallization, and melting temperature decreased linearly with increasing content of the functionalized PPG [31].

Weizman et al. [33] investigated the coating of PLA with modified whey protein (biodegradable) for improved gas barrier properties of biodegradable films. Results have demonstrated that the limited gas and vapor barrier properties of PLA films could be enhanced by incorporating whey protein isolate (WPI), a by-product of the cheese industry, as it possesses good oxygen, aroma, and oil barrier properties; however, its permeability to water vapor is high. The WPI coatings were blended with polyvinyl alcohol and pectin to improve the coated film properties. In addition, nanoclays were used to improve water-vapor barrier properties. The comparison of a neat poly(lactic acid) film versus poly(lactic acid) coated with WPI exhibited an improvement of about 90% in the oxygen barrier properties and about 27% in the water-vapor barrier properties.

1.4.1.2 PLA Fillers

Fillers are often added to biodegradable plastics to reduce cost. They are often inert and of mineral origin, thus posing no toxicity concerns at the levels found in biodegradable plastics. Typical fillers include calcium carbonate (most common), titanium dioxide, talc, and silicon dioxide.

Composite films made from poly(L/D-lactide) copolymer and hydroxyapatite filler were prepared and showed a lower level of degradation than the unfilled films. However, lactate release was higher for the filled films than for the unfilled films. These results implied that the fillers influenced the degradation mechanism by preventing the autocatalytic effect of the acidic end groups formed by hydrolysis of the polymer chains. Composites appeared to show an erosion type of degradation rather than the bulk degradation in the case of unfilled materials [34].

Kellersztein et al. investigated the effect of grafting wheat straw fibers with PCL via ring-opening polymerization on PLA reinforcement [35]. They demonstrated that the thermal stability of the wheat straw fibers improved after chemical modifications enabling higher processing temperatures. The flexural and tensile moduli were improved by 23% and 15%, respectively, compared with neat PLA, using 20 wt % modified fibers. An improvement of 20% in the impact strength was obtained using PCL-grafted fibers, as a result of entanglements and molecular interaction.

1.4.1.3 Biodegradable PLA Nanocomposites

PLA nanocomposites were prepared from polylactide (PLA) and nanoclays (NCs) (montmorillonites) modified with trimethyl octadecyl ammonium cation. All the nanocomposites exhibited an improvement in material properties such as storage modulus, flexural modulus, flexural strength, heat distortion temperature, and gas barrier properties compared with neat PLA. Moreover, the biodegradability of neat PLA and its respective nanocomposites was studied under composting conditions. The biodegradation rate increased significantly in the case of nanocomposites compared to neat PLA [36].

Lewitus et al. studied the effect of organically modified NCs at 5% loading on the properties of PLLA [37]. In the optimal case, the tensile modulus increased by 30%, the elongation increased by 40%, and the cold-crystallization temperature decreased by 15 °C, compared to neat PLLA. The properties improvement of PLLA films containing NCs demonstrated the possibility of extending the range of biodegradable film applications, especially in the field of packaging.

The properties of PCL or the PLLA bulk material are often modulated by the addition of plasticizing agents (oligomers, low-molecular-weight PCL, or PLLA), which leads to a decrease in stiffness. To eliminate this effect, PCL and organically modified NCs (Cloisite 30B) were prepared by in situ ring-opening polymerization of lactones. Since caprolactone possesses hydroxy groups, it may react with groups of the alkylammonium-modified NCs, resulting in tethered polyester NCs. Adding the PCL/NCs masterbatch into the PCL matrix increases stiffness and yield stress. Both properties increase with increasing inorganic content and intercalation/exfoliation. Selected samples with enhanced properties were studied concerning their degradation properties [38]. In another study, NCs containing biodegradable polymers exhibit enhanced biodegradability of the nanocomposites. It was postulated that incorporating organically modified NCs in PLA results in higher rates of enzymatic degradation due to the high concentration of accessible chain end groups. Another variable controlling the biodegradability of PLA is the crystallinity level, as the amorphous phase is easier to degrade than the crystalline phase. Finally, a series of biodegradable PLA/NCs composites prepared by melt extrusion exhibited a significant mechanical properties increase with a simultaneous improvement in biodegradability [39].

1.4.2 PHA Blends, Copolymers, and Additives

1.4.2.1 PHA Blends

In a natural state, PHB exists as an amorphous polymer that can be converted into a crystalline polymer by annealing. As PHB's density is around 1.4 g/cm^3, it sinks in water and is readily degraded by biogeochemical mechanisms [25]. The brittle nature of PHB leads to a very narrow processing window, which can be widened by incorporating HV (hydroxyvalerate) units into the PHB backbone, forming PHBV copolymers. These copolymers exhibit enhanced flexibility and tensile strength, reduced crystallinity, and biodegradability. A low HV content (~10%) leads to a hard and brittle copolymer resembling unplasticized PVC; a copolymer with mid-range HV content (10–25%) demonstrates a good balance of toughness, resembling PP; and a high HV content (25–40%) gives rise to soft and tough, PE-like properties.

Branching in PHB is another way to improve toughness, as in the case of soft polyethylene (LLDPE), due to a decrease in crystallinity by enhancing molecular entanglement. The benefits obtained are better processability, lower T_m, melt viscosity control, minimized thermal degradation, ease of processing, reduced crystallinity, strength, ductility, toughness, lower T_g, softness, and flexibility [17].

Due to their high crystallinity and lack of heat stability, the commercial use of isotactic PHB and copolymers of PHB has been limited [25]. PHBV copolymers, either plasticized or as neat resin, have been processed on conventional plastic equipment to produce films, bottles, and other injection-molded articles and successfully spun into fibers. Extensive physical aging contributes to an increased brittleness during storage. Blending the polymer with an appropriate polymeric additive can solve some problems. A simple blending procedure can prepare blends of synthetic PHB and natural bacterial isotactic PHBV containing 10 mol % 3HV units [40].

Blends of PHB with other biodegradable polymers, especially polycaprolactone (PCL), were investigated. The physical aging of the PHB was investigated for the homopolymer and its blends with PCL. It was found that blending does not affect the physical aging of PHB, while a negligible aging effect in the PCL phase was observed. Few examples of compatibilization of PHB/PCL blends via cross-linking in situ were studied [41].

PHB and other PHAs have also been blended with other biodegradable polymers. Blends of PHB and polycaprolactone (PCL) can be processed with conventional equipment, and the ratio of the two polymers determines the rate of degradation [24].

Studies on blends of poly(3-hydroxybutyrate-*co*-3-hydroxyvalerate) (PHBV) with cellulose esters such as cellulose acetate butyrate and cellulose acetate propionate revealed single glass transition temperatures for up to 50% PHBV content. For concentrations of PHBV higher than 50%, the crystallization of PHBV separated the two polymer phases, and different transitions were observed. The blending of lignin with PHB and PHBV exhibited a biodegradation identical to that of the unfilled PHB and PHBV. Crystallization was retarded by the presence of lignin, indicating that interaction takes place between PHB and lignin components. The glass transitions of PHB in the blends were shifted towards higher temperatures [42].

The compatibility and biodegradability of blends of PHB and PEO were investigated [25]. DSC demonstrated the miscibility of the blends, exhibiting a single T_g value for each composition studied. The T_g was dependent on the composition of the blend and varied with the molecular weight of the PEO. At a low level of PEO, the T_gs of the blends were within two degrees of one another, but as the PEO content increased, the T_g also increased. It was concluded that PHB and PEO are compatible in the melt and the amorphous state. The enzymatic degradation of PHB was confirmed via surface erosion; the degradation of PEO, via hydrolysis.

PHBV copolymers with 5–20% HV have been blended with CAB (cellulose acetate butyrate) and a polyester adipate plasticizer. Blends with above 30% CAB content and below 40% plasticizer are compatible. Increasing the CAB content of the blend (which needs to be over 50% for compatibility) improves the tensile strength and reduces the biodegradation rate. Additionally, the biodegradation rate seems to be affected by temperature and pH [25].

Poly(hydroxybutyrate-*co*-hydroxyvalerate) (P(HB-*co*-HV))/polyadipate (PEA) blends have been studied. Thermal analysis indicates that these blends are miscible until close to 30 wt % PEA content. An addition of PEA below 30 wt % decreases the crystallization induction time. Consequently, PEA can be used as a nucleating agent for crystallization (of P(HB-*co*-HV)). The elastic modulus and the activation energy for thermal decomposition decrease with PEA addition [43].

PLA and PHB are exceptionally compatible, and their blends can form a unique polymer alloy [28]. Blends containing PHB above 20 wt % are translucent, which is typical of semicrystalline polymers. Samples containing PHB below 20 wt % are transparent and clear, indicating very small or no crystalline phase domains.

1.4.2.2 PHA Additives

Nucleating agents or plasticizers shorten the molding processing cycles [44]. The low hydroxyvalerate grades with no plasticizers are difficult to process and require higher processing temperatures. Processing conditions are less critical with the higher hydroxyvalerate grades containing plasticizers [44].

The biodegradation rates of PHB containing plasticizers were studied using PEG, oxypropylated glycerol, and a commercial polyether (LAP). A 5–50% increase in plasticizer content resulted in a 4–25% and a 2–9% increase in elongation at break for LAP and PEG, respectively. The PHB composition with 10% PEG decomposed completely in soil during 30 days, while PHB containing 33% of the biologically stable LAP additive lost half of its mass over the same period [45].

1.4.3 PTMAT Blends

The PTMAT copolyester that is derived from diacids and glycols is entirely biodegradable and compostable. Blends of PTMAT with PBS and copolymers based on adipic acid, terephthalic acid, and 1,4-butanediol have been studied using a twin-screw extruder. The blends were immiscible, and the mechanical properties depended upon the crystallinity of the individual components and the blend composition. The tensile strength increased with increasing crystallinity. The degree of miscibility was found to depend on the intermolecular interactions [46].

1.5 Conclusions

Over the past few decades, increasing efforts have been directed to developing biologically degrading polymers in composting conditions. Scientific and industrial research has been focused on discovering and developing biodegradable polymers that are easily processable and exhibit good cost-effective properties and performance. Bioplastics can be organically recycled through composting according to international specifications. Biodegradability by composting represents a significant advantage in locations where organic waste including food packaging contaminated by food ends up in organic composting. In this case, organic recycling is preferred to mechanical recycling. Like other polymers, biodegradable polymers can be processed using conventional processes like injection molding, extrusion, thermoforming, and blow molding. Special attention should be directed to moisture removal from the polymer before processing to prevent hydrolytic degradation, which is the first stage in the biodegradation mechanism of biodegradable polymers. Thus, the processing operation should be performed under controlled humidity. During injection molding and extrusion, high processing temperatures may result in thermal degradation. Thus, biodegradable polymers should be compounded with appropriate plasticizers and blended with suitable polymers while processed at the lowest possible temperatures.

References for Chapter 1

[1] Nolan-ITU, Biodegradable Plastics – Developments and Environmental Impacts, Environment Australia, pp. 1–52, 2002.

[2] J.C. Middleton and A.J. Tipton, "Synthetic biodegradable polymers as medical devices," *Med. Plastics Biomater.*, vol. 3, pp. 30–39, 1998.

[3] D.M. Bigg, Controlling the Performance and Rate of Degradation of Polylactide Copolymer, ANTEC, p. 2816, 2003.

[4] F. Luo, A. Fortenberry, J. Ren, and Z. Qiang, "Recent Progress in Enhancing Poly(Lactic Acid) Stereocomplex Formation for Material Property Improvement," *Front. Chem.*, vol. 8, pp. 1–8, 2020.

[5] Y. Tokiwa and B.P. Calabia, "Biodegradability and biodegradation of poly(lactide)," *Appl. Microbiol. Biotechnol.*, vol. 72, pp. 244–251, 2006.

[6] S. Yeol Lee, J.W. Park, Y.T. Yoo, and S.S. Im, "Hydrolytic degradation behaviour and microstructural changes of poly(ester-co-amide)s," *Polym. Degrad. Stab.*, vol. 78, no. 1, pp. 63–71, 2002.

[7] T.D. Moshood, G. Nawanir, F. Mahmud, F. Mohamad, M.H. Ahmad, and A. AbdulGhani, "Biodegradable plastic applications towards sustainability: A recent innovations in the green product," *Clean. Eng. Technol.*, vol. 6, p. 100404, 2022.

[8] U. Witt, T. Einig, M. Yamamoto, I. Kleeberg, W.-D. Deckwer, and R.-J. Müller, "Biodegradation of aliphatic-aromatic copolyesters: Evaluation of the final biodegradability and ecotoxicological impact of degradation intermediates," *Chemosphere*, vol. 44, no. 2, pp. 289–299, 2001.

[9] A. Macintosh, A. Simpson, T. Neeman, and K. Dickson, "Plactic bag bans: Lessons from the Australian Capital Territory," *Resour. Conserv. Recycl.*, vol. 154, p. 104638, 2020.

[10] Z. Li, J. Yang, and X. Loh, "Polyhydroxyalkanoates: Opening doors for a sustainable future," *NPG Asia Mater.*, vol. 8, no. 4, p. e265, 2016.

[11] D.Y. Kim, H.W. Kim, M.G. Chung, and Y.H. Rhee, "Biosynthesis, modification, and biodegradation of bacterial medium-chain-length polyhydroxyalkanoates," *J. Microbiol.*, vol. 45, no. 2, pp. 87–97, 2007.

[12] R. van Tuil, P. Fowler, M. Lawther, and C.J. Weber, Biobased Packaging Materials for the Food Industry: Status and Perspectives, A European Concerted Action Report Book, November 2000.

[13] Y. Shaked, H. Dodiuk, S. Kenig, and S. McCarthy, "The effect of hyperbranched polymers on processing and thermal stability of biodegradable polyesters," *Polym. Eng. Sci.*, vol. 49, no. 3, pp. 559–566, 2009.

[14] Y. Shaked, H. Dodiuk, S. Kenig, and S. McCarthy, "Thermal stabilization of biodegradable poly hydroxy-butyrate in melt extrusion," *Poly. Eng. Sci.*, vol. 49, no. 3, pp. 559–566, 2008.

[15] I. Noda, S.B. Lindsey, and D. Caraway, Nodax™ Class PHA Copolymers: Their Properties and Applications. In: *Plastics from Bacteria*, G.Q. Chen (Ed.), Springer, Berlin-Heidelberg, pp. 237–255, 2010.

[16] P. Šmejkalová, V. Kužníková, J. Merna, and S. Hermanová, "Anaerobic digestion of aliphatic polyesters," *Water Sci. Technol.*, vol. 73, no. 10, pp. 2386–2393, 2016.

[17] N. Sridewi, K. Bhubalan, and K. Sudesh, "Degradation of commercially important polyhydroxyalkanoates in tropical mangrove ecosystem," *Polym. Degrad. Stab.*, vol. 91, no. 12, pp. 2931–2940, 2006.

[18] D.-M. Abou-Zeid, R.-J. Müller, and W.-D. Deckwer, "Degradation of natural and synthetic polyesters under anaerobic conditions," *Journal of Biotechnology*, vol. 86, no. 2, pp. 113–126, 2001.

[19] R. Leaversuch, "Biodegradable polyesters: Packaging goes 'green'," *Plast. Technol.*, vol. 48, no. 9, pp. 2–7, 2002.

[20] C. Bruns, B. de W.R. Gottschall, and B. de Wilde, Ecotoxicological trials with BAK 1095 (Bayer AG, Leverkusen) according to DIN V 54900, European Compost Network ECN e.V., Wolfsburg, September 2000.

[21] F.L. Buchholz, "Superabsorbent Polymers: An Idea Whose Time Has Come," *J. Chem. Educ.*, vol. 73, no. 6, p. 512, 1996.

[22] R.A. Gross and B. Kalra, "Biodegradable Polymers for the Environment," *Science*, vol. 297, pp. 803-807, 2002.

[23] R. Shalini and A. Singh, "Biobased Packaging Materials for the Food Industry," *J. Food Sci. & Technol. Nepal*, vol 5, pp. 16-20, 2009.

[24] M. Flieger, M. Kantorova, A. Prell, T. Rezanka, and J. Votruba, "Biodegradable plastics from renewable sources," *Folia Microbiol.*, vol. 48, pp. 27-44, 2003.

[25] G.F. Moore and S.M. Saunders, *Advances in Biodegradable Polymers* (Rapra Review Reports vol. 9), Rapra Publ., 1997.

[26] *S. Kumar, K.S. Thakur*, and *V. Joshi*, Bioplastics in Food Packaging (Ch. 37). In: *Postharvest Management of Fruits and Vegetables*, 2nd Ed., NIPA, pp. 766-777, 2000.

[27] S.A.T. van der Meer, J.R. de Wijn, and J.G.C. Wolke, "The influence of basic filler materials on the degradation of amorphous D- and L-lactide copolymer," *J. Mater. Sci.: Mater. Med.*, vol. 7, pp. 359-361, 1996.

[28] R. Khankrua, P.-A. Sommai, H. Hiroyuki, and S. Supakij, "Thermal and Mechanical Properties of Biodegradable Polyester/Silica Nanocomposites," *Energy Procedia*, vol. 34, pp. 705-713, 2013.

[29] M.N. Kim, A.R. Lee, K.H. Lee, I.J. Chin, and J.S. Yoon, "Biodegradability of poly(3-hydroxybutyrate) blended with poly(ethylene-co-vinyl acetate) or poly(ethylene oxide)," *Eur. Polym. J.*, vol. 35, no. 6, pp. 1153-1158, 1999.

[30] W. Ma, R.A. Gross, and S.P. McCarthy, Polylactic acid-based blends, United States Patent, 5,883,199A, 1999.

[31] C. Rozsa, P. Ortiz, V.P. Cyras, A. Vázquez, and N. Galego, "Poly(hydroxybutyrate-co-hydroxyvalerate)-polyadipate blends," *Int. J. Polym. Mater. Polym. Biomater.*, vol. 51, no. 7, pp. 619-631, 2002.

[32] N. Ljungberg and B. Wesslén, "The effects of plasticizers on the dynamic mechanical and thermal properties of poly(lactic acid)," *J. Appl. Polym. Sci.*, vol. 86, no. 5, pp. 1227-1234, 2002.

[33] O. Weizman, A. Dotan, Y. Nir, and A. Ophir, "Modified whey protein coatings for improved gas barrier properties of biodegradable films," *Polym. Adv. Technol.*, vol. 28, no. 2, pp. 261-270, 2017.

[34] S. Sinha Ray, K. Yamada, M. Okamoto, and K. Ueda, "New polylactide-layered silicate nanocomposites. 2. Concurrent improvements of material properties, biodegradability and melt rheology," *Polymer*, vol. 44, no. 3, pp. 857-866, 2002.

[35] I. Kellersztein, E. Amir, and A. Dotan, "Grafting of wheat straw fibers with poly (ε-caprolactone) via ring-opening polymerization for poly(lactic acid) reinforcement," *Polym. Adv. Technol.*, vol. 27, no. 5, pp. 657-664, 2016.

[36] O. Martin and L. Averous, "Poly(lactic acid): plasticization and properties of biodegradable multiphase systems," *Polymer*, vol. 42, no. 14, pp. 6209-6219, 2001.

[37] D. Lewitus, S. McCarthy, A. Ophir, and S. Kenig, "The Effect of Nanoclays on the Properties of PLLA-modified Polymers Part 1: Mechanical and Thermal Properties," *J. Polym. Environ.*, vol. 14, pp. 171-177, 2006.

[38] G. Laverde, Analysis of the Mechanical Properties of Biodegradable Films Made from Blends of Polylactic Acid (PLA) and Polyesters by Blown Film Extrusion, ANTEC, p. 2515, 1998.

[39] I. Bibers, V. Tupureina, A. Dzene, L. Savenkova, and M. Kalnins, "Biodegradable materials from plasticized PHB biomass," *Macromol. Symp.*, vol. 170, no. 1, pp. 61-72, 2001.

[40] J. Li, Y. He, and Y. Inoue, "Thermal and mechanical properties of biodegradable blends of poly(L-lactic acid) and lignin," *Polymer International*, vol. 52, no. 6, pp. 949-955, 2003.

[41] E. Chiellini, P. Cinelli, F. Chiellini, and S.H. Imam, "Environmentally Degradable Bio-Based Polymeric Blends and Composites," *Macromol. Biosci.*, vol. 4, no. 3, pp. 218-231, 2004.

[42] I. Ghosh, R.K. Jain, and W.G. Glasser, Blends of Biodegradable Thermoplastics with Lignin Esters (Ch. 17). In: *Lignin: Historical, Biological, and Materials Perspectives*, W.G. Glasser, R.A. Northey, and T.P. Schultz (Eds.), ACS Symposium Series Vol. 742, pp 331–350, 1999.

[43] J. John, R. Mani, and M. Bhattacharya, "Evaluation of compatibility and properties of biodegradable polyester blends," *J. Polym. Sci. Part A Polym. Chem.*, vol. 40, no. 12, pp. 2003–2014, 2002.

[44] R. Song, M. Murphy, C. Li, K. Ting, C. Soo, and Z. Zheng, "Current development of biodegradable polymeric materials for biomedical applications," *Drug Des. Devel. Ther.*, vol. 12, pp. 3117–3145, 2018.

[45] N. Ogata, T. Tatsushima, K. Nakane, K. Sasaki, and T. Ogihara, "Structure and physical properties of cellulose acetate/poly(L-lactide) blends," *J. Appl. Polym. Sci.*, vol. 85, no. 6, pp. 1219–1226, 2002.

[46] H. Abe, Y. Doi, and Y. Kumagai, "Synthesis and Characterization of Poly[(R,S)-3-hydroxybutyrate-b-6-hydroxyhexanoate] as a Compatibilizer for a Biodegradable Blend of Poly[(R)-3-hydroxybutyrate] and Poly(6-hydroxyhexanoate)," *Macromolecules*, vol. 27, no. 21, pp. 6012–6017, 1994. doi: 10.1021/ma00099a012

[47] H.-S. Kim, H.-J. Kim, J.-W. Lee, and I.-G. Choi, "Biodegradability of bio-flour filled biodegradable poly(butylene succinate) bio-composites in natural and compost soil," *Polymer Degradation and Stability, vol. 91, no. 5*, pp. 1117–1127, 2006.

2 Polyhydroxyalkanoates (PHAs): Advancements in Blends, Biocomposites, and Their Commercialization

Debarshi Nath, Ehsan Pesaranhajiabbas, Akhilesh K. Pal, Bandaru Ramarao, Manjusri Misra, and Amar K. Mohanty

This chapter explains the PHA blending process with other biopolymers and organic/inorganic fillers to develop biocomposites with balanced and desirable properties depending on their application, current market status, and biodegradability studies. PHA-based blends and composites, which extend the physical property space and the performance for applications as well as their use, have significant potential for further development. Other biopolymers such as polycaprolactone, polybutylene succinate (PBS), PBAT, and PBSA are widely used in conjunction with PHAs to balance/improve the properties and processing conditions. Furthermore, various types of fillers are also incorporated into the polymer/polymer blend matrix to enhance the modulus, barrier properties, and thermal resistance of the biocomposites.

2.1 Introduction

As a result of the globalization of manufacturing and its supply chains, single-use plastic products are being widely applied to increase manufacturing and product application efficiencies. Unfortunately, single-use plastics are discarded into landfills due to their non-biodegradable behavior or sometimes directly into the environment, contributing to a significant environmental burden worldwide. Fossil-based plastics also strain petroleum and natural gas supplies while putting extra fossil carbon into the atmosphere, contributing heavily to carbon pollution and leading to climate change consequences [1].

Plastics were first produced on a large scale industrially in the 1950s, and their low bulk density, excellent durability, outstanding processability, exceptional mechanical and barrier qualities, and low cost quickly made them indispensable for daily use and high-end applications. The adaptability of plastic products and materials has caused a sharp rise in the amount of plastic produced each year over the recent past, and this continues to increase at alarming rates [2]. Almost 6300 million tonnes of plastic waste have been produced since 1950, with annual plastic output exceeding 380 million tonnes and rising by 4% annually [3]. The application of polymer blends and composites in different sectors is shown in Figure 2.1. The commercial sectors with the highest worldwide plastic volume in 2015 were packaging (35.9%), construction (16.0%), textiles (14.5%), and consumer goods (10.3%). The worldwide output of plastics mainly comprises the following polymers: 36% polyethylene (PE), 21% polypropylene (PP), and 12% polyvinyl chloride (PVC), among others [4].

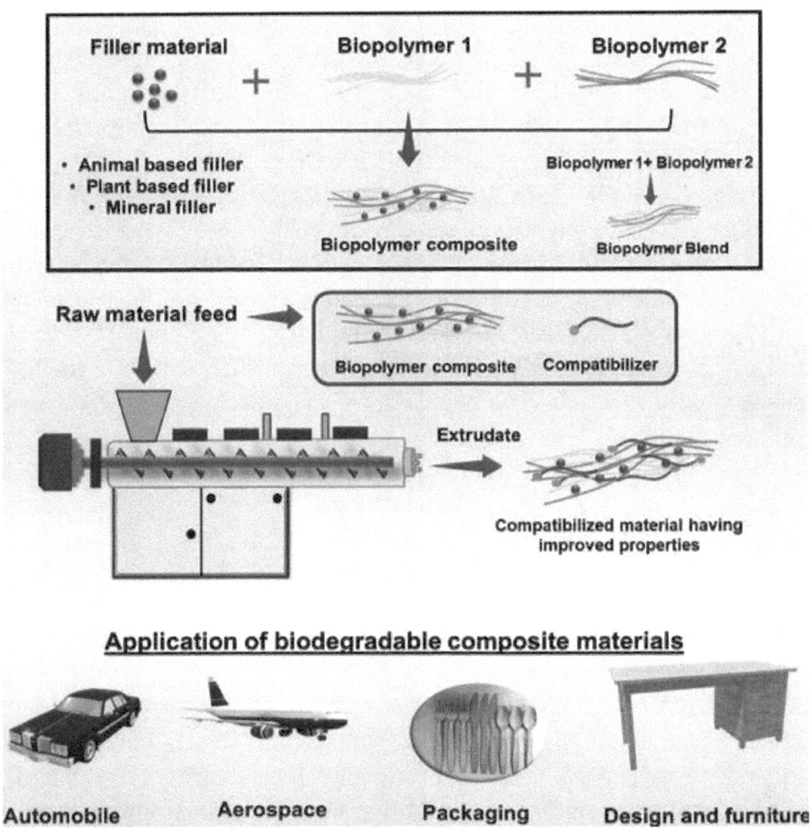

Figure 2.1 Application of polymer blends and composites in different sectors

Unfortunately, the same features that make plastics so desirable also generate plastic waste, a serious ecological concern if not managed effectively. According to current forecasts, 12 billion tonnes of plastic garbage will end up in landfills or even the natural habitat by 2050, compared to 4.9 billion tonnes in 2010 [5]. In addition, it is anticipated that 100 tonnes of plastic trash will have reached the marine environment between 2010 and 2025, where it will gradually disintegrate and break into microplastics [2]. As most commercial plastics come from a non-renewable source, the problem of plastic waste collection and treatment has recently focused much academic and industry effort towards more sustainable yet similarly performing options [6].

Bioplastics, or biodegradable plastics made from renewable resources, stand out as a possible alternative [6, 7]. Waste from biodegradable plastics may be disposed of in many ways, reducing the quantity of plastic in the environment. In contrast, the carbon footprint of bioderived bioplastics can be significantly reduced during the resource extraction phase [8]. Since CO_2 is absorbed for biomass growth during photosynthesis, the overall carbon footprint of bioplastic manufacturing is lower than that of petroleum-based plastic production [7, 9]. Production of biopolymers worldwide was 2.11 million tonnes in 2018, and the number is predicted to rise to 2.62 million tonnes by 2023 [7, 10]. In recent decades, rigorous research has led to the commercialization of several bioplastics of industrial importance, including PHAs [11], PBS [12], poly(lactic acid) (PLA) [13], and thermoplastic starch (TPS) [14], which are comparable to their non-renewable contemporaries [15]. Even though bioplastics have been studied for almost a century, their commercialization on a large scale remains in development.

PHAs are semi-crystalline, aliphatic polyesters synthesized and stored as a carbon and energy storage material in microbial cells [16]. Despite many desirable properties, the ability of PHA to be produced from renewable carbon resources, its biodegradability, and chemical variety make it a subject of intense commercial and academic attention [17]. PHAs are produced mainly by utilizing microbial resources in their synthesis. Increased PHA production takes place when microorganisms are subjected to unfavorable growth circumstances. This acts as a survival strategy for these microorganisms, where they store PHA granules, further supporting this synthesis by enhancing the production of the bioplastic [16].

According to the number of carbons in their monomeric components, PHAs are divided into two groups: short-chain-length PHAs (scl-PHA) and medium-chain-length PHAs (mcl-PHA). Regarding physical properties, scl-PHA is more crystalline and rigid, whereas mcl-PHA is less crystalline and exhibits more elasticity [17]. PHAs are distinguished from non-biodegradable plastics because microbes break down the polymer into fundamental components when exposed to soil, compost, or marine sediments. CO_2 and water are released during the PHA breakdown process under aerobic settings, whereas carbon dioxide and methane are produced under anaerobic conditions [18]. Unfortunately, owing in part to poor thermal and mechanical qualities, these polymers cannot yet surpass traditional plastics [19]. Polymer blending thus can act as a solution to the problem, whereby excellent dual qualities may be formed from separate polymer components, making polymer blends an essential field of research for polymeric material with multi-functional features and high performance [20].

Polymers (either petroleum-based or renewable resources-based) are seldom used by themselves. Biopolymers such as PBS, PBSA, PCL, and PBAT have emerged as the most promising candidate for blending with PHAs. The classification of polymers is presented in Figure 2.2. This is due to the desired mechanical properties of these co-polymers, such as their high toughness and elongation at break, and is complementary to PHAs. PBS is a key biodegradable polyester produced through the polycondensation of succinic acid and butanediol. In contrast, PBAT and PBSA are co-polymers containing two different randomly arranged co-monomers. In PBAT, the co-monomers are butylene adipate (obtained from adipic acid and 1,4-butanediol) and butylene terephthalate (terephthalic acid and 1,4-butanediol). In the case of PBSA, similar to PBAT, the co-monomers are butylene adipate and butylene succinate, created by the polycondensation reaction between 1,4-butanediol and succinic acid [21]. Similarly, PCL is a polyester prepared by the ring-opening polymerization of ε-caprolactone in the presence of a catalyst like stannous octanoate [22]. The degradation of biopolymers is influenced by their chemical composition and the conditions under which they are prepared [23]. In certain instances, biodegradation is caused by the enzymatic activity of microorganisms, such as bacteria, fungi, and algae, in the natural environment. Also, biodegradation happens as a result of a combined depolymerization process where the polymer chain is broken down by a nonenzymatic reaction (such as thermal degradation or chemical hydrolysis), and the microorganisms then metabolize the intermediates [23, 24]. Furthermore, in the case of PCL, it can be degraded by cutinases released from different phytopathogens due to its similarity with cutin [22].

Similarly, fibers are commonly added to polymeric materials to enhance their strength and durability. Vegetable fibers (such as flax, cotton, jute, and hemp) can be categorized as bast, leaf, or seed-hair fibers. Cellulose is the most abundant component extracted from plant fibers, and the prospects for cellulose fiber-reinforced polymers have emerged due to the renewed emphasis on sustainable raw resources [25, 26]. Reinforcement using fillers is especially crucial for polymers derived from renewable resources, as most of them suffer from lower modulus and lower melting temperatures [26]. Nonetheless, there are still numerous obstacles to overcome and opportunities to utilize fillers to increase the composite's performance and trigger its industrial production. Examples include issues with filler dispersion, reduction in filler content, and the interaction of filler with matrix [23].

Therefore, this chapter briefly summarizes the main features of PHAs along with their blends and biocomposites, followed by a thorough investigation of the synthesis and characterization techniques employed for the composites. A detailed discussion on the biodegradation of PHAs and their composites are also included. The chapter concludes with real-world examples of how these eco-friendly composites are used in industry.

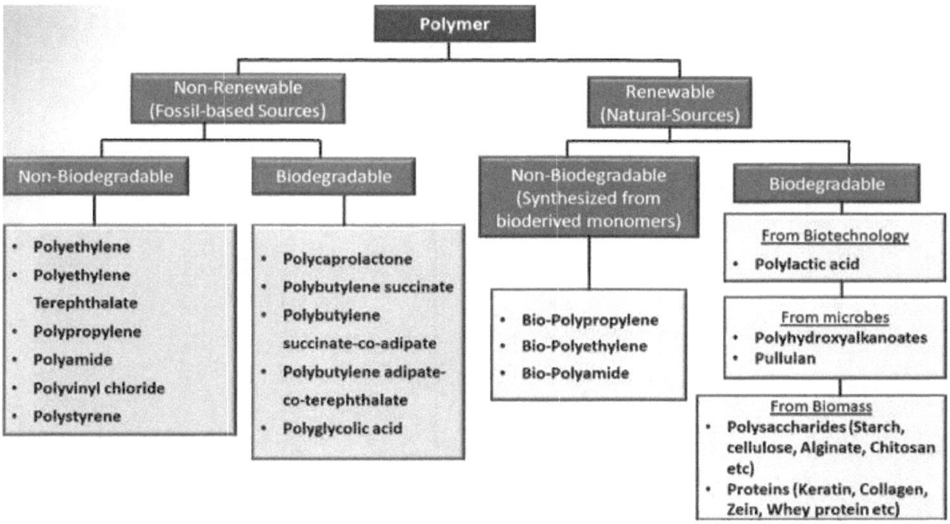

Figure 2.2 Classification of biodegradable and non-biodegradable polymers

2.2 PHA-Based Blends

Polyhydroxyalkanoates are polymers of the esters of hydroxyalkanoic acids (HAs), produced from lignocellulosic raw materials of a broad range, including starch, cellulose, molasses, and even carbon dioxide. Compared to other biodegradable polymers, PHA has superior oxygen and water vapor barrier properties, making it an excellent option for food packaging applications.

Another advantage of PHAs is that their mechanical performance can be tuned by introducing various co-monomers with different content. Hence, PHA biopolymers with various mechanical characteristics are now available for various applications. About 150 distinct varieties of HA monomers with unique chemical and physical properties have been identified based on their side chain structure and geometrical characteristics. Moreover, this number is expected to rise when more PHA varieties are developed by performing physical or chemical modification of naturally produced PHAs or by developing genetically modified organisms to create PHAs with specific functional groups as shown in Figure 2.3 [17]. For example, the mechanical attributes of PHA co-polymers such as mcl-PHA poly(3-hydroxyhexanoate-co-3-hydroxyoctanoate) P(3HHx-co-3HO) and scl-PHA [poly(3-hydroxybutyrate-co-3-hydroxyvalerate (PHBV)] have properties like polypropylene and polystyrene [27]. The chemical structures of various biodegradable polymers are shown in Figure 2.4.

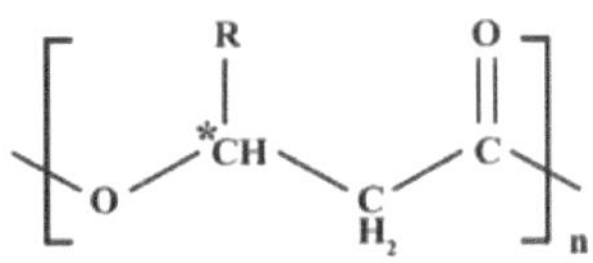

Poly(3-hydroxyalkanoate)

R group	Carbon no.	PHA polymer
methyl	C_4	Poly(3-hydroxybutyrate)
ethyl	C_5	Poly(3-hydroxyvalerate)
propyl	C_6	Poly(3-hydroxyhexanoate)
butyl	C_7	Poly(3-hydroxyheptanoate)
pentyl	C_8	Poly(3-hydroxyoctanoate)
hexyl	C_9	Poly(3-hydroxynonanoate)
heptyl	C_{10}	Poly(3-hydroxydecanoate)
octyl	C_{11}	Poly(3-hydroxyundecanoate)
nonyl	C_{12}	Poly(3-hydroxydodecanoate)
decyl	C_{13}	Poly(3-hydroxytridecanoate)
undecyl	C_{14}	Poly(3-hydroxytetradecanoate)
dodecyl	C_{15}	Poly(3-hydroxypentadecanoate)
tridecyl	C_{16}	Poly(3-hydroxyhexadecanoate)

Figure 2.3 Chemical structure for polyhydroxyalkanoate (PHA). It is the funct onal alkyl (R) group that determines the nomenclature and carbon number of PHA compounds. Chiral PHA-building-block centers are indicated with asterisks [17]. Reproduced from MDPI, 2014, open access

However, these polymers usually have a limited processing window due to their higher susceptibility toward thermal degradation during common melt-processing methods [28]. Furthermore, they are primarily rigid biopolymers with low flexibility. They are also more expensive than many other biodegradable polymers, limiting their industrial use [16]. Blending PHAs with other biodegradable polymers is one of the best ways to improve the competitive performance of PHAs with other biodegradable polymers. PHAs can be mixed with another polymer with complementary properties (as shown in Table 2.1), and the blend obtained can be tuned for the performance required for a specific application. For instance, PHAs show high brittleness, which can be undesirable. To mitigate its effect, a possible solution is to blend it with flexible biodegradable polymers such as PBAT or PCL to obtain more deformable blended bioplastics.

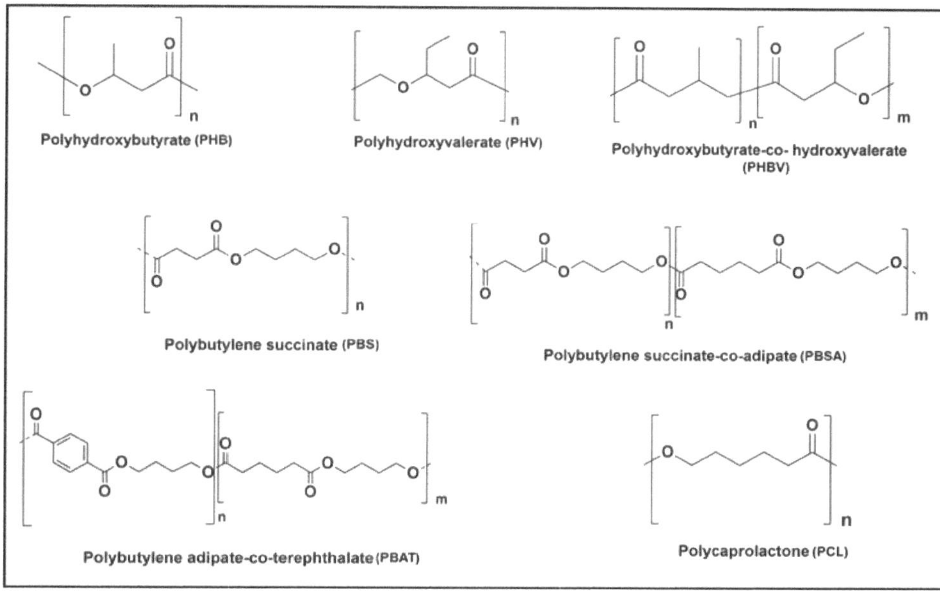

Figure 2.4 Chemical structures of PHB, PHV, PHBV, PBS, PBSA, PBAT, and PCL

Table 2.1 Mechanical Properties of Various Bioplastics

Polymer	Tensile strength (MPa)	Elongation at break (%)	Young's modulus (MPa)	Impact strength (J/m)	Ref.
PHBV	33.5	1.7	3260	20	[29]
PBAT	20	610	50	-	[30]
PBS	32.7	215	438	77	[31]
PBSA	17.0	207	310	409	[29]
PCL	17	314	450	-	[32]

The blending of polymers can be performed by two main approaches: blending in the solution state (solution casting) and blending in the melt state (melt blending). In solution casting, the neat polymers can be dissolved into a proper solvent and mixed into a uniform solution. The solvent is separated by evaporation or distillation under appropriate pressure conditions (vacuum or high pressure). However, blending in solution may not be practicable for industrial use due to several causes, even if it produces a homogenous polymer blend. First, complete removal of the solvent cannot be achieved, and the remaining solvent in the polymer blend can result in the degradation of the final material. Moreover, most of the solvents are toxic and costly, and if they cannot be recovered and reused at an adequate rate, this adds to the final cost of the developed material. Another approach is the melt blending of polymers, which are heated above their melting temperature. Then they are mixed by fluid mechanical forces to reach a uniform mixture. Despite blending in the solution state, melt blending is a continuous processing method, making it very attractive for industries. Furthermore, as no solvent is required for the melt blending of polymers, there is no concern related to the remaining solvent in the final manufactured material.

The performance of a polymer blend is dependent not only on the properties of its constituents but also on the interaction of these polymers in the blend. In other words, two polymers in the blend can be thermodynamically miscible or immiscible, as indicated by the free energy change of the mixing process as shown in Equation 2.1 [33]:

$$\Delta G = \Delta H - T\Delta S \tag{2.1}$$

where ΔG is the free Gibbs energy, ΔH is the enthalpy, and ΔS is the entropy of mixing of the polymer blend.

To be miscible and form a true solution, the free energy change for a polymer blend should be negative, and a single solution phase should be obtained. Conversely, a positive free energy change presents a barrier to solution formation, and the blend remains in two macroscopic or microscopic phases, each phase being rich in one of the polymer blend constituents. According to the literature, most PHA-based blends turn out to be immiscible, with a two-phase morphology. The following section will discuss such immiscible (i. e., two-phase) blends, and the effect of different factors that can influence the properties of the final material.

2.2.1 Non-Compatibilized Blends of PHAs and Biodegradable Polymers

Literature reports on the processing and properties of PHAs blended with other biodegradable polymers, including PBAT, PBS, PBSA, PCL, and PLA, have been published, and some of them are given here [34–38]. As mentioned earlier, most PHA blends with other biodegradable polymers usually are immiscible blends with a two-phase morphology. Therefore, their performance highly depends on the interaction between phases, also known as compatibility. To reduce the high brittleness of PHAs, they are often blended with more flexible biodegradable polymers to impart flexibility.

PBAT is an example co-blending polymer with sufficiently high deformability to compensate for the brittleness of PHA. Zhang and co-workers studied the blends of poly(3-hydroxybutyrate-co-3-hydroxyhexanoate) (PHBH) and PBAT [39]. Blends with a wide range of compositions (PBAT content of 20, 40, 50, 60, and 80 wt%) were prepared by a twin-screw extruder and then blown into thin films. The microstructure showed a two-phase morphology, implying the immiscibility of these two co-polyesters. Another clue for their miscibility came from the thermal analysis, where the blends showed two different glass transition temperatures (T_g) corresponding to each constituent. Despite this immiscibility, the processability and mechanical properties of produced films made from these blends showed significant changes compared to the native pure components. Neat PHBH was too brittle to be processed by film blowing; but adding only 20 wt% of PBAT made it possible to blow films with moderate elongation at a break of 103% in the machine direction. By increasing the content of PBAT to 40 wt%, the deformability of films in both machine and transverse direction increased significantly (240% and 483%, respectively), indicating the capability of the PBAT to improve the deformability of brittle PHBH. The effect of PBAT on the processability of PHBV films by made film-blowing process was investigated by Cunha and co-workers [40]. Observations confirmed that PHBV blends with PBAT had better bubble stability with less draw resonance in the bubble and lower fluctuation in film dimensions compared to neat PHBV films.

The effect on the morphology and mechanical properties of PHBV was studied after including PBAT by Zytner and co-workers [41]. Blends with different content of PBAT (30, 50, and 70 wt%) were melt-blended and injected using a micro-compounder and injection molder. PHBV/PBAT blends indicated a two-phase morphology characteristic of immiscible polymer blends, as shown in Figure 2.5. However, a clear transition was observed in the morphology of the blends with different content of the PBAT. For example, in a blend containing 30 wt% of PBAT, a droplet-matrix morphology was observed with PBAT particles dispersed with the PHBV matrix. By increasing the PBAT content to 50 wt%, a connected network of PBAT was formed, revealing a transition from droplet-matrix to co-continuous morphology. Such a transition in the morphology significantly affected the mechanical performance of the blends, as revealed by tensile and impact tests. Accordingly, PHBV showed high tensile strength and modulus of 36.6 and 3281 MPa, respectively. In contrast, PHBV showed low percent elongation at break and an impact strength of 2.5% and 20.4 J/m, respectively. Blending this biopolymer with 30 wt% of PBAT slightly improved the percent elongation at break and impact strength to 3.8% and 28 J/m, respectively. However, by increasing the PBAT content to 50 wt%, a remarkable enhancement was observed in these properties, as the percent elongation at break and impact strength were increased to 138% and 293 J/m, along with the decrease in tensile strength and tensile modulus to 27.5 and 1563 MPa, respectively.

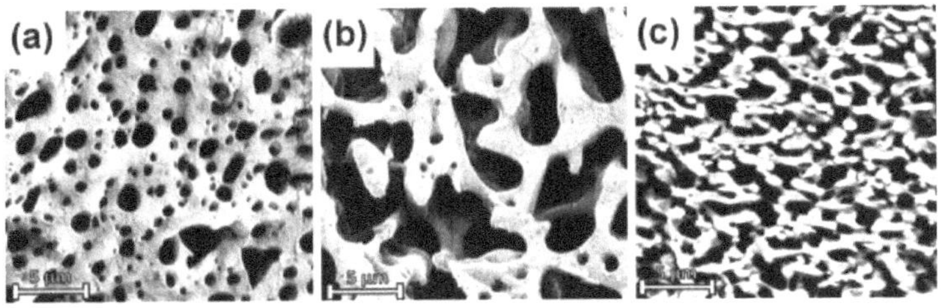

Figure 2.5 SEM analysis on the fracture surface of different blends, showing a transition in the morphology of blends from droplet-matrix in PHBV/PBAT (70/30 wt%) blend (a) to co-continuous in PHBV/PBAT (50/50 wt%) blend (b) and again back to droplet-matrix in PHBV/PBAT (30/70 wt%) blend (c) [41]. Copyright American Chemical Society, 2023

PBSA and PCL are other biodegradable polymers with high toughness that can be blended with PHA biopolymers to modify their deformability. In a study by Le Delliou and co-workers, PHBV was melt blended with various contents of PBSA [42]. Results confirmed the immiscibility of two biopolymers, evidenced by the two-phase morphology of the blends. At low content of PBSA, a droplet-matrix morphology was detected, while a co-continuous morphology was formed in the blend with equal constituents content. The thermal analysis also showed that the crystallization of both PHBV and PBSA was accelerated in the presence of other biopolymers. However, blending with PBSA did not show a significant effect on the deformability of PHBV, as a slight enhancement in elongation at the break of blends was reported, mainly due to a non-efficient stress transfer between the two phases. The effect of melt blending of poly(3-hydroxybutyrate) (PHB) with PBS and PBSA on the characteristic of this biodegradable polymer was investigated by Righetti and co-workers [43]. For both blends, a transition from the brittle fracture in neat PHB to a more ductile one in blends was reported, as evidenced by the appearance of necking in tensile test samples during testing. Moreover, a consistent improvement in percent elongation at break was indicated by increasing either PBS or PBSA. This transition in fracture behavior of PHB after blending with PBS and PBSA was also supported by the micrographs obtained from the fracture surface of the samples. Accordingly, while the PHB fracture surface was smooth, implying its brittle behavior, the blends showed a rougher fracture surface confirming their more ductile behavior than neat PHB.

In research by Jost and Miesbauer, the cast extrusion method was used to produce sheets from PHBV blended with PBAT and PBS [44]. The blends were first prepared by melt-blending in a twin-screw extruder, and then sheets with a thickness of 80 μm were produced by a flat-film extrusion process. According to the result, blending PHBV with PBS or PBAT reduced the Young's modulus and tensile strength of this biopolymer, accompanied by a slight improvement in its elongation at break. Moreover, both blends showed inferior oxygen and water vapor barrier properties compared to neat PHBV, with a greater decrease in PHBV/PBS blends.

2.2.2 Compatibilized PHA-Based Blends

Based on the research on the blends of PHAs with other biodegradable polymers, most of these blends show the characteristic of immiscible blends with two phases co-existing in the microstructure of the blend. Therefore, the compatibility and interaction between these two phases are of great importance and can determine the final performance of the blend. In other words, in the absence of good interaction between the two phases, the interfacial area between these two phases can act as a weak point and result in the degradation of the properties of the blend [45]. Accordingly, improving the compatibility between PHA polymer and the other biodegradable polymer in the blend is critical to achieving the best performance. There are some approaches available in the literature for improving the compatibility in immiscible polymer blends, which results in the formation of a strong interface between two phases contributing to the finer morphology and higher mechanical properties of the blend. Generally, these approaches can be divided into two main groups: ex-situ compatibilization and in-situ compatibilization.

Ex-situ compatibilization involves adding pre-made high-molecular-weight block or graft co-polymer to blended polymers as a compatibilizer. Ex-situ compatibilization involves two steps: synthesizing the compatibilizer, and then adding it during melt blending to the blended polymers. Each block or graft in the compatibilizer has an affinity to interact with the components in the blends. Accordingly, when located at the interface of two phases in an immiscible polymer blend, the compatibilizer can enhance the interaction between these two phases and contribute to the enhanced performance of the blend. The main advantage of ex-situ compatibilization is that the structure of the compatibilizer can be designed and controlled as it is made in a separate step.

In-situ compatibilization is another approach for improving the compatibility of two phases in immiscible blends. This approach involves reactive melt blending (reactive extrusion) of two immiscible blends in the presence of a compatibilizer containing reactive functional groups such as epoxy, anhydride, or isocyanates. During reactive extrusion and in the presence of such a compatibilizer, a chemical reaction occurs between these reactive functional groups and other functional groups in the biodegradable polymers, such as hydroxyl and carboxyl groups, resulting in the formation of in-situ graft or block co-polymers. When located at the interface of two polymers, the formed in-situ co-polymer functions as a compatibilizer and enhances the compatibility of two phases and, consequently, the performance of the blend. Reviewing the literature, it can be noticed that in-situ compatibilization by reactive extrusion is the main approach investigated in PHA-based blends with other biodegradable polymers.

A hyperbranched ethylenediamine triazine polymer (HBETP) was synthesized and used as a compatibilizer in PHBV/PBAT (70/30 wt%) to improve the compatibility of the blend [46]. An improvement in the toughness of the material, indicated by increased percent elongation at break and higher impact strength, was observed. A thermo-mechanical analysis (DMA) showed the lower difference between T_g of the two polymers in the compatibilized blends. Besides, with this hyperbranched polymer, fine morphology with less debonding between phases was reported. All these trends confirmed the compatibilization effect of the added hyperbranched polymer. According to the proposed mechanism by the author, this hyperbranched polymer contained amino groups in its chemical structure, which are capable of making both hydrogen bonds and chemical bonds with hydroxyl and carboxyl groups in PHBV and PBAT. Such interactions could result in better compatibility between PHBV and PBAT phases, contributing to the improvement in the mechanical performance of the blend.

Chain extenders, usually used for improving the thermal stability of the polymers during melt processing, can also be utilized to enhance compatibility if they can trigger a reaction between two blended polymers by their functional groups. Among these types of materials, those containing epoxy functional groups have been extensively used as compatibilizing agents in biodegradable polyester blends. In a study by Feijoo and co-workers, different amounts of an epoxy-functionalized chain extender were used in PHBV/bio-based PBSA blends [29]. The microstructural analysis revealed a reduction in the size of the dispersed phase and an increase in the interfacial surface between two biopolyesters after including a chain extender. The thermal analysis also showed that the crystallization ability of PHBV was reduced by the incorporation of the chain extender, implying that the chain extender effectively improved the linking of polymer macromolecules, which restricted their mobility and crystallization ability. After adding a chain extender, the rheological analysis also demonstrated the enhanced complex viscosity and storage modulus of the blends, mainly due to better interaction between two phases in compatibilized blends. However, the mechanical analysis indicated a slight improvement in blends with a specific concentration ratio of two polymers, as evidenced by the increase in elongation at the break of PHBV/PBSA (70/30 wt%) blend after adding 0.5 parts per hundred resin (phr) of the chain extender. The authors concluded that although there was clear evidence showing the improved interaction between two phases in these immiscible blends, the compatibilization effect of the chain extender with the epoxy functionalized group was not high enough to make a significant improvement in the mechanical behavior of PHBV/PBSA blends.

Reactive chemicals such as peroxides are another type of compatibilizing agent studied in PHA-based blends for improving the compatibility of two materials. They can trigger a chemical reaction between two polymers during melt processing by making free radicals, which form a co-polymer of these two constituents to serve as compatibilizers. This strategy was used in PHB/PBS and PHBV/PBS blends using dicumyl peroxide (DCP) [47]. Significant modification in percent elongation at break was reported for PHBV/PBS (80/20 wt%) blends after incorporating various content of DCP. Accordingly, the elongation at the break of the blend increased from only 8% in the neat blend to 400% in the blend containing 0.5 wt% of DCP. However, the improvement in elongation at break was not significant in the case of PHB/PBS blends. Micrographs obtained from SEM and transmission electron microscopy (TEM) also confirmed a remarkable change in the microstructure of the blends, as the size of the PBS dispersed phase significantly reduced after incorporation of DCP, showing enhanced interfacial interaction between two phases in the presence of compatibilizer, as shown in Figure 2.6. The authors attributed these effects to the formation of a co-polymer of PHBV and PBS as a result of a free-radical reaction between these two bio polyesters triggered by the reactive DCP chemical.

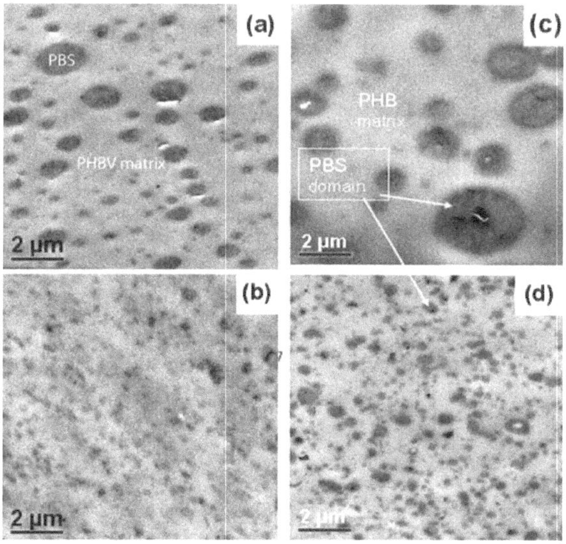

Figure 2.6 Effect of DCP on the morphology of PHB/PBS and PHBV/PBS blends; (a) PHBV/PBS blend without DCP; (b) PHBV/PBS containing 0.5 wt% of DCP; (c) PHB/PBS blend without DCP; (d) PHB/PBS containing 0.5 wt% of DCP [47]: Copyright Wiley, 2012 (License Number: 5567090624417)

DCP was similarly utilized in blends of P3HB and PBAT [48]. Comparing the morphology of non-compatibilized and compatibilized blends, the authors noticed more deformation in the PBAT phase in the former. They attributed this change in morphology to the stronger interaction between PHB and PBAT in the presence of DCP. In research done by Garcia and co-workers, different amounts of DCP were utilized in P3HB/PCL blends as a compatibilizer [49]. Based on the mechanical analysis, the DCP was an effective compatibilizer in this immiscible blend, as it enhanced impact strength and elongation at the break of the material. In contrast, the compatibilized blend retained its tensile strength. The highest improvement in toughness was reported in the blend compatibilized with the highest level of DCP (1 wt%), where elongation at break and impact strength were enhanced from 11.1% and 2 kJ/m^2 in the non-compatibilized blend to 21.3% and 6.3 kJ/m^2 in compatibilized one, respectively. Comparison of morphology between non-compatibilized and compatibilized blends revealed more evidence of improvement in the compatibility of two phases by incorporating DCP, evidenced by no obvious interface between the two phases and finer morphology in the compatibilized blend. The images from the atomic force microscope (AFM) also showed the smaller size of PCL-rich domains and more homogeneous dispersion of PCL particles within PHB matrix after adding DCP, giving another clue of enhanced compatibilization as shown in Figure 2.7(a–d). The authors correlated these improvements in mechanical properties [Figure 2.7(e)] and compatibility in compatibilized blends to forming PHB-PCL co-polymers in the presence of DCP as a free-radical initiator during the melt-blending process.

The role of two organic peroxides, DCP and di-(2-*tert*-butylperoxyisopropyl)benzene (BIB), in the compatibilization of PHB/PCL blends was investigated [50]. Although both reactive chemicals successfully improved the mechanical properties of the blend, BIB was more effective. According to the tensile test, the neat blend revealed elongation at a break of 125%; this value was enhanced to 278% by incorporating one part per weight of DCP, while only 0.5 part per weight of BIB improved the elongation of the blend to 305%. A rheological study also confirmed that at the same level of two peroxides, BIB decreased the melt flow index of the blend more than DCP. The authors discussed that the presence of more than one peroxy group in the chemical structure of BIB make this chemical more reactive so that it can make more free-radicals than DCP, and more co-polymer formation/cross-linking can occur in the blends compatibilized with this reactive chemical, contributing to a higher improvement in the mechanical performance of the blend.

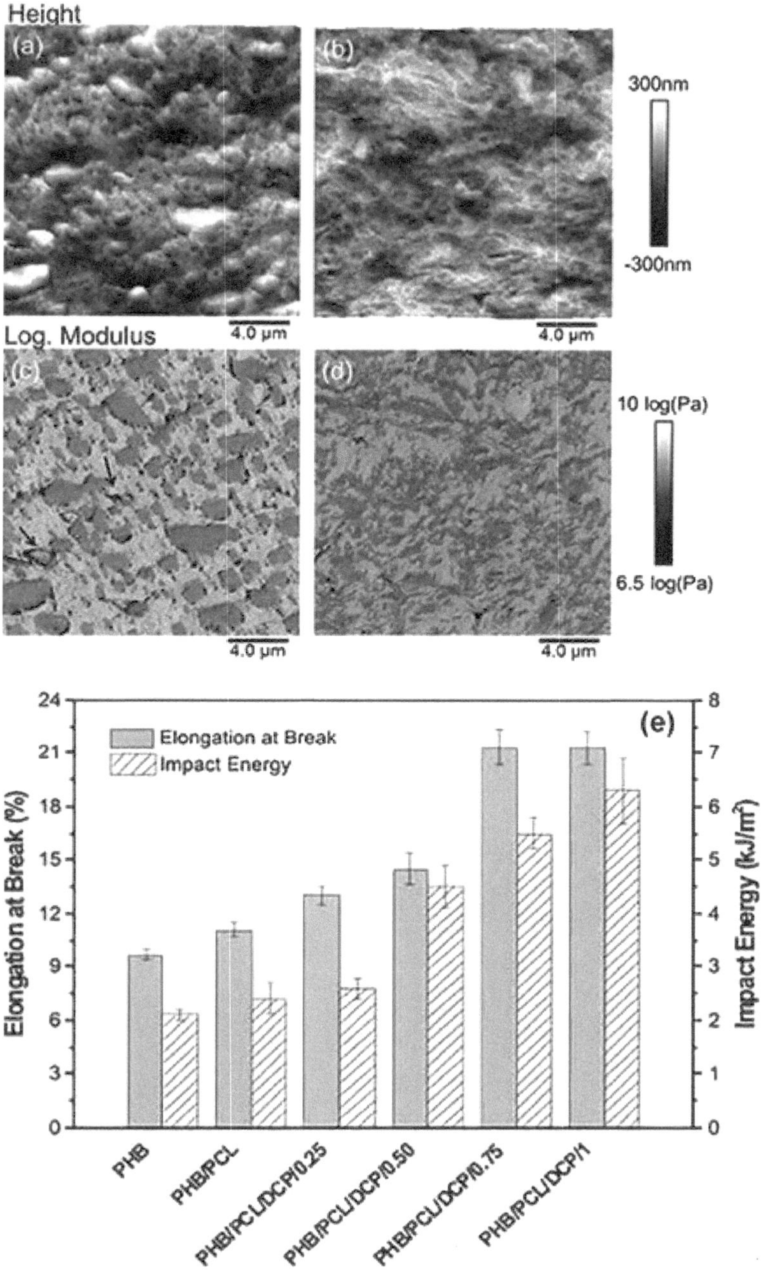

Figure 2.7 Effect of DCP on morphology and mechanical performance of PHB/PCL (75/25 wt%) blend; (a) and (c) AFM surface topography (height channel) and elastic modulus image map of un-compatibilized blend; (b) and (d) AFM surface topography (height channel) and elastic modulus image map of compatibilized blend with 1 wt% of DCP; (e) Modification of impact strength and elongation at break of blend after compatibilization by different amounts of DCP [49]. Copyright Elsevier, 2017 (License Number: 5567100156037)

Zytner and coworkers used a reactive extrusion approach in the presence of two reactive chemicals, namely DCP as free radical initiator and 3,5-tri-2-propenyl-1,3,5-triazine-2,4,6(1H,3H,5H)-trione (TAIC) as a cross-linker agent in binary blends of PHBV and PCL [51]. When just one of these reactive chemicals was added, no significant modification could be observed in the mechanical properties of the blends. However, adding both a free-radical initiator and cross-linker contributed to an improvement in the toughness of the blend, demonstrated by the increase in impact strength and percent elongation at break. The outstanding change was also revealed in the rheological behavior of the blends, as the neat blend and blends containing a sole reactive chemical showed Newtonian behavior over a long period, while the blend containing both peroxide and TAIC indicated a strong shear-thinning behavior, implying the presence of strong interaction and chain branching in the blend. Moreover, a huge difference between the crystallization behavior of the neat blend and the compatibilized blend was demonstrated by optical microscopy, as shown in Figure 2.8. In the neat blends, irregular and imperfect spherulites of PHBV were formed, with nucleation mainly occurring at the clear interface of PHBV and PCL. However, after compatibilization with peroxide and TAIC, uniform crystals of PHBV were formed without sharp interphase boundaries and without clear interference from PCL crystals.

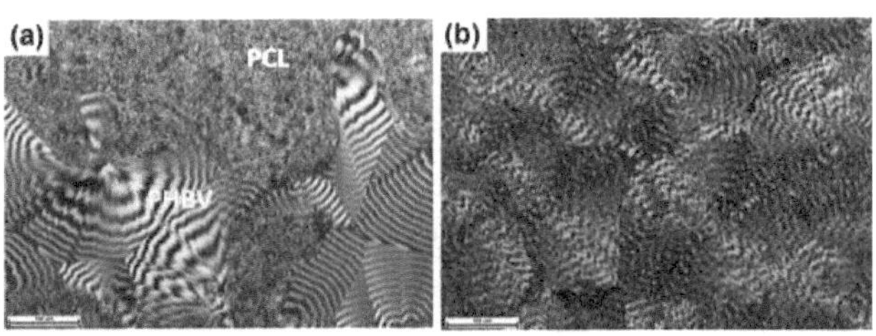

Figure 2.8 Effect of compatibilization on the crystallization of PHBV/PCL blend; (a) un-compatibilized PHBV/PCL (80/20 wt%); (b) compatibilized PHBV/PCL (80/20 wt%) blend with peroxide and TAIC [51]. Copyright American Chemical Society, 2023

Another type of compatibilizing agent of interest is the polymers grafted by reactive chemicals such as maleic anhydride on the backbone of a polymer chain. The grafted material can function as a compatibilizer by reacting with one of the components in the blend with its backbone, and reacting with the other component with its grafted reactive group. Such a function of the grafted material can enhance the interfacial interaction and reduce the interfacial tension between two polymers in the blend, contributing to the improvement in the compatibility of the final material. Kennouche and co-workers used this idea to improve the compatibility of PHBV/PBS blends [52]. In this research, they synthesized a compatibilizer by grafting maleic anhydride on the backbone of PHBV (MA-g-PHBV). Although the

FTIR spectrum of the compatibilized blends did not show any new type of interaction between components and compatibilizer, the compatibilizer revealed its effectiveness in improving the interaction between PHBV and PBS, as evidenced by SEM micrographs. The images from the microstructure clearly showed that after incorporation of only 5 wt% of MA-g-PHBV, the size of PBS dispersed particles within the PHBV matrix became finer with more homogeneous size distribution, as a result of reduced interfacial tension between two phases.

2.3 PHA-Based Biocomposites and Nanocomposites

2.3.1 Filler-Based Biocomposites and Nanocomposites

Polymer matrix composites have traditionally utilized fillers as reinforcement. They perform many tasks, including lowering the ultimate cost of the material because they are often less expensive than polymers. They might improve the thermal stability, barrier qualities, stiffness, impact resistance, and processability of the polymer matrix, among other attributes. A wide range of fillers with various chemical structures, forms, and sizes are available for strengthening polymers. Mixing PHA with PBAT, PBS, or PBSA can create a biodegradable polymer with balanced qualities that can be utilized in many applications. Some blends may not meet all requirements for some particular applications, and additional modifications are necessary to get the required attributes. For this reason, novel biocomposites might be synthesized by including appropriate fillers in these mixtures. PHBV has been the subject of extensive research within the PHA family due to its commercialization, superior thermal resistance, and good mechanical properties [53]. However, owing to the high cost of PHBV (~US$2.81/kg) compared to traditional polymers, incorporating low-cost fillers is one way to develop competitively priced materials with various applications [19].

In the preparation of biocomposites, a widely used technique named the masterbatch dilution method is performed to improve the filler dispersion in the polymer matrix. The masterbatch dilution approach involves preparing a strongly filled masterbatch and then diluting it with a neat polymer. A better dispersion results from the high shear stress of the second dilution step on the masterbatch [54]. Pal and coworkers [55] prepared a PHBV/PBAT-based composite blend incorporated with organo-modified nanoclay at different concentrations. For better exfoliation of the filler into the matrix, the masterbatch method was used to prevent agglomeration of the nanoclay, thereby improving the composite properties. The PHBV/PBAT/

nanoclay was melt-blended to form composite pellets, which were then used to develop cast-extruded and compression-molded films. Compared to compression-molded films, the cast-extruded PHBV/PBAT films exhibited superior barrier characteristics (oxygen and water vapor) due to the chain orientation that occurred during the high-speed stretching. There was an enhancement of ~70 and ~79% for water vapor and oxygen permeability, respectively, after 1.2% nanoclay inclusion for cast films, which suggests that the nanoclay has been well dispersed and that the matrix and filler have a high interfacial adhesion.

Different physical and chemical matrix-filler compatibilization can increase the mechanical properties and dispersion of fillers in the composites by enhancing the matrix-filler interfacial bonding. Among them, maleic anhydride (MA) graft copolymerization has been widely used [56]. By altering the matrix-fiber surface energy through matrix grafting, resulting in improved filer wettability and as a result, fiber dispersion has improved. In a study, Rodriguez et al. [19] developed PHBV and biobased PBS (BioPBS) composites by incorporating talc into the blends. The properties of the blends were improved by adding the compatibilizer prepared by maleation of 80% BioPBS/20% PHBV by adding 5% MA in the presence of 1% initiator. The mechanism of MA grafting into PHBV to produce MA-g-PHBV is provided in Figure 2.9. The processing conditions decompose the initiator to free radicals. These free radicals catalyze the grafting of MA molecules onto polymer chains by removing hydrogen atoms from the polymer backbone via nucleophilic reaction. Talc was also added to the mix to improve the processability and barrier properties further. The authors observed that the water vapor and oxygen properties decreased by about ~77% and ~75%, respectively, due to the synergistic effect of talc and MA grafting. Moreover, the composite with coupling agents exhibited higher impact, tensile and flexural strength for the blends. The interfacial adhesion of the composites was improved by the application of 1% MA-BioPBS/PHBV-blend coupling agents (Figure 2.10), which led to a consistent increase in the tensile strength and barrier properties.

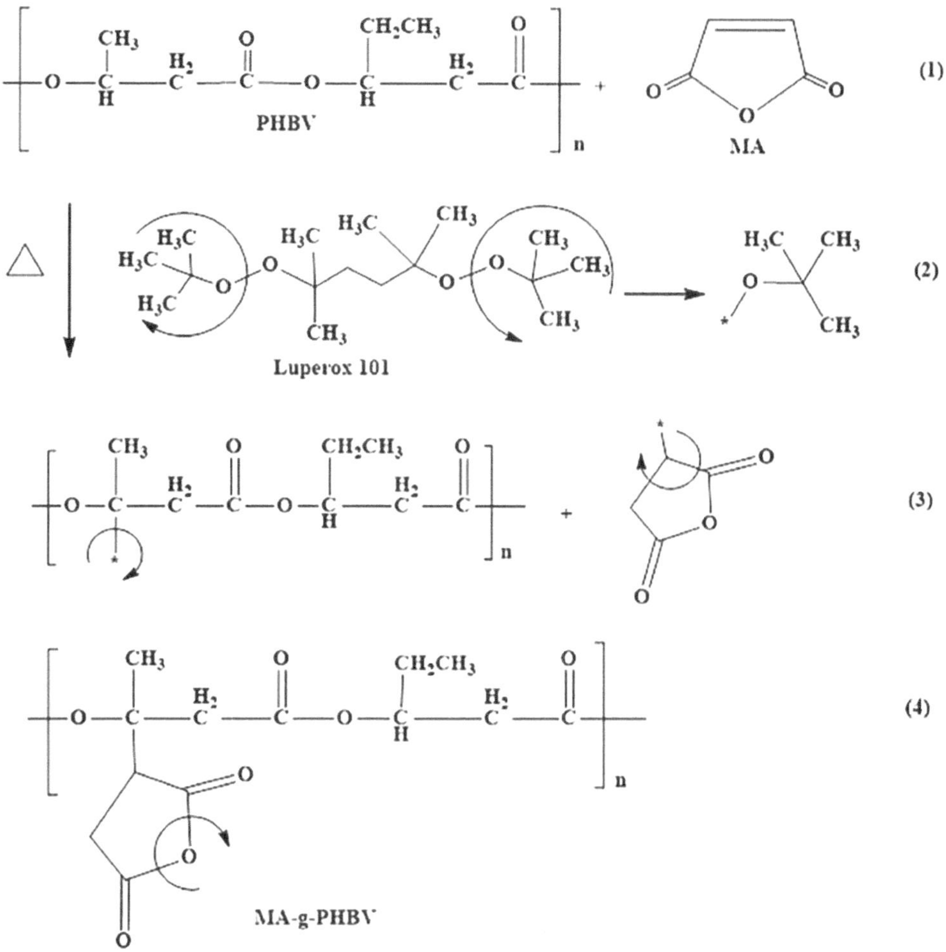

Figure 2.9 Schematic representation of MA-grafting mechanism into PHBV chain [19]. Reproduced from Elsevier, 2021, open access (License Number: 5566820330626)

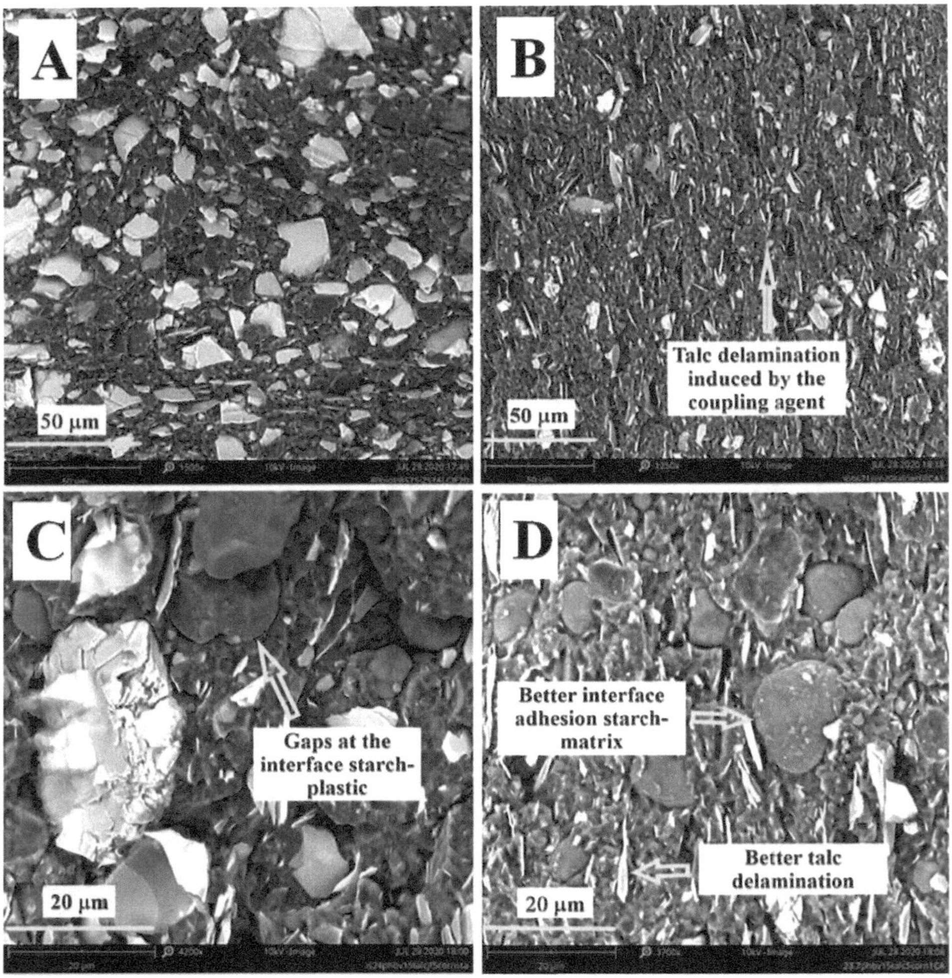

Figure 2.10 BioPBS/Talc and BioPBS/Talc/MA-g-PHBV are presented in Label A and B, respectively. The addition of maleic anhydride (MA) improved the exfoliation of talc particles, reducing the particle size to 3 µm from 6 µm [19]. After the addition of talc in the composite, a gap between the interfaces can be observed (Label C), which became visibly less after the incorporation of MA (Label D) [19]. Reproduced from Elsevier, 2021, open access (License Number: 5566820330626).

Nature-derived fillers can also be used to enhance composite properties due to their dispersibility, thermal stability, and high stiffness. Therefore, a biocomposite based on PHBV/PBAT (50/50 wt%) blend has been prepared by Pawar et al.[57], by incorporating vermiculite (1–10 wt%) or graphene nanosheets (0.1–5 wt%) by melt-compounding. The addition of fillers (especially graphene) significantly improved the modulus of the biocomposites due to the effective stress transfer ability of the fillers, with 5% graphene showing an improvement of ~30% compared to neat PHBV.

Similarly, 5% vermiculite and graphene addition enhanced the elongation at the break by 350 and 225%, respectively, compared to neat PHBV, indicating that vermiculite composites were more ductile than graphene composites. It was noticed that the filler phase was concentrated in the PBAT phase. In addition, vermiculite platelets delaminated more consistently than graphene. The unusual filler placement in the soft matrix surrounded by the hard matrix provided it with greater mechanical qualities compared to PHBV. Vermiculite-containing blends, in contrast to pure polymers and PHBV/PBAT blend and PHBV/PBAT/graphene, showed a quicker rate of enzymatic breakdown, which was attributed to the increase of the hydrophilic nature of the composite due to the presence of vermiculite.

In another study by Bittmann et al. [58], the authors used different concentrations of PHBV/PBAT and incorporated 3% organomodified montmorillonite (OMMT). Thermogravimetric study shows that the thermal stability of PHBV is improved by the addition of OMMT and 50% or more PBAT to PHBV. The addition of nanoclay to the polymer mixture delays breakdown, giving the nanocomposites marginally greater thermal stability than the pristine polymers and their corresponding blends. The percent crystallinity reduces after the blending of PBAT to PHBV (41 to 4%) and the composites PHBV/MMT (44 to 6%). However, after the addition of OMMT to PHBV/PBAT blends, they exhibited a higher degree of crystallinity throughout each composition investigated. This is because nanoclay fillers serve as nucleation locations and encourage the growth of more polymer crystals compared to neat polymer. However, the presence of clay agglomerates simultaneously causes restrictions on the motions of polymer chains, delaying crystallization, and this delay results in the formation of smaller and less perfect crystals.

Antimicrobial properties are of immense importance in biocomposite food packaging films, since they can prevent food contamination and microbial growth. Food spoilage is a major issue affecting food safety, and hence certain additives (antimicrobials and antioxidants) are incorporated into the packaging material to enhance the shelf life of the packaged products [59]. In this work, the effects of mixing PHBV/PBAT and adding functionalized montmorillonite (OMMT) to the blends are examined in relation to changes in morphology and thermal properties changes due to various environmental factors such as weathering, antimicrobial attack, and water absorption [60] Antimicrobial properties were imparted to the composites by the addition of industrial antimicrobial agents (Irgaguard B1315 and F300) and propolis agents (Propolis succumb extract and 5% SiO_2 added propolis powder). Studies on thermal behavior following water absorption indicate that the thermal stability of all samples significantly decreases after moisture absorption. Materials that contain the industrial antibacterial ingredient show the greatest decline. This drop is thought to be the result of chain scission brought on by hydrolysis. Moreover, the antimicrobial test against *Staphylococcus aureus* and *Escherichia coli* after 24 h of exposure revealed that composites with propolis did not exhibit antimicrobial activity. One possible reason for the inactivity of propolis agents might be due to their encapsulation in the blend matrix suppressing its activity. On the other hand, the antibacterial action was exhibited by the industrial antimicrobial additive, and a halo was observed near the bacterial growth, showing the efficiency of antimicrobial additives in bacterial growth.

2.3.2 Lignocellulosic Filler- and Fiber-Reinforced Biocomposites

Lignocellulosic fibers are well-known among industrial manufacturers because of their adaptability, extensive availability, and contribution to the idea of environmental sustainability. Biocomposites are made by incorporating lignocellulosic fiber with different polymer compositions [61]. These renewable fillers are typically inexpensive and abundant and are produced primarily from agricultural residues and trash. Lignocellulosic fibers and fillers are derived from various plant parts (root, seed, leaf). The classification of natural fillers is presented in Figure 2.11. These materials are hierarchically structured natural composites. Due to their availability and environmental friendliness, lignocellulosic fiber composite manufacturing increases throughout such applications. In addition, lignocellulosic fiber composites have other benefits, such as good formability, low cost, and renewability. Lignocellulosic fibers include cellulose as their primary cell structure, and they also contain hemicelluloses, pectin, lignin, and water-soluble materials [61]. As lignocellulosic fillers and fibers are renewable, inexpensive, biodegradable, and less corrosive to processing machinery than synthetic fibers, they have lately attracted much interest in formulating biocomposites. They also have great specific mechanical qualities due to their low density and excellent mechanical properties.

Despite their potential to improve biodegradable polymer performance, hydrophilicity complicates use of these fibers in a polymer matrix. Since most biodegradable polymers are hydrophobic, inadequate fiber – matrix adhesion in biocomposites can lead to inefficient stress transfer between constituents and, in turn, a deterioration of biocomposite mechanical properties. Another commonly used PHA, named PHB, is a semi-crystalline, biodegradable and biocompatible polymer produced by bacterial fermentation of sugarcane [62]. PHB has properties similar to a polymer such as polypropylene but its high cost, brittleness, low impact resistance, and poor thermal stability after processing limit its use. Therefore, in a study by Beber et al.[62], the authors developed PHB/PBAT-based biocomposites at different polymer concentrations by incorporating 20 wt% babassus natural filler. The babassu palm trees are endemic to the northern, northeastern, and central parts of Brazil. Babassu filler inclusion decreased the strength and elongation at the break of the blends by a maximum of 32.3% and 66.3%, respectively. The inferior mechanical performance of after-filler inclusion might be attributable to the fact that babassu filler absorbed water before/during processing. Water aging induces a noticeable decrease in tensile characteristics due mostly to fiber degradation-induced damage and fiber-matrix interface deterioration. Hence, this property deterioration after fiber addition calls for the use of a compatibilizer to improve the interaction between the matrix and the filler.

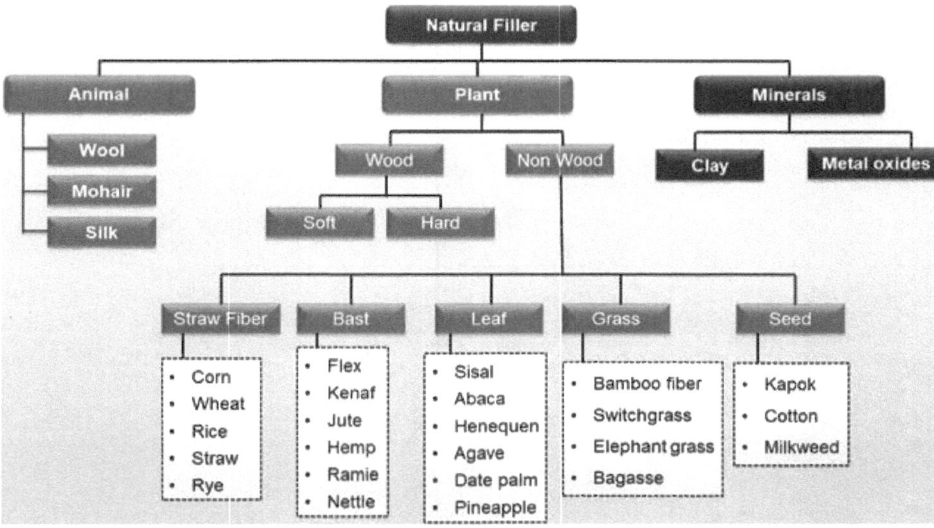

Figure 2.11 Classification of natural fibers

Hence, many studies have been devoted to enhancing the compatibility of the filler and matrix and among them. Surface treatment of fibers and compatibilizers in biocomposites are two main methods for resolving this issue. Therefore, Nagarajan et al., in a study [63], developed a PHBV/PBAT biodegradable matrix system loaded with 20–40 wt% switchgrass. Poly diphenylmethane diisocyanate (PMDI), a compatibilizer, was added at 0.1–1 phr to the matrix to address the hydrophobic – hydrophilic imbalance between the matrix and fiber. The authors reported that compatibilized PHBV/PBAT/switchgrass composites significantly outperformed their virgin counterparts in terms of mechanical properties. Chemically, a urethane-linked intermediate is formed when the hydroxyl group of the fibers reacts with the isocyanate functional group of PMDI. The carbonyl group reaction from the above reaction has two lone pairs. One lone pair form a hydrogen bonding with an intermediate, while the other lone pair of electrons may bond with the hydroxyl group on the lignocellulosic fibers (Figure 2.12). The enhanced interfacial adhesion between the matrix and the fiber in the compatible composites was further supported by the SEM micrographs (Figure 2.13).

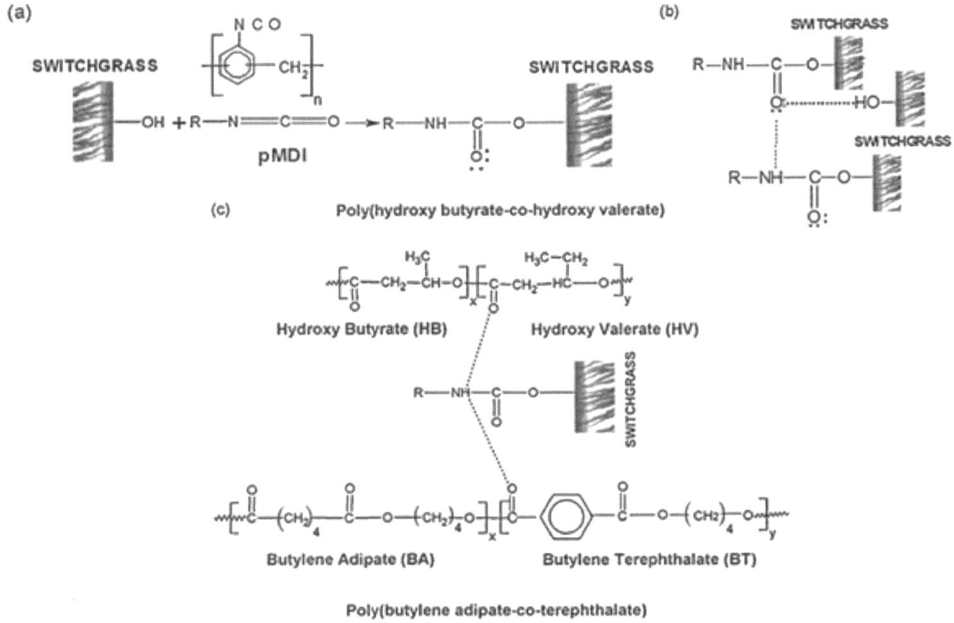

Figure 2.12 Mechanism of compatibilization between PHBV, PBAT, and switchgrass in the presence of PMDI [63], Copyright Elsevier, 2013 (License Number: 5566820701389)

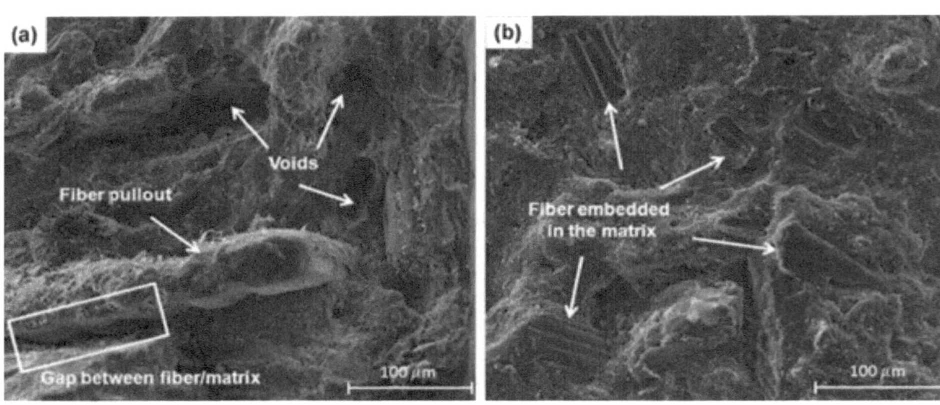

Figure 2.13 SEM images showing (a) PHBV/PBAT/switchgrass (b) PMDI compatibilized PHBV/PBAT/switchgrass composites. Sample (b) exhibited better interfacial adhesion compared to non-compatibilized composites due to the presence of PMDI [63]. Copyright Elsevier, 2013 (License Number: 5566820701389)

In another study by Zarrinbakhsh and coworkers [64], the authors successfully produced a green biocomposite of PBHV/PBS and distillers' dried grains with solubles (DDGS) using PMDI as a compatibilizer. The interaction between PHBV, PBS, PMDI, and DDGS is shown in Figure 2.14. DDGS is a low-cost by-product of the corn ethanol business, which is produced when corn is converted into about equal amounts of ethanol, CO_2, and DDGS. The DDGS was used as received and also after treatment. The DDGS treatment included washing 100 g of DDGS in a 1.5 L beaker followed by filtering through a filter paper. PHBV/PBS and DDGS were compounded in the presence of PMDI. The deterioration onset temperature of DDGS increased noticeably from 140 to 235 °C after surface treatment of DDGS, according to thermogravimetric analysis. The green composite with treated and compatibilized DDGS demonstrated better heat deflection temperature (by ~21%) and increased modulus (~28%). This was attributed to the better interfacial adhesion between the polymer matrix and DDGS by using PMDI as a compatibilizer, which was also confirmed by the SEM images.

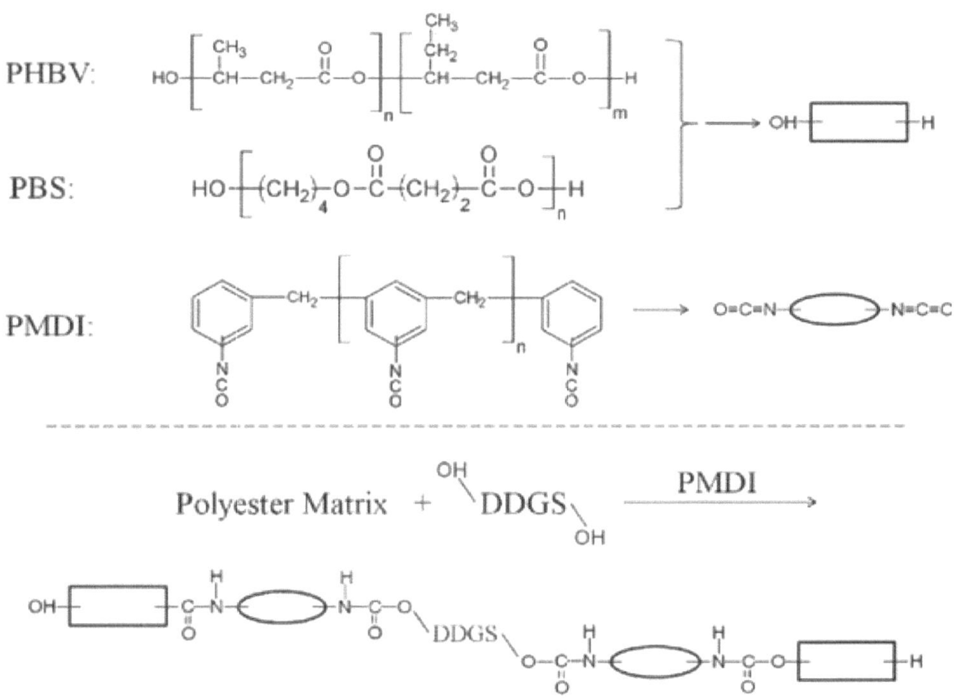

Figure 2.14 Diagram depicting the hypothesized reactions between the PHBV, PBS, PMDI, and DDGS [64], Copyright Wiley, 2011 (License Number: 5566821020259)

The findings of this study indicate that DDGS has the potential to be used successfully to develop a kind of biodegradable green composite that may be used for applications requiring more stiffness and better temperature performance. However, the poor impact and elongation at break may be a limiting factor for its use in real-world applications. Hence, a high-impact DDGS-based biocomposite was also developed from commercially available premixed PHBV/PBAT (55/45) polymer blend by adding PMDI as a compatibilizer and commercial corn oil as a processing aid, and their impact on the biocomposite properties was also analyzed by Zarrinbakhsh et al. [65]. The presence of compatibilizer and lubricant promoted cross-linking in the biopolymer matrix resulting in improved modulus, strength, and elongation at break. Moreover, after DDGS inclusion, the dynamic mechanical analysis demonstrated an increment in the storage modulus of the polymer matrix. The storage modulus of the biocomposites increased and decreased with further additions of PMDI and corn oil, respectively. In PMDI-compatibilized biocomposites, PBAT and PHBV, T_g values shifted towards each other, suggesting compatibilization and cross-linking at the interface. SEM images of the DDGS-matrix interface and the PHBV/PBAT interface within the matrix itself both showed increased interfacial adhesion. In addition, the impact strength of the biocomposite was significantly enhanced by ~184% because of the synergistic interactions of 1% PMDI and 3% corn oil.

In this study, Sarasini et al. [66] looked at the potential of using coffee silverskin (CSS), one of the most common coffee byproducts, as a reinforcing agent in PHBV/PBAT blends, and used compatibilizers to improve the adhesion between the components. The CSS fiber was incorporated into the polymer blend matrix at 10–30 wt%, either as-received or after extraction of antioxidant, and in the presence of two compatibilizers, namely melanized linseed oil and (3-aminopropyl)triethoxysilane (silane), and their impact was also studied. The obtained findings verified that extracted CSS and silane-CSS offered the optimal balance of strength and ductility. Silane increased interfacial adhesion, allowing for effective stress transmission between the polymer and the CSS filler resulting in filler breakage. The improved filler wettability and dispersion obtained by compatibilization may have contributed to the higher ductility by lowering the presence of filler clusters that may serve as stress concentration locations. In contrast, melanized linseed oil only had a moderate compatibilization impact on CSS because the antioxidant extraction process reduced the number of hydroxyl groups on CSS, indicating that the silane treatment had more of an impact than the plasticizer inclusion.

In a study by Gupta et al., PHBV was incorporated with PBAT and hemp powder residue as a functional additive to improve the cost structure while also optimizing the characteristics of PHBV [67]. The compatibility between the biocomposite constituents was improved by the maleation of PBAT. Maleation of PBAT (m-PBAT) was achieved by the radical-induced interaction of PBAT with maleic anhydride (MA) during a reactive extrusion process. The authors found that the biocomposite consisting of 60 wt% hemp powder with PHBV/PBAT (80/20) has similar HDT, impact strength, and tensile properties to pristine PHBV and is cost-effective over other traditional plastics. Hence the presence of m-PBAT enhanced the interfacial interaction between the PHBV/PBAT and hemp powder. The increased tensile characteristics and other properties were a result of the chemical grafting of hemp powder with m-PBAT, where it was distributed effectively in the PHBV/PBAT matrix. After 60 days, compost completely disintegrated the PHBV sample, while the optimized sample started disappearing after 75 days. Due to the chemical bonding caused by the integration of m-PBAT, the biocomposite specimens in compost may have a significantly longer composability timeline.

The PHBV/PBAT blend with fiber can be further toughened by the addition of a commercial elastomer through reactive blending, as demonstrated by Zhang et al. [53]. In the study, a ternary blend of PHBV/PBAT and epoxidized natural rubber (ENR) was designed as a high-toughening matrix for miscanthus fiber-based composites in the presence of reactive compatibilizer DCP (0.3 wt%). The DCP acted as a reaction initiator, causing a free-radical reaction between the phases that resulted in the formation of the grafting co-polymer and a partly cross-linked network between both phases, greatly improving the compatibility of the polyester blends. When combined with DCP, ENR served as both a coupling agent and an impact modifier in the multiphase blends and composites. The natural fiber was added at 10–20 wt%, and their properties were analyzed. Because of the incorporation of miscanthus fibers into the matrix, its stiffness, heat resistance, and toughness were greatly improved. Even after the addition of 20 wt% miscanthus, the maximum impact strength of the material was still 240.5 J/m due to the presence of DCP. High impact resistance in these composites is likely due to the strong bond between the matrix and the fiber, further supported by scanning electron microscopy and thermomechanical findings. An overview on the research conducted on PHBV-based biocomposites in the presence/absence of compatibilizer are shown in Table 2.2.

Table 2.2 Overview of Studies on Biocomposites of PHBV Blends with Fillers and Lignocellulosic Fibers

Fillers/ fibers used	Polymer matrix	Compatibilizer	Impact of additives on biocomposite properties	Ref.
Miscanthus	PHBV/ PBAT/ epoxidized natural rubber	Dicumyl peroxide	▪ Enhancement in stiffness and thermal resistance ▪ Better interfacial adhesion between the phases in the biocomposite	[53]
Hemp	PHBV/PBAT	Maleic anhydride grafted PHBV	▪ Comparable heat deflection temperature, tensile properties and impact strength to pristine PHBV	[67]
Switchgrass	PHBV/PBAT	Diphenylmethane diisocyanate	▪ Improvement in strength, modulus and heat deflection temperature of the biocomposites ▪ Remarkable adhesion between matrix and fiber	[63]

Table 2.2 Overview of Studies on Biocomposites of PHBV Blends with Fillers and Lignocellulosic Fibers *(continuation)*

Fillers/ fibers used	Polymer matrix	Compatibilizer	Impact of additives on biocomposite properties	Ref.
Babassu mesocarp	PHB/PBAT	-	▪ Tensile strength and elongation reduced after filler addition	[62]
Halloysite	PHBV/PBS	Maleic anhydride grafted PHBV	▪ Enhanced thermal stability and fire reaction compared to neat PHBV ▪ Improved dispersion of PBS nodules in the PHBV matrix	[52]
Talc	PHBV/PBS	Maleic anhydride grafted PHBV	▪ Improved tensile and flexural strength ▪ Enhancement in barrier properties ▪ Enhance interfacial adhesion between the composite's constituents	[19]
Vermiculite, graphene nanosheets	PHBV/PBAT	-	▪ Tensile modulus and strain at break are significantly increased compared to neat PHBV ▪ Vermiculite incorporated blend exhibited higher biodegradation rate (~ 1500%) after 8 days	[57]
Functionalized montmorillonite	PHBV/PBAT	-	▪ Thermal stability of the PHBV increased ▪ The degree of crystallinity increased after filler incorporation ▪ Clay and clay agglomerates restrict polymer chain mobility, slowing crystallization.	[57]

■ 2.4 Biodegradation of PHA-Based Materials

Considerable amounts of discarded plastic materials are present as debris in landfills and natural habitats because of the durability of the polymers involved. Considering our diminishing fossil fuel reserves and limited capacity for waste disposal to landfills, this linear consumption of hydrocarbons in packaging and other short-lived applications of plastic is not sustainable. Various solutions exist to shift the paradigm towards sustainability, such as reduction in material usage, recyclability after end-of-life, improved recycling capacity, advancement of bio-based reservoirs, policies to minimize littering, the implementation of green chemistry life-cycle analyses, revised risk assessment tactics, and most importantly, industrial-scale biodegradation of used plastics in various environments such as composting, soil biodegradation, and marine biodegradation [68]. Such actions will be successful only after the collective activities of researchers, industrialists, policymakers, and the public.

Alternative polymers, advertised as biodegradable, show a potential solution to reduce this environmental damage. Though we consider these deliberations supportive and forward-oriented, we documented several recently published research articles that came under insufficient knowledge of plastic biodegradation principles. Such types of published research articles can be divided into two classes. The first class comprises publications that claim to evaluate the percent biodegradation of plastics without providing essential experimental data (e.g., data on carbon dioxide produced during microbial integration of available carbon in plastics). The second class comprises publications that raise questions on the biodegradation of certified biodegradable polymers on the basis of insufficient data on these polymers in modified environments other than the mentioned ASTM standards. The researchers are concerned that such publications could lead to misconceptions and misdirect the conclusive argument on the prominent part of biodegradable polymers in reducing plastic pollution, and eventually adversely impact the research development on biodegradable plastics [69].

It is well understood that the environment is not a dumping ground for generated plastic waste. On the other hand, the development of biodegradable polymers has not been performed with the ultimate aim of dumping the polymer into the environment after use. However, plastic wastes can be handled beneficially by the property of biodegradability, which originates from individual industrial sectors such as food packaging. Packaging is considered one of the most significant sectors for both cases, i.e., using plastics and generating waste [70]. Polymers consumed in packaging are intended to disintegrate/biodegrade in various environments, such as industrial composts and anaerobic digesters. In that case, the compostable plastics consumed for various applications can be redirected from landfills at their end-of-life and open dumps to handle composting systems. Likewise, we have confidence in the utilization of biodegradable plastics instead of conventional plastics, an environmentally supportive solution for certain applications that need the use of plastics directly in the environment.

Because of the dependence of biodegradation on polymer structure and environmental conditions, a broad range of trials is essential to fully describe the biodegradability of a given polymer [71]. International organizations have defined standardized testing procedures to ensure a global standard of plastic biodegradability.

The American Society for Testing and Materials (ASTM) and the International Organization for Standardization (ISO) define and regulate several standard tests used to assess biodegradable polymers and encompass various environmental factors adequately. The factors, including temperature, pH level, humidity, oxygen, and microorganisms (in compost, soil, and seawater), are mainly responsible for the biodegradation of these polymers. The characteristics of the biodegradable polymer and the biodegradability of the consequent composites can be altered in comparison to the basic polymer by the inclusion of natural fiber. Just as with pure polymer materials, the actual biodegradability of polymer composites (in terms of rate of degradation and toxicity for safe disposal in environmental conditions) must be evaluated using specific standardized testing methods. These tests include a wide scope of conditions to covering all types of plastics and the several environmental settings they could be exposed to as plastic waste. Thus, with incessant development and revolution in the biopolymer industry, these standard test methods for evaluating the biodegradability of polymeric materials are critical to the plastics industry's future, and will be essential to diminish the global problem of plastic pollution.

2.4.1 Biodegradation Studies on PHA-Based Blends and Composites

PHA has several useful properties as a biodegradable polymer, which include photo-activation to promote pollutant oxidation, complete disintegration within a short time, and a biodegradation path similar to starch (90% of the microbes that can break down starch can also break down short-chain PHAs). There are two types of bacteria involved in biodegradation: those that physically break down PHAs and those that consume the by-products of the degradation (valeric acid, butyric acid, etc.) [72]. Overall, many different types of bacteria are responsible for the biodegradation of short and medium PHAs. These bacteria come from the following genera: *Corynebacterium*, *Enterobacter*, *Mycobacterium*, *Nocardia*, *Gracilibacillus*, *Klebsiella*, *Micrococcus*, *Actinomyces*, *Alcaligenes*, *Arthrobacter*, *Aspergillus*, *Bacillus*, *Clostridium*, *Comamonas*, and others [72]. Fungi are capable of degrading PHA and have been shown to be even more efficient degraders of PHA than bacteria. In aerobic and anaerobic settings, *Ascomycota*, *Basidiomycetes*, *Deuteromycetes*, and *Zygomycotina* from the Kingdom of fungi degrade PHA effectively [72]. Therefore, it can be concluded that PHA can be biodegraded in both aerobic and anaerobic settings, with several studies performed that evaluate the biodegradability and composability of PHA in marine and soil environments.

2.4.1.1 Biodegradation Studies on PHAs and Their Blends

In a study by Muniyasamy et al. [73], the authors evaluated the biodegradability of PHBV by partially following ASTM D5338-15 (industrial composting) and ASTM D5338-12 (soil burial test) standards. Polymers with a heterochain backbone, like PHBV, degrade chemically by hydrolysis or enzyme-catalyzed hydrolysis. Hydrolytic enzymes released by microorganisms are crucial to hydrolytic biodegradation because they influence the breakdown of polymer chains with a high molecular weight into oligomers and monomers with a low molecular weight. Based on these facts, typical biodegradation tests measure CO_2 emitted from finished polymer. Direct proof of the degree and pace of biodegradation may be obtained by measuring the released CO_2 and comparing it to the theoretical amount based on the carbon content in the original plastic sample [73]. CO_2 emissions throughout the test show that the material has joined the natural carbon cycle rather than merely decomposing into biostable tiny pieces that are hard to identify. Therefore, upon testing PHBV, it exhibited a biodegradation rate of almost 90% under composting conditions, while under soil burial test, the degradation rate of PHBV was 35%. The authors concluded that the most influential factors for the eventual biodegradation of these biobased polymers were the various environmental conditions in soil and compost.

Similarly, another study by Arauz and co-workers [74] measured the biodegradability of PHA when subjected to soil biodegradation following ASTM D5988-03. The authors compared the kinetics of carbon dioxide generation between PHA and control PHB samples and observed that PHA degraded more slowly than PHB. The lag phase, which indicates a delay in the start of biodegradation, was much higher for PHA by approximately 8 days. Nonetheless, the calculated CO_2 generation maximum values for both polymers were comparable. The observed delay in the onset of biodegradation for PHA may be attributed to variations in the specificity of the enzymes involved in biodegradation or the availability of carbon as a result of surface structure changes in the samples. The authors finally concluded that the novel PHA film exhibited a weight loss of 75% after 80 days. The biodegradability of PHBV/PCL mulch films was developed by Othman et al. [75] and their biodegradability was evaluated by the soil burial test in accordance with the European standard EN 17033:2018. The PHBV/PCL films exhibited a weight loss of around 30% in 5 weeks, which testifies to the high biodegradability of the developed mulch films. Moreover, the neat PHBV and neat PCL films demonstrated a weight loss of 45 and 40%, respectively, in 5 weeks which is higher than for PHBV/PCL film. The probable reason for this behavior might be due to the increase of molecular weight of the mulch film because of the mixed ratio of PHBV and PCL, which slowed down the degradation rate. In another study by Kim et al. [76], the authors synthesized poly(3-hydroxybutyrate) using marine yeast (*Pichia* sp. TSLS24). The developed PHA exhibited outstanding biodegradability after 28 days, with a biodegradability of 28%. The findings showed that PHA generated by the strain TSLS24 might be easily broken down by the environment's indigenous bacteria in a saline environment, though a lot of other factors such as high pressure, salinity level, low temperature, and currents also affect this process.

2.4.1.2 Biodegradation Studies on PHA-Based Biocomposites

Studies on the biodegradation of biocomposites that adhere to any ASTM standard are frequently contrasted with studies on virgin polymers. Natural fillers in PHA-based biocomposites have the potential to lower the cost of PHAs and improve the water diffusion rate and maximize water absorption while promoting their biodegradation [77]. Several studies have demonstrated that adding fiber improves the biodegradability of PHBV [78]. In a study by Muniyasami and coworkers [79], PHBV was incorporated with flax fiber, and their biodegradability was tested by the CO_2 evolution method under composting conditions. Moreover, the flax fiber was treated with alginic acid to improve the thermo-mechanical properties and stability of the composite system. From the biodegradation study, it was observed that the PHBV/flax-alginic acid had the highest biodegradation of 88%, followed by PHBV/flax (85%) and PHBV (63%) in 100 days. Also, the alginic acid-treated composite had the earliest exponential phase, resulting in faster degradation compared to other samples. The main factor favoring the rapid biodegradation of PHBV biocomposites could be the poor dimensional stability and high moisture absorption of flax fibers. Additionally, PHBV/flax-alginic acid degraded more quickly than expected because of the properties of alginic acid, which is water-soluble and easily degradable. Similar to the previous study, PHBV was incorporated with palm fiber and silane-treated palm fiber, and their biodegradability was tested in a soil burial test [80]. The weight loss during the tested period indicated that degradation increased at higher palm fiber and silane-palm fiber content. At 40% fiber content, the composite degraded by ~85% over 60 days, which was the highest among the tested samples. It is widely known that increased water absorption can expand the region vulnerable to microbial attack, which in turn causes ester groups on the PHA backbone to hydrolyze and speed up biodegradation.

While untoughened PHBV/natural fiber composites have been studied for their biodegradability, there is surprisingly little written about the biodegradability of toughened PHBV composites. In a study by Zaidi et al. [78], a soil burial test was performed for neat PHBV, PHBV/PBAT/flax, and PHBV/ENR/flax biocomposites. Under the testing conditions. PHBV lost a maximum of 0.5% of its mass due to biodegradation throughout the course of 112 days. However, for the composites, the biodegradation rate was much higher, with toughened composites exhibiting even better results. After 112 days, the PHBV/ENR/flax, PHBV/PBAT/flax, and PHBV/flax had degraded by 17%, 9%, and 6%, respectively. By comparing ENR- and PBAT-toughened samples, filamentous formations were more common and intricate on the former. The filamentous formations on ENR-toughened samples may have been caused by actinomycetes, an order of actinobacteria that attacks natural rubber, which might be the reason for the faster biodegradation of PHBV/ENR/flax samples.

PHAs are among the most widely recognized marine-biodegradable biopolymers, making them ideal for minimizing the long-term effects of ocean plastic trash. In a study by Meereboer et al. [81], PHBV was incorporated with miscanthus and DDGS to determine if proteinaceous fillers and natural fibers can increase the marine biodegradability of PHA. The ASTM standard D7991-15 was used for this study for neat PHBV, PHBV/miscanthus, and PHBV/DDGS. The proteinaceous PHBV/DDGS exhibited a maximum biodegradation of ~40% compared to neat PHBV; moreover, the PHBV/DDGS and PHBV/miscanthus were biodegraded in 295 and 412 days, respectively, indicating that DDGS has better biodegradability than miscanthus. It has been hypothesized that due to the presence of lignin in miscanthus, it is more challenging to bio-assimilate, except in a fermentative environment. Moreover, protein offers a good nitrogen supply that may have already undergone some breakdown due to composite processing and may be utilized for local cell growth and development that promotes DDGS biodegradation and assimilation.

2.5 Commercial Applications of PHAs

Among the several bioplastics used to combat plastic pollution, PHAs have rapidly gained popularity for commercial applications. Nevertheless, research and development is only one stage in the product development process, and when theoretical scientific knowledge is put into practical use, only then does an actual product emerge [82]. Consequently, the manufacturing process brings a new product to the market. Shopping bags, rubbish bags, cutlery, mulch film, etc., are some of the numerous goods developed with bioplastics. Several companies have started producing and developing bioplastic and its products on a large scale to fill the void left by the ban on plastic by different governments.

PHA is made from renewable materials and is biocompatible and biodegradable, and has applications in packaging, medical devices, and drug delivery. Many new industries have been formed at the pilot and industrial scales to drive the widespread industrialization of PHAs. China is home to seven PHA manufacturers, the oldest of which, TianAn, has operated for almost two decades. Recently established companies like PhaBuilder, Medpha, COFCO, and Bluepha manufacture low-cost PHA. GreenBio and Ecomann are other companies established in China [83]. North American countries such as US and Canada have also established industries for PHA production. In the US, Fullcycle and Newlight are the companies that use mixed organic waste and greenhouse gas for inexpensive manufacturing of PHAs.

In contrast, Danimer and RWDC used recombinant *Ralstonia eutropha* cultured in fatty acids for PHAs production [83]. Similarly, Bosk and TerraVerdae Bioworks, established in Canada, develop PHA-based resins for various applications. Other PHA manufacturers are exploring novel approaches to using various low-cost waste substrates, forest waste, kitchen waste, wood and paper scraps, shale gas, and agricultural by-products.

Due to the improved manufacturing capabilities for PHAs and other bioplastic resins, bioplastic-based products are becoming more common in the market nowadays, and they are reasonably priced, making them attractive to consumers as well. Commercial examples of PHA-based agricultural mulch are Nodax and Mirel [84]. Nodax would decompose anaerobically, making it suitable for coating urea fertilizers as well as insecticides and herbicides to be used in rice fields [85]. Other forms of Nodax include films, latex, fibers or nonwovens, and foams. Hygiene wipes and tampon applicators, among other flushable that might break down in septic systems, can be made with this polymer. Moreover, they may be used to create packaging, biodegradable bags, lids, carpets, and surgical and medical apparel [85]. The Food and Drug Administration (FDA) has given its approval to a PHA co-polymer, "Metabolix", which is a combination of PHB and P(3HB-co-3HO).

BIOPOL, a product of Metabolix, is used for coating paper and paperboards, and due to its antistatic properties, can also be used for electronic packaging. BIOPOL fiber makes fishing nets, crab cages, and nets for cultivation [85]. It is used to create packaging materials with similar qualities to petrochemical-based polymers [85, 86]. Furthermore, PHA may be utilized in the production of diapers. Biofilms used as diaper back sheets made of P(3HB) or P(3HB-co-3HV) are described in two US patents, No. 4826493 and No. 4880592 [86]. DegraPol is a urethane product developed from P(3HB)-diol and α,ω-dihydroxy-poly(ε-caprolactone-block-diethylenegly-col-block-ε-caprolactone). DegraPol also demonstrated interesting applications in bone-healing methods like chondrocyte transplantation and autologous osteoblast [85].

Bioplastic blend products are also widely popular for packaging applications due to their increasing manufacturing potential resulting in cost-effectiveness and a variety of products. TIPA, a company producing compostable packaging films, was established to tackle the environmental issue that plastic packaging poses and provides solutions for food and fashion industries which require a high volume of packaging material [87]. BioPak is an Australia-based company that also manufactures a wide variety of compostable containers and lids, cutlery, packaging materials, etc. [88]. Mater-Bi offers bioplastic blends that can be used in different applications, from mulch films to carrier bags. These bioplastic-based products can offer solutions to everyday products with reduced environmental impact.

2.6 Conclusion

The unrestricted usage of traditional plastics has caused severe environmental issues caused by plastic wastes and microplastic formation. The use of biopolymers is one of the options to prevent further aggravation of these issues. Therefore, this chapter has described the work done on blending biodegradable polymers like PHA with other bioplastics in order to overcome the various shortcomings of PHA that inhibit its commercialization. Research in this area has demonstrated that combining biopolymers has paved the way for a novel material with the required balancing qualities. The addition of filler materials to these blends is also an approach to developing biocomposites. Natural fibers are of primary interest due to their functionalization potential, improved fiber – matrix adhesion, and enhanced modulus and barrier properties. They can also be used in controlling waste generation because of their biodegradability and ability to reduce production costs for biocomposites. The sustainability and innovation of polymers and composites made from renewable resources depend on research efforts, particularly in the areas of surface modifications, compatibilizing methodologies, and advanced processing techniques. Nonetheless, there is still more work to be done in the development of renewable resources-based polymeric blends and composites that are able to replace traditional plastics.

References for Chapter 2

[1] S. Kakadellis and Z. M. Harris, "Don't scrap the waste: The need for broader system boundaries in bioplastic food packaging life-cycle assessment – A critical review," *J Clean Prod*, vol. 274, pp. 122831, Nov. 2020, doi: 10.1016/J.JCLEPRO.2020.122831

[2] G. Fredi and A. Dorigato, "Recycling of bioplastic waste: A review", *Adv Ind Eng Polym Res*, vol. 4, no. 3, pp. 159–177, Jul. 2021, doi: 10.1016/J.AIEPR.2021.06.006

[3] J. G. Rosenboom, R. Langer, and G. Traverso, "Bioplastics for a circular economy", *Nat Rev Mater*, vol. 7, no. 2, pp. 117–137, Jan. 2022, doi: 10.1038/s41578-021-00407-8

[4] R. Geyer, J. R. Jambeck, and K. L. Law, "Production, use, and fate of all plastics ever made", *Sci Adv*, vol. 3, no. 7, Jul. 2017, doi: 10.1126/sciadv.1700782

[5] K. D. Cox, G. A. Covernton, H. L. Davies, J. F. Dower, F. Juanes, and S. E. Dudas, "Human consumption of microplastics", *Environ Sci Technol*, vol. 53, no. 12, pp. 7068–7074, Jun. 2019, doi: 10.1021/ACS.EST.9B01517

[6] G. Guidotti, M. Soccio, N. Lotti, M. Gazzano, V. Siracusa, and A. Munari, "Poly(propylene 2,5-thiophenedicarboxylate) vs. poly(propylene 2,5-furandicarboxylate): Two examples of high gas barrier bio-based polyesters", *Polymers*, vol. 10, no. 7, pp. 785, Jul. 2018, doi: 10.3390/POLYM10070785

[7] F. M. Lamberti, L. A. Román-Ramírez, and J. Wood, "Recycling of bioplastics: Routes and benefits", *J Polym Environ*, vol. 28, no. 10, pp. 2551–2571, Jun. 2020, doi: 10.1007/S10924-020-01795-8

[8] E. White, R. Bassilakis, and S. Nogués, "The bioplastics innovation landscape, players and market opportunities – From the plastics present to a sustainable future", Clarivate, 2020

[9] Y. Zhu, C. Romain, and C. K. Williams, "Sustainable polymers from renewable resources", *Nature*, vol. 540, no. 7633, pp. 354–362, Dec. 2016, doi: 10.1038/nature21001

[10] B. European, "Bioplastics facts and figures", *European Bioplastics Association*, 2019

[11] Y.M. Corre, S. Bruzaud, J.L. Audic, and Y. Grohens, "Morphology and functional properties of commercial polyhydroxyalkanoates: A comprehensive and comparative study", *Polym Test*, vol. 31, no. 2, pp. 226–235, Apr. 2012, doi: 10.1016/J.POLYMERTESTING.2011.11.002

[12] O. Platnieks, S. Gaidukovs, V. Kumar Thakur, A. Barkane, and S. Beluns, "Bio-based poly (butylene succinate): Recent progress, challenges and future opportunities", *Eur Polym J*, vol. 161, pp. 110855, Dec. 2021, doi: 10.1016/J.EURPOLYMJ.2021.110855

[13] E. Castro-Aguirre, F. Iñiguez-Franco, H. Samsudin, X. Fang, and R. Auras, "Poly(lactic acid) – Mass production, processing, industrial applications, and end of life", *Adv Drug Deliv Rev*, vol. 107, pp. 333–366, Dec. 2016, doi: 10.1016/J.ADDR.2016.03.010

[14] A. Surendren, A.K. Mohanty, Q. Liu, and M. Misra, "A review of biodegradable thermoplastic starches, their blends and composites: recent developments and opportunities for single-use plastic packaging alternatives", *Green Chemistry*, vol. 24, no. 22, pp. 8606–8636, Nov. 2022, doi: 10.1039/D2GC02169B

[15] V. Siracusa and I. Blanco, "Bio-polyethylene (Bio-PE), bio-polypropylene (Bio-PP) and bio-poly(ethylene terephthalate) (Bio-PET): Recent developments in bio-based polymers analogous to petroleum-derived ones for packaging and engineering applications", *Polymers*, vol. 12, no. 8, pp. 1641, Jul. 2020, doi: 10.3390/POLYM12081641

[16] V. Kumar, R. Sehgal, and R. Gupta, "Blends and composites of polyhydroxyalkanoates (PHAs) and their applications", *Eur Polym J*, vol. 161, pp. 110824, Dec. 2021, doi: 10.1016/J.EURPOLYMJ.2021.110824

[17] G.Y.A. Tan, C.L. Chen, L. Li, L. Ge, L. Wang, I.M.N. Razaad, Y. Li, L. Zhao, Y. Mo, and J.Y. Wang, "Start a research on biopolymer polyhydroxyalkanoate (PHA): A review", *Polymers*, vol. 6, no. 3, pp. 706–754, Mar. 2014, doi: 10.3390/POLYM6030706

[18] E. Rudnik, "Biodegradability Testing of Compostable Polymer Materials", *Handbook of Biopolymers and Biodegradable Plastics: Properties, Processing and Applications*, pp. 213–263, Jan. 2013, doi: 10.1016/B978-1-4557-2834-3.00011-2

[19] A. Rodriguez-Uribe, T. Wang, A.K. Pal, F. Wu, A.K. Mohanty, and M. Misra, "Injection moldable hybrid sustainable composites of BioPBS and PHBV reinforced with talc and starch as potential alternatives to single-use plastic packaging", *Compos Part C: Open Access*, vol. 6, pp. 100201, Oct. 2021, doi: 10.1016/J.JCOMC.2021.100201

[20] X. dong Qi, J.H. Yang, N. Zhang, T. Huang, Z.W. Zhou, I. Kühnert, P. Pötschke, and Y. Wang, "Selective localization of carbon nanotubes and its effect on the structure and properties of polymer blends", *Prog Polym Sci*, vol. 123, pp. 101471, Dec. 2021, doi: 10.1016/J.PROGPOLYMSCI.2021.101471

[21] S. Coiai, M.L. di Lorenzo, P. Cinelli, M.C. Righetti, and E. Passaglia, "Binary green blends of poly(lactic acid) with poly(butylene adipate-co-butylene terephthalate) and poly(butylene succinate-co-butylene adipate) and their nanocomposites", *Polymers*, vol. 13, no. 15, pp. 2489, Jul. 2021, doi: 10.3390/POLYM13152489

[22] N.R. Nair, V.C. Sekhar, K.M. Nampoothiri, and A. Pandey, "Biodegradation of biopolymers", *Curr Dev Biotechnol Bioeng: Prod, Isol Purif Ind Prod*, pp. 739–755, Jan. 2017, doi: 10.1016/B978-0-444-63662-1.00032-4

[23] F.v. Ferreira, L.S. Cividanes, R.F. Gouveia, and L.M.F. Lona, "An overview on properties and applications of poly(butylene adipate-co-terephthalate)–PBAT based composites", *Polym Eng Sci*, vol. 59, no. s2, pp. E7–E15, Mar. 2019, doi: 10.1002/PEN.24770

[24] S.K. Awasthi, M. Kumar, V. Kumar, S. Sarsaiya, P. Anerao, P. Ghosh, L. Singh, H. Liu, Z. Zhang, and M.K. Awasthi, "A comprehensive review on recent advancements in biodegradation and sustainable management of biopolymers", *Environmental Pollution*, vol. 307, pp. 119600, Aug. 2022, doi: 10.1016/J.ENVPOL.2022.119600

[25] A.K. Bledzki and J. Gassan, "Composites reinforced with cellulose based fibres", *Prog Polym Sci*, vol. 24, no. 2, pp. 221–274, May 1999, doi: 10.1016/S0079-6700(98)00018-5

[26] L. Yu, K. Dean, and L. Li, "Polymer blends and composites from renewable resources", *Prog Polym Sci*, vol. 31, no. 6, pp. 576–602, Jun. 2006, doi: 10.1016/J.PROGPOLYMSCI.2006.03.002

[27] J.J. Andrew and H.N. Dhakal, "Sustainable biobased composites for advanced applications: recent trends and future opportunities – A critical review", *Compos Part C: Open Access*, vol. 7, pp. 100220, Mar. 2022, doi: 10.1016/J.JCOMC.2021.100220

[28] Y. Ke, X.Y. Zhang, S. Ramakrishna, L.M. He, and G. Wu, "Reactive blends based on polyhydroxyalkanoates: Preparation and biomedical application", *Mater Sci Eng: C*, vol. 70, pp. 1107–1119, 2017, doi: 10.1016/j.msec.2016.03.114

[29] P. Feijoo, A.K. Mohanty, A. Rodriguez-Uribe, J. Gámez-Pérez, L. Cabedo, and M. Misra, "Biodegradable blends from bacterial biopolyester PHBV and bio-based PBSA: Study of the effect of chain extender on the thermal, mechanical and morphological properties", *Int J Biol Macromol*, vol. 225, pp. 1291–1305, Jan. 2023, doi: 10.1016/J.IJBIOMAC.2022.11.188

[30] S. Torres, R. Navia, R. Campbell Murdy, P. Cooke, M. Misra, and A.K. Mohanty, "Green composites from residual microalgae biomass and poly(butylene adipate-co-terephthalate): Processing and plasticization", *ACS Sustain Chem Eng*, vol. 3, no. 4, pp. 614–624, Apr. 2015, doi: doi.org/10.1021/sc500753h

[31] A. Anstey, S. Muniyasamy, M.M. Reddy, M. Misra, and A. Mohanty, "Processability and biodegradability evaluation of composites from poly(butylene succinate) (PBS) bioplastic and biofuel co-products from Ontario", *J Polym Environ*, vol. 22, no. 2, pp. 209–218, Feb. 2014, doi: 10.1007/S10924-013-0633-8

[32] B. Adams, M. Abdelwahab, M. Misra, and A.K. Mohanty, "Injection-molded bioblends from lignin and biodegradable polymers: Processing and performance evaluation", *J Polym Environ*, vol. 26, no. 6, pp. 2360–2373, Jun. 2018, doi: 10.1007/S10924-017-1132-0

[33] K. Pajula, M. Taskinen, V.P. Lehto, J. Ketolainen, and O. Korhonen, "Predicting the formation and stability of amorphous small molecule binary mixtures from computationally determined Flory-Huggins interaction parameter and phase diagram", *Mol Pharm*, vol. 7, no. 3, pp. 795–804, Jun. 2010, doi: 10.1021/MP900304P

[34] A. Javadi, A.J. Kramschuster, S. Pilla, J. Lee, S. Gong, and L. Turng, "Processing and characterization of microcellular PHBV/PBAT blends", *Polym Eng Sci*, vol. 50, no. 7, pp. 1440–1448, 2010, doi: 10.1016/j.compositesa.2010.04.002

[35] M. Nishida, K. Yasuda, and M. Nishida, "Correlative analysis between morphology and mechanical properties of poly-3-hydroxybutyrate (PHB) blended with polycarprolactone (PCL) using solid-state NMR", *Polym Test*, vol. 91, pp. 106780, 2020, doi: 10.1016/j.polymertesting.2020.106780

[36] Z. Qiu, T. Ikehara, and T. Nishi, "Poly (hydroxybutyrate)/poly (butylene succinate) blends: miscibility and nonisothermal crystallization", Polymer, vol. 44, no. 8, pp. 2503–2508, 2003, doi: 10.1016/S0032-3861(03)00150-2

[37]. H. Zhao, Z. Cui, X. Sun, L.-S. Turng, and X. Peng, "Morphology and properties of injection molded solid and microcellular polylactic acid/polyhydroxybutyrate-valerate (PLA/PHBV) blends", *Ind Eng Chem Res*, vol. 52, no. 7, pp. 2569–2581, 2013, doi: 10.1021/ie301573y

[38] H. Zhao, S. Yu, Y. Zhang, and G. Zhao, "Mechanical properties and structure of injection molded poly (hydroxybutyrate-co-hydroxyvalerate)/poly (butylene adipate-co-terephthalate)(PHBV/PBAT) blends", *J Appl Polym Sci*, pp. e53880, doi: 10.1002/app.53880

[39] M. Zhang, X. Diao, Y. Jin, and Y. Weng, "Preparation and characterization of biodegradable blends of poly (3-hydroxybutyrate-co-3-hydroxyhexanoate) and poly (butylene adipate-co-terephthalate)", *Journal of Polymer Engineering*, vol. 36, no. 5, pp. 473–480, 2016, doi: 10.1515/polyeng-2015-0076

[40] M. Cunha, B. Fernandes, J.A. Covas, A.A. Vicente, and L. Hilliou, "Film blowing of PHBV blends and PHBV-based multilayers for the production of biodegradable packages", *J Appl Polym Sci*, vol. 133, no. 2, Jan. 2016, doi: 10.1002/APP.42165

[41] P. Zytner, A.K. Pal, F. Wu, A. Rodriguez-Uribe, A.K. Mohanty, and M. Misra, "Morphology and performance relationship studies on poly (3-hydroxybutyrate-co-3-hydroxyvalerate)/poly (butylene adipate-co-terephthalate)-based biodegradable blends", *ACS Omega*, vol. 8, no. 2, pp. 1946–1956, 2023, doi: doi.org/10.1021/acsomega.2c04770

[42] B. Le Delliou, O. Vitrac, M. Castro, S. Bruzaud, and S. Domenek, "Characterization of a new bio-based and biodegradable blends of poly(3-hydroxybutyrate-co-3-hydroxyvalerate) and poly(butylene-co-succinate-co-adipate)", *J Appl Polym Sci*, vol. 139, no. 19, pp. 52124, May 2022, doi: 10.1002/APP.52124

[43] M.C. Righetti, P. Cinelli, L. Aliotta, E. Bianchi, F. Tricoli, M. Seggiani, and A. Lazzeri, "Immiscible PHB/PBS and PHB/PBSA blends: Morphology, phase composition and modelling of elastic modulus", *Polym Int*, vol. 71, no. 1, pp. 47–56, 2022, doi: 10.1002/pi.6282

[44] V. Jost and O. Miesbauer, "Effect of different biopolymers and polymers on the mechanical and permeation properties of extruded PHBV cast films", *J Appl Polym Sci*, vol. 135, no. 15, pp. 46153, 2018, doi: 10.1002/app.46153

[45] R. Muthuraj, M. Misra, and A.K. Mohanty, "Biodegradable compatibilized polymer blends for packaging applications: A literature review", *J Appl Polym Sci*, vol. 135, no. 24, pp. 45726, 2018, doi: 10.1002/app.45726

[46] Y. Jin, E. Wang, Y. Weng, S. Men, Y. Dong, Y. Sima, and Z. Huang, "The investigation of the toughening mechanism of PHBV/PBAT with a novel hyperbranched ethylenediamine triazine polymer based modifier: the formation of the transition layer and the microcrosslinking structure", *J Polym Environ*, vol. 26, pp. 4158–4167, 2018, doi.org/10.1007/s10924-018-1286-4

[47] P. Ma, D.G. Hristova-Bogaerds, P.J. Lemstra, Y. Zhang, and S. Wang, "Toughening of PHBV/PBS and PHB/PBS blends via in situ compatibilization using dicumyl peroxide as a free-radical grafting initiator", *Macromol Mater Eng*, vol. 297, no. 5, pp. 402–410, 2012, *https://doi.org/10.1002/mame.201100224*

[48] M. Larsson, O. Markbo, and P. Jannasch, "Melt processability and thermomechanical properties of blends based on polyhydroxyalkanoates and poly (butylene adipate-co-terephthalate)", *RSC Adv*, vol. 6, no. 50, pp. 44354–44363, 2016, doi.org/10.1039/C6RA06282B

[49] D. Garcia-Garcia, E. Rayón, A. Carbonell-Verdu, J. López-Martínez, and R. Balart, "Improvement of the compatibility between poly (3-hydroxybutyrate) and poly (ε-caprolactone) by reactive extrusion with dicumyl peroxide", *Eur Polym J*, vol. 86, pp. 41–57, 2017, doi.org/10.1016/j.eurpolymj.2016.11.018

[50] M. Przybysz, M. Marć, M. Klein, M.R. Saeb, and K. Formela, "Structural, mechanical and thermal behavior assessments of PCL/PHB blends reactively compatibilized with organic peroxides", *Polym Test*, vol. 67, pp. 513–521, 2018, doi: 10.1016/j.polymertesting.2018.03.014

[51] P. Zytner, F. Wu, M. Misra, and A.K. Mohanty, "Toughening of biodegradable poly (3-hydroxybutyrate-co-3-hydroxyvalerate)/poly (ε-caprolactone) blends by in situ reactive compatibilization", *ACS Omega*, vol. 5, no. 25, pp. 14900–14910, 2020, doi: 10.1021/acsomega.9b04379

[52] S. Kennouche, N. Le Moigne, M. Kaci, J.C. Quantin, A.S. Caro-Bretelle, C. Delaite, and J.M. Lopez-Cuesta, "Morphological characterization and thermal properties of compatibilized poly (3-hydroxybutyrate-co-3-hydroxyvalerate)(PHBV)/poly (butylene succinate)(PBS)/halloysite ternary nanocomposites", *Eur Polym J*, vol. 75, pp. 142–162, 2016, doi: 10.1016/j.eurpolymj.2015.12.009

[53] K. Zhang, M. Misra, and A.K. Mohanty, "Toughened sustainable green composites from poly(3-hydroxybutyrate-co-3-hydroxyvalerate) based ternary blends and miscanthus biofiber", *ACS Sustain Chem Eng*, vol. 2, no. 10, pp. 2345–2354, Oct. 2014, doi: 10.1021/SC500353V

[54] B.J. Rashmi, K. Prashantha, M.F. Lacrampe, and P. Krawczak, "Scalable production of multifunctional bio-based polyamide 11/graphene nanocomposites by melt extrusion processes via masterbatch approach", *Adv Polym Technol*, vol. 37, no. 4, pp. 1067–1075, Jun. 2018, doi: 10.1002/ADV.21757

[55] A.K. Pal, F. Wu, M. Misra, and A.K. Mohanty, "Reactive extrusion of sustainable PHBV/PBAT-based nanocomposite films with organically modified nanoclay for packaging applications: Compression moulding vs. cast film extrusion", *Compos B Eng*, vol. 198, pp. 108141, Oct. 2020, doi: 10.1016/J.COMPOSITESB.2020.108141

[56] W.V. Srubar, S. Pilla, Z.C. Wright, C.A. Ryan, J.P. Greene, C.W. Frank, and S.L. Billington, "Mechanisms and impact of fiber – matrix compatibilization techniques on the material characterization of PHBV/oak wood flour engineered biobased composites", *Compos Sci Technol*, vol. 72, no. 6, pp. 708–715, Mar. 2012, doi: 10.1016/J.COMPSCITECH.2012.01.021.

[57] S.P. Pawar, A. Misra, S. Bose, K. Chatterjee, and V. Mittal, "Enzymatically degradable and flexible bio-nanocomposites derived from PHBV and PBAT blend: Assessing thermal, morphological, mechanical, and biodegradation properties", *Colloid Polym Sci*, vol. 293, no. 10, pp. 2921–2930, Oct. 2015, doi: 10.1007/S00396-015-3700-Y

[58] B. Bittmann, R. Bouza, L. Barral, M. Castro-Lopez, and S. Dopico-Garcia, "Morphology and thermal behavior of poly (3-hydroxybutyrate-co-3-hydroxyvalerate)/poly(butylene adipate-co-terephthalate)/clay nanocomposites", *Polym Compos*, vol. 36, no. 11, pp. 2051–2058, Nov. 2015, doi: 10.1002/PC.23115

[59] D. Nath, R. Santhosh, J. Ahmed, and Preetam Sarkar, "Optical, mechanical, structural, and antimicrobial properties of tamarind kernel powder, halloysite, and cinnamaldehyde nanocomposite films", *J Food Process Eng*, pp. e14065, May 2022, doi: 10.1111/JFPE.14065

[60] B. Bittmann, R. Bouza, L. Barral, R. Bellas, and A. Cid, "Effect of environmental factors on poly(3-hydroxybutyrate-co–3-hydroxyvalerate)/poly(butylene adipate-co-terephthalate)/montmorillonite nanocomposites with antimicrobial agents", *Polym Compos*, vol. 39, no. 3, pp. 915–923, Mar. 2018, doi: 10.1002/PC.24018

[61] M.R.M. Asyraf, M.R. Ishak, M.N.F. Norrrahim, N.M. Nurazzi, S.S. Shazleen, R.A. Ilyas, M. Rafidah, and M.R. Razman, "Recent advances of thermal properties of sugar palm lignocellulosic fibre reinforced polymer composites", *Int J Biol Macromol*, vol. 193, pp. 1587–1599, Dec. 2021, doi: 10.1016/J.IJBIOMAC.2021.10.221

[62] V.C. Beber, S. De Barros, M.D. Banea, M. Brede, L.H. De Carvalho, R. Hoffmann, A.R.M. Costa, E.B. Bezerra, I.D. Silva, K. Haag, and K. Koschek, "Effect of Babassu Natural filler on PBAT/PHB biodegradable blends: An investigation of thermal, mechanical, and morphological behavior", *Materials*, vol. 11, no. 5, pp. 820, May 2018, doi: 10.3390/MA11050820

[63] V. Nagarajan, M. Misra, and A.K. Mohanty, "New engineered biocomposites from poly(3-hydroxybutyrate-co-3-hydroxyvalerate) (PHBV)/poly(butylene adipate-co-terephthalate) (PBAT) blends and switchgrass: Fabrication and performance evaluation", *Ind Crops Prod*, vol. 42, no. 1, pp. 461–468, Mar. 2013, doi: 10.1016/J.INDCROP.2012.05.042

[64] N. Zarrinbakhsh, M. Misra, and A.K. Mohanty, "Biodegradable green composites from distiller's dried grains with solubles (DDGS) and a polyhydroxy(butyrate-co-valerate) (PHBV)-based bioplastic", *Macromol Mater Eng*, vol. 296, no. 11, pp. 1035–1045, Nov. 2011, doi: 10.1002/MAME.201100039

[65] N. Zarrinbakhsh, A.K. Mohanty, and M. Misra, "Improving the interfacial adhesion in a new renewable resource-based biocomposites from biofuel coproduct and biodegradable plastic", J *Mater Sci*, vol. 48, no. 17, pp. 6025–6038, Sep. 2013, doi: 10.1007/S10853-013-7399-1

[66] F. Sarasini, F. Luzi, F. Dominici, G. Maffei, A. Iannone, A. Zuorro, R. Lavecchia, L. Torre, A. Carbonell-Verdu, R. Balart, and D. Puglia, "Effect of different compatibilizers on sustainable composites based on a PHBV/PBAT matrix filled with coffee silverskin", *Polymers*, vol. 10, no. 11, pp. 1256, Nov. 2018, doi: 10.3390/POLYM10111256

[67] A. Gupta, L. Lolic, and T.H. Mekonnen, "Reactive extrusion of highly filled, compatibilized, and sustainable PHBV/PBAT – Hemp residue biocomposite", *Compos Part A: Appl Sci Manuf*, vol. 156, pp. 106885, May 2022, doi: 10.1016/J.COMPOSITESA.2022.106885

[68] R.C. Thompson, C.J. Moore, F.S. Vom Saal, and S.H. Swan, "Plastics, the environment and human health: current consensus and future trends", *Philos Trans R Soc B*, vol. 364, pp. 2153–2166, 2009, doi: 10.1098/rstb.2009.0053.

[69] M.T. Zumstein, A. Schintlmeister, T.F. Nelson, R. Baumgartner, D. Woebken, M. Wagner, H.P.E. Kohler, K. McNeill, and M. Sander, "Biodegradation of synthetic polymers in soils: Tracking carbon into CO_2 and microbial biomass", *Sci Adv*, vol. 4, no. 7, Jul. 2018, doi: 10.1126/SCIADV.AAS9024

[70] S. Agarwal, "Biodegradable polymers: Present opportunities and challenges in providing a microplastic-free environment", *Macromol Chem Phys*, vol. 221, no. 6, pp. 2000017–2000024, Mar. 2020, doi: 10.1002/macp.202000017

[71] A. Krzan, S. Hemjinda, S. Miertus, A. Corti, and E. Chiellini, "Standardization and certification in the area of environmentally degradable plastics", *Polym Degrad Stab*, vol. 91, pp. 2819–2833, 2006, doi: 10.1016/j.polymdegradstab.2006.04.034

[72] K.W. Meereboer, M. Misra, and A.K. Mohanty, "Review of recent advances in the biodegradability of polyhydroxyalkanoate (PHA) bioplastics and their composites", *Green Chemistry*, vol. 22, no. 17, pp. 5519–5558, Aug. 2020, doi: 10.1039/D0GC01647K.

[73] S. Muniyasamy, O. Ofosu, M.J. John, and R.D. Anandjiwala, "Mineralization of poly(lactic acid) (PLA), poly(3-hydroxybutyrate-co-valerate) (PHBV) and PLA/PHBV blend in compost and soil environments", *J Renew Mater*, vol. 4, no. 2, pp. 133–145, Jan. 2016, doi: 10.7569/JRM.2016.634104

[74] A.O. Pérez-Arauz, Aguilar-Rabiela, A. Vargas-Torres, A.I. Rodríguez-Hernández, N. Chavarría-Hernández, B. Vergara-Porras, and M.R. López-Cuellar, "Production and characterization of biodegradable films of a novel polyhydroxyalkanoate (PHA) synthesized from peanut oil", *Food Packag Shelf Life*, vol. 20, pp. 100297, Jun. 2019, doi: 10.1016/J.FPSL.2019.01.001

[75] N.A.F. Othman, S. Selambakkannu, and N. Seko, "Biodegradable dual-layer polyhydroxyalkanoate (PHA)/polycaprolactone (PCL) mulch film for agriculture: Preparation and characterization", *Energ Nexus*, vol. 8, pp. 100137, Dec. 2022, doi: 10.1016/J.NEXUS.2022.100137

[76] N.T.T. Thu, L.H. Hoang, P.K. Cuong, N. Viet-Linh, T.T.H. Nga, D.D. Kim, Y.K. Leong, and L.T. Nhi-Cong, "Evaluation of polyhydroxyalkanoate (PHA) synthesis by *Pichia* sp. TSLS24 yeast isolated in Vietnam", *Scientific Reports*, vol. 13, no. 1, pp. 1–14, Feb. 2023, doi: 10.1038/s41598-023-28220-z.

[77] J.D. Badia, T. Kittikorn, E. Strömberg, L. Santonja-Blasco, A. Martínez-Felipe, A. Ribes-Greus, M. Ek, and S. Karlsson, "Water absorption and hydrothermal performance of PHBV/sisal biocomposites", *Polym Degrad Stab*, vol. 108, pp. 166–174, Oct. 2014, doi: 10.1016/J.POLYMDEGRADSTAB.2014.04.012

[78] Z. Zaidi, D. Mawad, and A. Crosky, "Soil biodegradation of unidirectional polyhydroxybutyrate-co-valerate (PHBV) biocomposites toughened with polybutylene-adipate-co-terephthalate (PBAT) and epoxidized natural rubber (ENR)", *Front Mater*, vol. 6, pp. 275, Nov. 2019, doi: 10.3389/FMATS.2019.00275

[79] S. Muniyasamy, O. Ofosu, B. Thulasinathan, A.S.T. Rajan, S.M. Ramu, S. Soorangkattan, J.B. Muthuramalingam, and A. Alagarsamy, "Thermal-chemical and biodegradation behaviour of alginic acid treated flax fibres poly(hydroxybutyrate-co-valerate) PHBV green composites in compost medium", *Biocatal Agric Biotechnol*, vol. 22, pp. 101394, Nov. 2019, doi: 10.1016/J.BCAB.2019.101394.

[80] C.S. Wu, H.T. Liao, and Y.X. Cai, "Characterisation, biodegradability and application of palm fibre-reinforced polyhydroxyalkanoate composites", *Polym Degrad Stab*, vol. 140, pp. 55–63, Jun. 2017, doi: 10.1016/J.POLYMDEGRADSTAB.2017.04.016

[81] K.W. Meereboer, A.K. Pal, E.O. Cisneros-López, M. Misra, and A.K. Mohanty, "The effect of natural fillers on the marine biodegradation behaviour of poly(3-hydroxybutyrate-co-3-hydroxyvalerate) (PHBV)", *Scientific Reports*, vol. 11, no. 1, pp. 1–11, Jan. 2021, doi: 10.1038/s41598-020-78122-7

[82] J. Jian, Z. Xiangbin, and H. Xianbo, "An overview on synthesis, properties and applications of poly(butylene-adipate-co-terephthalate)-PBAT", *Adv Ind Eng Polym Res* vol. 3, no. 1, pp. 19–26, Jan. 2020, doi: 10.1016/J.AIEPR.2020.01.001

[83] D. Tan, Y. Wang, Y. Tong, and G.Q. Chen, "Grand challenges for industrializing polyhydroxyalkanoates (PHAs)", *Trends Biotechnol*, vol. 39, no. 9, pp. 953–963, Sep. 2021, doi: 10.1016/J.TIBTECH.2020.11.010

[84] K. Saravanan, M. Umesh, and P. Kathirvel, "Microbial polyhydroxyalkanoates (PHAs): A review on biosynthesis, properties, fermentation strategies and its prospective applications for sustainable future", *J Polym Environ*, vol. 30, no. 12, pp. 4903–4935, Sep. 2022, doi: 10.1007/S10924-022-02562-7

[85] S. Philip, T. Keshavarz, and I. Roy, "Polyhydroxyalkanoates: Biodegradable polymers with a range of applications", *J Chem Technol Biotechnol*, vol. 82, no. 3, pp. 233–247, Mar. 2007, doi: 10.1002/JCTB.1667

[86] F. Muneer, I. Rasul, F. Azeem, M.H. Siddique, M. Zubair, and H. Nadeem, "Microbial polyhydroxyalkanoates (PHAs): Efficient replacement of synthetic polymers", *Polym Environ*, vol. 28, no. 9, pp. 2301–2323, Jun. 2020, doi: 10.1007/S10924-020-01772-1

[87] "Compostable, eco-friendly packaging solutions | TIPA", *https://tipa-corp.com/* (accessed Mar. 21, 2023)

[88] "Market leaders in eco-friendly food packaging | BioPak Australia", *https://www.biopak.com/au/* (accessed Mar. 21, 2023)

3

Polyhydroxyalkanoates as Biodegradable Biopolymers for a Circular Economy: Characteristics, Challenges, and Recent Advances

Chiara Magnani, Paloma Cabesa Segura, Baptiste Leroy, Giada Lo Re, Ruddy Wattiez, and Jean-Marie Raquez[1]

PHAs are viewed as real circular plastic materials, as they can be even produced from wastes, without depleting sources of food-based raw materials. The shortcomings of PHAs are related to their cost, processing, and performance. PHAs are characterized by low thermal stability during melt-processing with too slow crystallization kinetics. They are commonly brittle with a low elongation at break, subject to aging, and possess insufficient barrier properties for packaging applications.

To overcome these drawbacks, different strategies have been adopted. Reduction of production cost is achieved by using industrial wastes as carbon sources, new bacteria strains and mixed strains have been studied, and innovative purification techniques have been adopted. The upgrading of mechanical properties is attained by melt-blending and incorporation of (nano)fillers, plasticizers, or reinforcing and nucleating agents. To meet the objective of a circularity, efforts are directed toward combining biodegradable additives such as natural fibers and nanocrystals, as will be described in this chapter.

[1] G. L. R. acknowledges the Chalmers Areas of Advance, Genie and KAW Biocomposites No. V-2019-0041 for financial support. Funding from the Region Wallon and European Comission in the frame of the ARC 2017–2021 Program (PHASYN Project) and from European Union's Horizon 2020 Research and Innovation Programme (BIODEST project) under the Marie Sklodowska-Curie Grant Agreement No. 778092 are gratefully acknowledged. JMR is a Senior FRS-FNRS Research Associate.

3.1 Introduction to PHAs

Plastic materials have revolutionized our daily lives, but their increased ubiquity coupled with the inappropriate management of plastic waste have impacted our environment, polluting soils and seas, and becoming one of the greatest emergencies of our century [1]. 348 million tonnes of plastics were produced in 2017 [2] reaching 368 million tonnes in 2020 [3], with production estimated to increase of about 4% each year [4]. It has been estimated that in 2015 only 9% of the plastic produced got recycled and 12% got incinerated, while 79% got dumped in landfills or in the natural environment [5, 6]. In Europe, recycled plastic is estimated to be 30% of the total production [5], but this percentage includes the waste exported to other countries for recycling [7–9].

Among the solutions proposed to solve plastic pollution, the replacement of fossil-derived plastics with bioplastics is by far the most promising, and interest in them has steadily grown in recent years. Many of them are both bio-sourced and biodegradable, and could partly or even completely encompass the circularity concept, because the products resulting from their biodegradation or recycling can be used to produce new polymers, thereby closing the loop. Among them, the most well-known are thermoplastic starch, poly(lactic acid) (PLA), and polyhydroxyalkanoates (PHAs).

However, at the moment the production of bioplastics constitutes only 1% of the total amount of plastics produced per year [1, 3]. Focusing only on biopolyesters, PHAs can be considered as the "rising stars" in this category of materials. Compared to PLA, the most industrialized biopolyesters in the market, PHAs offer several advantages:

- They can be microbially produced through any low-cost substrate resulting from waste such as sludges and biomass residues.
- They cover a broad spectrum of thermal and mechanical properties thanks to the variation in their chemical structure (a function of the substrate used), which make them exploitable in many fields.
- They are readily biodegradable in any natural environment (home compost, soil, marine conditions), including the possibility to be chemically recycled.

These different features enable PHAs to be viewed as truly circular plastics, as they can be even produced from waste, without depleting sources of food production, and contributing to "close the loop". However, they are still struggling to conquer the bioplastic market, and the reasons for this are related to their cost, processing, and performance. The most limiting factor is their high cost compared to fossil-based plastics (as well as to PLA), mainly due to their bacterial production and more particularly the purification process involved. Regarding the issues about their processability and performances, they have a low thermal stability on melt-processing and also slow crystallization kinetics, which often impede their properties and do not reach the minimal performance of commodity plastics. For example, they are brittle with a low elongation at break, they age easily, and they do not satisfy the barrier properties required for packaging applications.

To overcome these drawbacks, various strategies have been adopted. To reduce the production costs, industrial waste is being increasingly employed as carbon sources, new bacteria strains or mixed strains are being studied, and innovative purification techniques have been adopted. The most widely adopted strategy to control the main properties of PHAs has been the melt-blending approach, in which (nano) fillers, plasticizers or reinforcing and nucleating agents are added. To fulfill the objective of a circular economy, the best biodegradable additives such as natural fibers and nanocrystals need to be found, and efforts towards this end will be described in the forthcoming paragraphs [1].

3.1.1 Production

Polyhydroxyalkanoates (PHAs) are a group of bio-sourced polyesters naturally produced by microorganisms mainly when they are under stress conditions and in the presence of a carbon source in excess. Among other functions [30], they constitute a kind of energy stock, that can be consumed in the absence of other nutrients [31]. The most-studied PHA is the homopolymer poly(hydroxybutyrate), which was identified for the first time in the 1920s in the bacterium *Bacillus* sp. [32, 33].

Since then, many types of bacteria have been found capable of producing polyhydroxyalkanoates. Lu et al. [34] comprehensively summarized all the strains natively able to accumulate PHAs. Among the most common microorganisms, we can mention *Cupriavidus necator*, also known as *Ralstonia eutropha*, a bacterium from the genus *Pseudomonas*, and *Azotobacter vinelandii* [35]. The cited microorganisms are part of the group of bacteria that accumulate PHAs when a major nutrient is limited [36]. A second group of bacteria does not require any limiting element to produce PHAs during growth. Engineered bacteria are part of this group, and a paragraph is dedicated to them below.

In general, PHAs are accumulated in the form of granules inside the bacterial cell walls [31] (Figure 3.1), which get coated in a layer of proteins and phospholipids [37], although the presence of the latter is in doubt [38, 39].

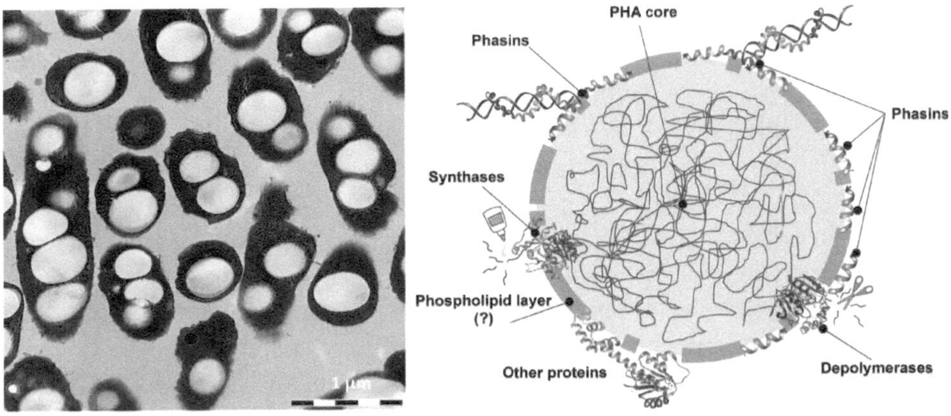

Figure 3.1 Left: Transmission electron micrograph of granules in the cell of *Cupriavidus necator* [39]. Right: PHA granule composition [40]

Biosynthesis of PHAs can occur through various pathways depending on the bacterium and its enzymes, but also on the external conditions and the carbon source used [41]. Indeed, if the substrate consists of a sugar or a fatty acid, the metabolic pathway will be different, and the composition of the final polyester could change [42]. In general, sugars and fatty acids are transformed into the precursors acetyl coenzyme A (AcetylCoA) or acyl coenzyme A (AcylCoA) as the main building monomers, as shown in Figure 3.2.

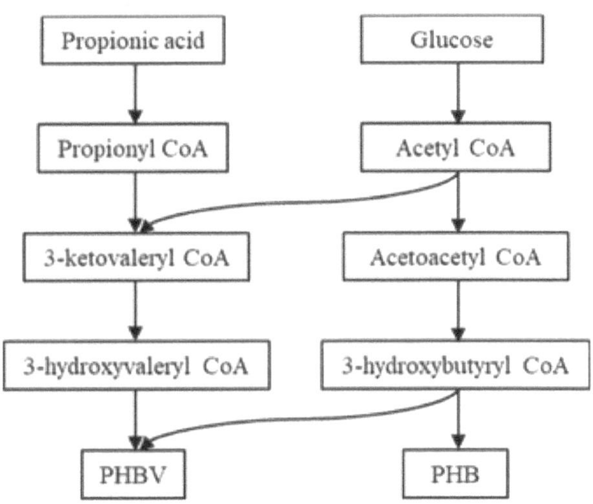

Figure 3.2 Biosynthetic pathway to obtain PHBV and PHB [44]

Biosynthetic pathways for the formation of PHA involve specific enzymes for each step that start from a molecule of AcetylCoA linked to the monomer (*R*)-3-hydroxybutyryl-CoA, which is then polymerized to form a type of PHA, known as poly(hydroxybutyrate) (PHB) [43].

In the case of the copolymer poly(hydroxybutyrate-co-hydroxyvalerate) (PHBV), if we take propionic acid as a substrate, a molecule of 3-ketovaleryl-CoA is formed from propionyl-CoA and acetyl-CoA, and then is transformed into 3-hydroxyvaleryl-CoA [44]. Then a specific enzyme condenses the different monomers synthesized, and its specificity determines the final composition of the polyester. Valeric acid is another precursor for the hydroxyvalerate building block together with propionic acid, while acetic and butyric acid, together with simple sugars, are generally metabolized as hydroxybutyrate monomers. As previously mentioned, the external conditions affect the polymer synthesized – for example, a variation in the micronutrients or in the temperature can influence the polymer chain length [45].

Despite their biological origin, PHAs obtained from renewable feedstocks are not necessarily sustainable. For example, the use of fertilizers, pesticides, or the huge quantity of water and (often non-renewable) energy necessary to grow crop fields is not considered sustainable, because they result in marine and terrestrial eutrophication, gas emissions and land occupation, competing with the food chain. In this sense, efforts must be made to reduce to a minimum the use of toxic substances and to turn to renewable energies instead of fossil-derived ones. The land use can be reduced by employing non-food biomass like lignocellulose [22]. Moreover, PHAs have a huge potential compared to other bioplastics since they can be obtained from agricultural, industrial, and household waste. For example, Nitkiewicz et al. used a LCA tool to compare PHAs produced from crude vegetable oil, exhausted oil, and glycerol as a side-product of biofuel production [23]. They found that the use of crude oil had a huge impact due to land occupation and consumption of energy and water, while the production from waste or side-products dramatically reduced the environmental impact. In this view, the integration of PHA production into biorefineries is the best scenario, in which industrial waste is used to produce high-value products, helping to close the loop.

3.1.2 Structural Aspects

PHAs have a common chemical structure, as shown in Figure 3.3. According to the number of carbons in the monomer, they are classified as:

- Short chain length PHA (scl-PHA): R from C_1 to C_5
- Medium chain length PHA (mcl-PHA): R from C_6 to C_{14}
- Long chain length PHA (lcl-PHA): R greater than C_{14}.

Among them, the homopolymer poly(hydroxybutyrate) (PHB) is the most studied and the easiest to obtain.

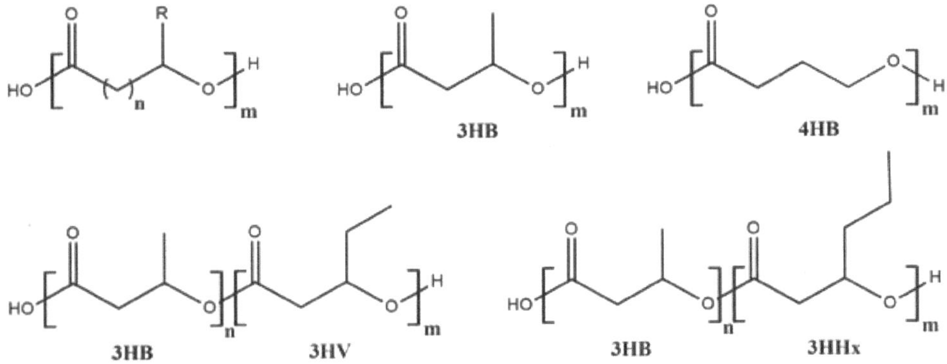

Figure 3.3 General chemical structure of PHAs; chemical structure of PHB and related copolyesters

The capacity of bacteria to incorporate different carbon sources allows the synthesis of diverse monomers other than hydroxybutyrate itself, giving a broad range of copolymers. The most widespread ones both in research and industry are poly(hydroxybutyrate-co-hydroxyvalerate) (PHBV) and poly(hydroxybutyrate-co-hydroxy-hexanoate) (PHBHx), followed by the copolyester poly(3-hydroxybutyrate-co-4-hydroxybutyrate) (PHB4HB). The combination of co-monomers and their ratio strongly influences the properties of the resulting polyester, ranging from highly crystalline thermoplastic polymers to amorphous and elastomer-like ones. The crystallization kinetics of PHAs is of particular interest because it affects all the properties of PHAs. Due to their microbial origin, PHAs are particularly pure, so in general the nuclei density from which the polymer starts to crystallize is very low, as shown in Figure 3.4, compared to commodity plastics. Therefore, they are characterized by the formation of very large spherulites that can reach millimeter dimensions and that contain imperfections and cracks (Figure 3.4). In PHB this microstructure is coupled with a high crystallinity (around 60%), which can reach up to 80% [35], resulting in a polyester with a high Young's modulus (around 3–4 GPa) and a low elongation at break [35]. Moreover, its melting point is around 174 °C, which makes the processing window quite narrow, because thermal degradation processes start at around 200 °C.

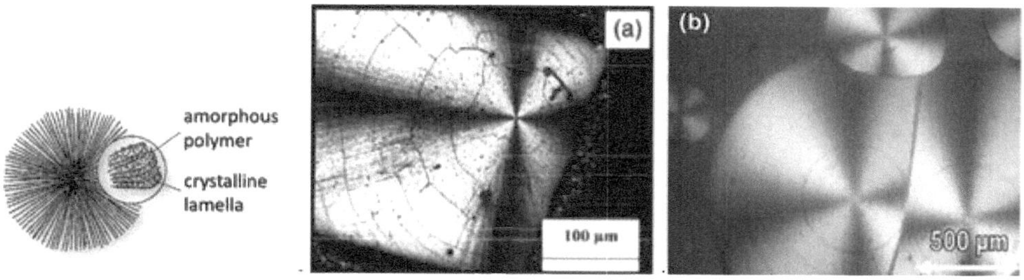

Figure 3.4 Left: schematic representation of a spherulite. Center: PHB spherulite with typical cracks [46]. Right: PHB crystallization at 80 °C from the melt state [47]

When a co-monomer is incorporated randomly in PHB, the side chain R determines the capacity of the copolymer chains to fold. As a consequence, the co-monomer ratio influences the crystallization kinetics and the morphology of the spherulites formed. Consequently, the short side chain of HV co-monomer does not perturb the chain folding too much, and it is able to co-crystallize in the HB crystalline lattice [48]. However, when the crystallization rate slows down [49–51], the crystallinity is reduced. Interestingly, beyond a certain percentage of HV the crystalline lattice passes from an HB to an HV lattice, a phenomenon called isodimorphism [48]. From a macroscopic viewpoint, the melting point of PHBV decreases as the HV mol% rises, and it has a glass transition near or below 0 °C due to the higher mobility of the chains. Overall, it results in a thermoplastic polymer with slightly improved elongation at break and higher impact strength compared to PHB [48, 52, 53]. The side chain of the hydroxy-hexanoate monomer is longer compared to HV and HHx is excluded from the HB crystalline lattice, thus its incorporation along the PHB chain can prevent the crystallization of the polymer [53–55]. This results in an amorphous polymer with an elongation at break of 400% and 850% with 10 and 17 mol% of HHx, respectively [54]. Interestingly, the co-monomer 4HB has the same effect and the homopolymer P4HB is elastomeric, with the elongation at break reaching 1000%, a very low melting point, and a glass transition around 54 °C and -48 °C, respectively [48, 56].

PHAs are known to have promising barrier properties to water, oxygen, and odors for packaging applications [57], even compared to many biosourced polymers [58]. In particular, they have barrier properties to oxygen better than polypropylene and polyethylene, with an oxygen transmission rate (OTR) in the range 23–29 $cm^3/m^2/$ day, and better barrier properties to water than polypropylene, with a water vapor transmission rate (WVTR) in the range 5–19 g/m^2/day [31] The permeability to water and gases is deeply influenced by the crystallinity, among other factors [58, 59]. Indeed, PHB and PHBV with low amount of HV are hydrophobic and highly crystalline, so their permeability to water and oxygen is low. Shogren et al. demonstrated that when the HV percentage is increased from 6 to 18 mol%, the crystallinity was lower and the mobility of the chain in the amorphous region (as well as the free volume) increased, resulting in a permeability to water two times higher [60]. The higher permeability of copolymers such as PHBH and P3HB4HB is ascribed to their much lower crystallinity [61, 62].

The variation of crystallinity in PHA copolymers is a factor also influencing their biodegradability [54]. It is known that the degradation starts from the erosion of the surface, so the rate of biodegradation is related to the capacity of microorganisms to form a biofilm on the surface of the object [63,64]. It has been proved that films of copolymers such as PHBV, PHBHx, and P3HB4HB have a biodegradation rate higher than PHB, both in soil and sea water [65, 66]. In particular, the crystallinity of the copolymer increased over time, proving that the amorphous phase was the first one to be degraded. Indeed, the less crystalline surface together with a less smooth surface (due to the presence of pores) favors the activity of the microorganisms. The rate followed the order PHB/4HB > PHBHx > PHBV > PHB, which is inversely proportional to the crystallinity index [67]. Also the percentage of the co-monomer makes a difference. Indeed, it was found that PHBH containing 12% HHx degrades faster than PHBH with 5% HHx [68].

On the other hand, the chemical structure of the polymer influences the biodegradability on its own. The reactive center of the enzyme PHA-depolymerase can have a certain specificity. Different bacteria have different capabilities to degrade specific polyesters. Doi et al. used P(3HB) depolymerase from *Alcaligenes faecalis*, which is specific for the HB monomer. Therefore, longer side chains were attacked with difficulty by the enzyme, and as a result it hardly hydrolyzed the ester bonds HV - HV, HHx - HHx, HV - HB, or HHx - HB, resulting in the presence of dimers and oligomers in the medium, as already demonstrated [54]. The bacterial composition of a medium is therefore important for successful biodegradation. In general, if a bacterium can polymerize a certain copolymer, it can also degrade it.

3.1.3 Applications

The global production of bioplastics in 2020 was 2.11 million tonnes, of which 1.227 million tonnes was biodegradable [3]. Among them, PLA and starch blends are leading the market while PHAs constituted only 1.7% of the total amount. However, their production is estimated to dramatically increase in the next five years, as shown in Figure 3.5, thanks to the restrictions imposed on use of fossil-derived plastics, and the rising awareness of society regarding environmental issues.

Although research and investment in PHAs have made the price drop from 9 €/kg (in 2002) [32, 69], today it is still 2–6 times higher than commodity plastics such as PP and PET [2, 13], and also higher than other biopolymers such as PLA, as shown in Table 3.1.

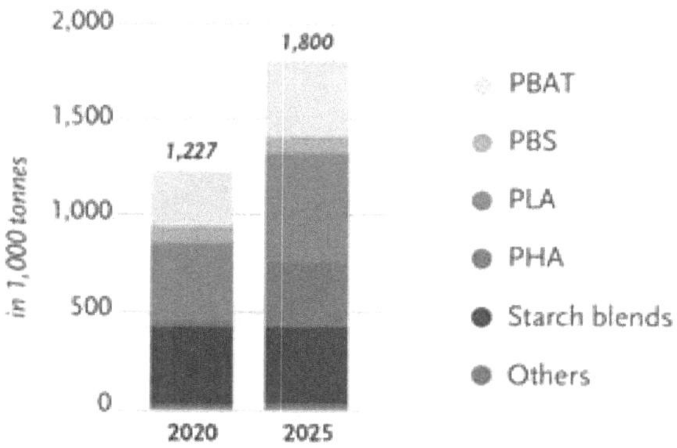

Figure 3.5 Comparison of production of biodegradable bioplastics between 2020 (actual) and 2025 (estimated) [3]

According to the analysis made by Lettner et al., the cost, together with the marketing activities, is the main factor limiting the uptake of PHAs in the market [70]. Indeed, the market reflects the prices of biodegradable polymers, with PLA and starch constituting the larger slice of industrial production. According to their analysis, the production process is the main driver of the high cost of PHAs [70].

To tackle this problem, many companies have developed (or are trying to develop) new technologies using cheaper carbon sources like organic waste, which are transformed in volatile fatty acids to feed the bacteria. Table 3.2 contains an updated list of the companies producing PHAs, which indicates the country, the commercial name, as well as the type of PHA produced and the carbon source utilized.

Table 3.1 Average Cost of Bioplastics Compared to Petrochemical-Based Plastics [6, 71] (applied exchange factor $/lbs= 2.205 €/kg)

	Average cost of material (€/kg)
Biodegradable plastics	
PLA	~2.21
Starch - polymer blends	2.21-4.42
PHAs	4.42-13.23
Commodity petrochemical plastics	
PP	1.87-2.09
PE	1.43-1.76
PET	1.87-1.98
PS	2.21-2.65
PVC	1.87-1.98

As the market for PHAs is quite young, it is still under development. Some companies have chosen to sell their patents (like Procter & Gamble did with Meridian, which is now merged with Danimer Scientific). The patent of Metabolix, after the closure of the implant shared with ADM, went to the Korean CJ Bio corporation [4]. Other companies went bankrupt, like the Italian Bio-on, which produced PHAs microspheres for cosmetic applications until the end of 2019 [72]. In their place, new start-ups have emerged, and today there are about 15 companies producing PHAs, mainly situated in US, Canada, and China, with a few spreads in Europe.

Table 3.2 Companies Producing PHAs (adapted from Shogren et al.[4] and Bugnicourt et al. [31])

Company	Trade name	Materials	Location	Origin
Danimer Scientific	Nodax	PHBV	U.S.	Vegetable oils
Newlight Technologies	AirCarbon	PHB	U.S.	CH_4, CO_2 (from agricultural waste)
Full Cycle Bioplastics		PHA	U.S.	Organic waste
Tepha	TephaFLEX	P4HB	U.S.	
Mango Materials		PHA	U.S.	Waste, CH_4
Kaneka	Green Planet	PHBH	Japan/Belgium	Plant oils
CJ Bio	CJ PHA	P(3,4-HB)	South Korea	
GoodFellow		PHB, PHBV	U.K.	
BioMatera		PHA	Canada	
Tianjin	SoGreen	P(3,4-HB)	China	
TianAn	Enmat	PHB, PHBV	China	Dextrose (from corn or cassava)
Bluepha		PHA	China	Crops, kitchen waste, others
PolyFerm Canada		PHA elastomers	Canada	Sugars, vegetable oils
Terraverdae		PHA	Canada	Waste, others
PHB Industrial	Biocycle	PHB, PHBV	Brazil	Sugar (from sugar cane)
Biomer	Biomer	PHB	Germany	

Many collaborations between companies producing PHAs and other brands are under development. Bacardi set an agreement with Danimer Scientific to use their Nodax PHBV to produce their spirit bottles with a 100% biodegradable polymer, eliminating completely the use of commodity plastics in all of their products by the end of 2023 [73]. Danimer created a compostable sneak plastic bag in partnership with Pepsico, for which they won the Plastics Industry Association's "Innovation in Bioplastics" award in 2018.

Newlight Technologies collaborates with Dell to produce their PC packaging with Sprint to produce cellphone cases, and an agreement has been concluded with Ikea as well [4].

The partnerships between large corporations and companies producing PHAs indicate that also from an industrial point of view the transition to more sustainable materials is perceived as inevitable and the market for PHAs is growing fast. Price decreases and the increased awareness of society will raise the profile of PHAs in the market, resulting in a gradual substitution for commodity plastics like PP and PE.

Since the thermal and mechanical properties of PHAs and PHBV change as a function of the copolymer composition, the variety of applications is quite broad, and includes packaging, agriculture, and biomedicine but also consumer goods, textiles, electronics, and adhesives. The actual market sectors in which bioplastics (including PHAs) are applied are shown in Figure 3.6.

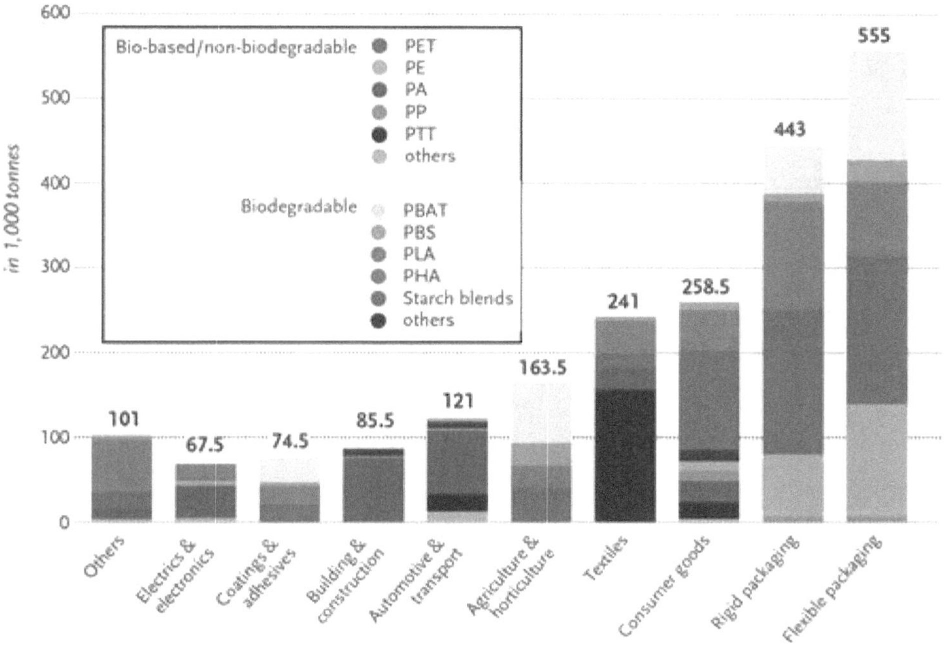

Figure 3.6 Global production capacities of bioplastics in 2020, by market sector [3]

PHBV was industrially applied for the first time by ICI and Wella to produce shampoo bottles in the 1980s [4,74]. Indeed, like the majority of bioplastics, the biggest part of global production is destined to be packaging (flexible or rigid), as confirmed by the partnerships of PHA producers. As mentioned earlier, PHAs have good barrier properties to water and oxygen (even better than PLA [31]), and together with their biocompatibility this makes them a good biodegradable alternative for packaging in contact with food [75]. For example, they have been studied as biodegradable cheese coatings [76, 77]. PHAs have been also employed as disposable absorbent products like diapers and sanitary napkins [78].

The biomedical sector constitutes another widespread application for PHAs linked with their biocompatibility [71]. They can be used for drug delivery and tissue engineering [79] or for suture points thanks to their resorbability. For example, the American company Tepha produces P4HB filaments to make sutures, or medical resorbable textiles and films [80]. They are also used for cosmetics [81], for their properties of oil-absorption.

In agriculture, PHAs are already used as mulch films and nets to protect crops [82], and composites with fillers of soy proteins are potentially marketable as agricultural growbags [82, 83]. For example, the Canadian company Terraverdae produces films and pots for agricultural applications [75]. The company PolyFerm Canada produces medium-chain length PHA copolymers specifically for markets in adhesives and sealants, plastic additives, inks and toners, paints and coatings, and medical devices [84].

3.1.4 Drawbacks

There are many reasons why bioplastics in general (and PHAs in particular) are not yet leading the market over fossil-based plastics. First of all, the cost of PHAs is far higher than commodity plastics and higher even than other bioplastics such as PLA, as mentioned earlier. The high cost is due to:

- the cost of feedstocks used for the bacteria fermentation
- the low yield of polymer produced compared to the amount of biomass
- the energy consumption of the production process
- the cost of the extraction and purification processes.

Linked to the production of PHAs is the reproducibility of the process, which is crucial. Bacteria are sensitive to different culture factors like pH, temperature, agitation, and of course to the composition of the carbon source and the ratio between the other nutrients such as phosphorus, nitrogen, etc. A slight change in these conditions could significantly affect both the yield of PHAs stored in the cells and the chemical composition of the final polyester. A careful examination of the literature shows that the declared percentage of valerate in PHBV of the same grade is often different. This variability is a limitation for the exploitation of PHAs. First, the thermal and mechanical properties change, so the requirements for an established application could not be always respected. Secondly, if the properties change, the processing conditions should be also adapted, implying an additional cost for the companies.

The second main reason impeding the widespread use of PHAs is that they do not perform like commodity plastics. In particular, PHB is brittle, as shown in the previous section. The brittleness is due to the formation of very large and crack-prone spherulites caused by the poor density of nuclei. Furthermore, the Tg is below room temperature, so it has a tendency to crystallize, and voids are created along the borders of spherulites, contributing to the brittleness. This fragility therefore reduces the exploitability of PHAs as an effective substitute for commodity plastics. In particular, the impact resistance is a required property for application in packaging, and the limited elongation at break is also an issue for packaging, textile and agriculture applications.

The slow crystallization is also a disadvantage for the processability. Indeed, a longer time to mold the polymer is necessary, which means a higher energy consumption. These drawbacks are only partially reduced by incorporating the co-monomer valerate. PHBV is the easiest PHB copolymer to obtain, but the side chain is not long enough to deeply influence the thermal and mechanical properties of PHB. Indeed, PHBV with a low percentage of HV still has high crystallinity coupled with a slow crystallization rate, and only the elongation at break is slightly improved.

Still related to the processability, the window of temperature is quite narrow due to the high melting temperature and the low thermal stability, since PHAs start to degrade at around 200 °C. Thus, the molecular weight decreases drastically every time that the polymer is processed. With a relatively high percentage of HV the melting point shifts to a lower temperature. However, most of the PHBV that is commercially available contains a low amount of the co-monomer. The main mechanism of degradation for PHB and PHBV is the random chain scission with the formation of crotonic acid, as shown in Figure 3.7 [31, 85]. As a consequence, the crystallization is also affected, shifting to lower temperature with slower kinetics.

Many efforts have been made both by academic and industrial researchers to reduce the cost of PHAs and to tackle the limitations related to PHA performances. The next section proposes a more detailed analysis of the potential solutions, and the solutions already adopted for each PHAs drawback.

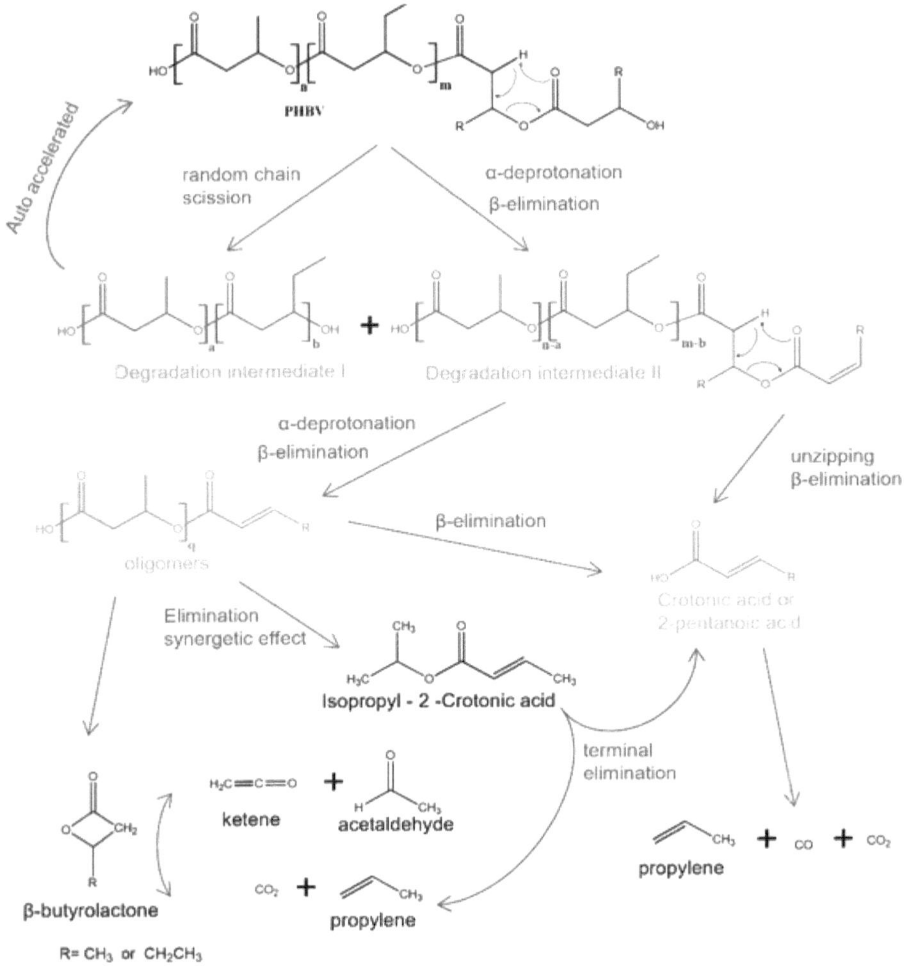

Figure 3.7 Thermal degradation of PHBV [85]

3.2 Reducing Production Costs of PHAs

To reduce the cost of PHAs, various strategies have been adopted tackling both the upstream and downstream processes. Concerning the upstream process, efforts have been pushed to find new and cheap carbon sources, and to increase the yield of PHA produced compared to the amount of non-PHA biomass. Photosynthetic bacteria and mixed cultures have also been tested to reduce the cost. The association of PHA production with other by-products was also found to be an interesting strategy. Concerning the downstream process, efforts have been made to reduce the cost of PHA extraction and purification.

3.2.1 Cheaper Carbon Sources

The easiest way to obtain PHAs from biomass fermentation is the implementation of pure carbon sources such as glucose and sucrose and organic acids like butyric, acetic, propionic, and valeric acids [86]. However, it has been estimated that 50% of the cost of PHA production is related to the use of expensive pure carbon sources as feedstocks [87]. This aspect made scientists focusing on finding cheap and renewable feedstocks also motivated by the goal of achieving circularity. Every year, vast quantities of waste is produced from agriculture and food industries, but also from household daily activities. Their use as feedstocks could partially solve the problem of (expensive) waste treatment processes [86]. In Figure 3.8, carbon substrates from renewable resources that have been already tested to obtain PHAs are summarized. They can be wastewaters rich in volatile fatty acids, exhausted cooking oil, or food waste [57], but they can also come from the production of dairy products like cheese whey (which contains up to 20% lactose), from agro-waste like molasses, or from biodiesel production coproducts like glycerol. Other substrates tested are algae [88], starch, wheat, and rice bran [87]. Edible plant oils were formerly widely used, but due to their increased cost, they are no longer considered as cheap carbon substrates [89].

Pre-treatment is needed for some renewable feedstocks before being consumed as a direct carbon source. Whey and lignocellulosic materials are hydrolyzed into simple sugars, wastewaters are transformed into volatile fatty acids through fermentation, while fats and oils are trans-esterified.

According to the renewable carbon source and the bacterial strain employed, PHAs of different chemical structure are obtained. Molasses from sugar cane or sugar beet are by-products derived from sugar production industries, and they are rich in simple sugars like sucrose, fructose, and xylitol. To give an example, *Azotobacter salinestris* produced PHBV when fed with molasses [90], while *A. vilenandii* gave PHB in the presence of beet molasses [91].

From glycerol (crude or purified), PHB is the main polyester obtained from most bacterial strains [48], however Koller et al. were able to obtain PHBV with 8–10% valerate from a wild osmophilic bacterium [92]. The same strain produced PHBV, also starting from hydrolyzed whey permeate. Ashby et al. coupled glycerol and levulinic acid (from the paper industries) as substrates for a strain of *Pseudomonas oleovorans*, and they were able to obtain an array of PHBV with different HV content and properties [93].

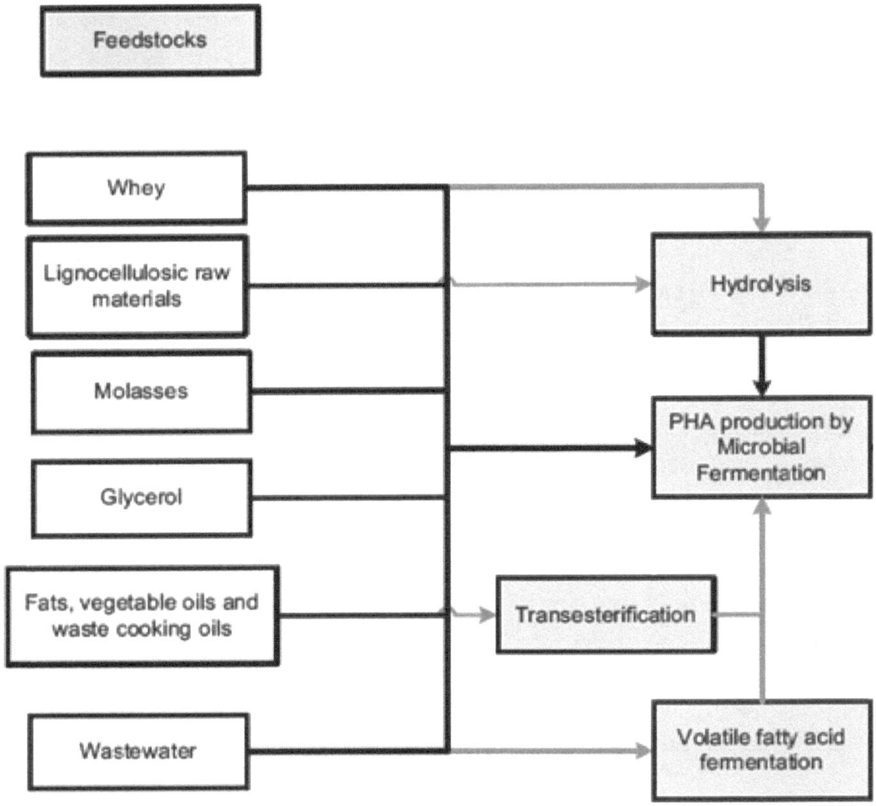

Figure 3.8 Renewable feedstocks (modified from Du et al. [89])

Haba et al. were able to obtain mcl-PHA from a specific strain of *Pseudomonas aeruginosa* employing as feedstocks waste cooking oil and free fatty acids deriving from soybean oil production [45]. From oils and fats, mcl- and lcl-PHAs are usually obtained [45]. However, despite the long carbon chain, *Cupriavidus necator* can produce scl-PHAs, consuming spent palm oil as the only carbon source [94].

Cheese whey contains up to 20% lactose, and after hydrolysis into galactose and glucose it was employed as a sole carbon source to obtain PHB from *Pseudomonas hydrogenovora* [95]. From the same substrate, in a previous study by the same group, a highly osmophilic bacterium was able to synthetize PHBV with 10% valerate, although the productivity was very low [92].

Wastewater derived from domestic waste, animal industries, paper, clothes or food production are very expensive to treat [89]. They contain high concentrations of inorganic and organic residues that are harmful for the environment, but they constitute a rich substrate for PHA-producing bacteria. *Azotobacter vinelandii*, for example, converted swine wastewater into PHBV with valerate up to 8 mol% [96], while a specific strain of *A. chroococcum* produced PHB when fed just with wastewater from olive oil mills, or PHBV when a small concentration of valerate was added [97]. *Enterobacter aerogenes* was found to produce PHB from domestic waste [98]. Due to their complex composition, not many studies have been done with real wastewater in the presence of a single bacterial strain.

The use of renewable feedstocks is not limited to academic research, but it is already spreading in industry as well. The company Full Cycle Bioplastics produces its PHAs starting from organic waste coming from food waste, cellulosic and dirty paper or woody biomass, agricultural by-products and wastewater [99]. With their technology they claim to produce PHAs with the most competitive price on the market compared to fossil-derived plastics. Their PHAs can be used to make different items from packaging and plastic bags to fibers and utensils. Bluepha also claims a low-cost production process in which alternative carbon sources like crops and kitchen waste are used to feed their reprogrammed microorganisms [100].

A limitation in the exploitation of renewable carbon sources comes from the limited capacity of the bacterium to metabolize them, beside the fact that waste often contains a mix of carbon substrates [101], and not the single substrates commonly used for bacterial cultures. Moreover, the presence of impurities and solid materials decrease PHA yield and productivity. Other types of bacteria like photosynthetic ones, with a more flexible metabolism, are considered as a promising alternative. Other solutions are the use of mixed microbial cultures (MMCs) or the engineering of bacteria [102]. The following sections are focused on these three potential alternatives.

3.2.2 Photosynthetic Microorganisms

Photosynthetic microorganisms have the capacity to produce their energy from sunlight and atmospheric CO_2 through photosynthesis. As a result, they do not need any other energy source for growing, so they can contribute to the growth of the circular bioeconomy. The CO_2 is incorporated in metabolites like proteins, amino acids, and lipids, but also biopolymers. Indeed, microorganisms like cyanobacteria [32] and purple bacteria [103, 104] are Gram-negative bacteria able to produce polyhydroxyalkanoates. They can proliferate in many different environmental conditions, and they are able to assimilate carbon sources coming from domestic, agricultural, and industrial waste. As already mentioned above, the production of bioplastics from water waste as feedstocks could reduce the cost of PHAs and at the same time the use of substrates that compete for food production.

Although the productivity of photosynthetic microorganisms is lower than commonly used bacteria, they present some advantages if PHA production is integrated into a biotechnological system. On one hand, the biomass residues can be used for animal feeding [105, 106] thanks to the rich nutrient content, so the wastewater obtained from animal farming contributes to bacterial growth, and then the biomass is reintegrated in the cycle. On the other hand, the production of by-products is a great strategy to reduce the cost of PHA production, as well as to contribute to the circular economy. Photosynthetic bacteria contain pigments necessary for their photosynthesis that can be recovered and used in cosmetics or other industrial sectors, such as food or pharmaceuticals [32, 107, 108].

Interestingly, these photosynthetic bacteria can produce PHAs without any depletion of any nutrient like (as in the most common cases), but just under redox stress [28, 109]. This means that the production of PHA occurs during the cell growth, so we have a single step instead of a two-steps process ("feast and famine"). Moreover, it is easier to control the stress by modulating the light intensity instead of preparing a second type of medium with another nutrient limitation, especially if we work with waste. In particular, purple non-sulfur bacteria are known for the versatility of their metabolism [109] and they can assimilate a mixture of carbon sources to obtain PHA copolymers [101]. Moreover, since they grow in anaerobic conditions, aeration is not necessary.

3.2.3 Mixed Cultures

Single-culture bacteria are usually employed to produce PHAs, and need sterile conditions and specific carbon sources - two requirements that increase the cost of the final polyester [110]. Mixed microbial cultures (MMCs) have been found to be promising in order to reduce the cost of PHA production [111]. The microbial mix contains bacteria able to accumulate PHAs and bacteria that cannot. The mix is enriched in PHAs, producing bacteria through alternating "feast" phases (with an excess of carbon source) and "famine" phases (characterized by a depletion of nutrients) [110, 112]. Indeed, the accumulated PHAs are used by the bacterial PHA producers to survive during the famine phase. In a second step, the PHAs are accumulated in the biomass.

The advantage of MMCs is that they do not need sterile conditions, thus cheap carbon sources from wastewater or other industrial waste can be used in place of pure and more expensive feedstocks [102, 113]. According to the results obtained from different studies, the amount of PHAs obtained with MMCs using various cheap feedstocks was comparable to that of single bacterial strains [110]. However, looking across all the studies done with MMCs, the cell density reached is very low compared to single-strain cultures, so the productivity is not yet at the same level of single cultures [57, 111].

Photosynthetic mixed cultures (PMCs) are a possible solution to reduce the cost of PHA production [110]. The use of different carbon sources at the same time is possible and leads to the production of PHAs with different chemical composition [114]. Like MMCs, PMCs can be fed with waste-derived carbon sources like volatile fatty acids (also in a mix) [110], but they do not need aeration, which is considered the most expensive part of MMC technology in term of energy consumption [110, 115]. However, PHA accumulation is not yet comparable to MMC and single-strain culture, so further studies are required [110, 116, 117].

Activated sludge has also been used to obtain PHAs, for example from wastewater derived from the paper [118] or sugar [119] industries.

3.2.4 Recombinant Bacteria

Other factors in addition to the carbon substrate and the downstream processes drive the high cost of PHAs, and they are responsible for the low PHA productivity (defined as amount of PHA per unit volume per unit time), the low content of PHAs compared to the total biomass, and the low yield relative to the carbon substrate [120]. Among them, PHA content is the most critical one because if it is very low it implies a huge amount of dried biomass is needed to obtain a reasonable amount of PHAs [36]. Thus, the energy consumption increases as well as the cost of the downstream processes.

The engineering of microorganisms is one of the techniques adopted to increase the amount of PHAs obtained and to improve efficiency. This strategy is also adopted to genetically modify a PHA producer in order to obtain different types of PHAs [121, 122]. Different approaches exist [120, 123] but the most common approach consists in genetically modifying a bacterium that usually does not produce PHAs by introducing the gene of another bacterial strain which expresses the enzyme producing PHAs [35].

The most studied and exploited genetically modified bacterium is recombinant *Escherichia coli* [120, 124]. Lee et al. produced *E. coli* expressing *Ralstonia eutropha* genes achieving a concentration of PHB above 100 g/L when fed with cheese whey. PHB accumulation of other common and wild bacteria strains did not reach 10 g/L, deemed to be a minimum threshold to be cost-effective.

Although PHB is the main polyester obtained from recombinant bacteria, PHBV, PHBH and other copolymers were also obtained from recombinant *E. coli* by harvesting different genes for PHA synthase from different bacteria and by introducing co-substrates [120]. However, the concentration and productivity of PHAs are reduced in these cases.

Beside the increased PHA amount, genetically modified bacteria have also other advantages. Since recombinant bacteria do not produce PHAs, in principle they do not express depolymerase enzymes, which can degrade the accumulated PHAs

[120]. Moreover, with natural PHA-producers it is often hard to lyse their cell walls, as recombinant bacteria become less sensitive due to the high accumulation of PHA grains [120, 125].

Another important advantage is the possibility to exploit many cheap and renewable carbon sources that bacteria like *E. coli* are able to digest differently from PHA producers, which often need pure and expensive carbon sources [120].

3.2.5 Valorization of By-Products

One strategy adopted to reduce the cost of PHAs and also to contribute to a circular economy is coupling bacterial fermentation with the production of valuable by-products like organic acids, amino acids, lipids, and others [71]. For example, photosynthetic microorganisms like cyanobacteria [32] and purple bacteria contain pigments to harvest the light necessary for the photosynthetic process. *Rhodospirillum rubrum*, a non-sulfur purple bacteria, contains mainly bacteriochlorophyll and carotenoids like spirilloxanthin [126]. Pigments are largely exploited in food, cosmetics, textiles, and pharmaceutical industries. Moreover, their naturally derived pigments are more appreciated by consumers compared to synthetic ones [107].

Succinate has been produced by engineered *E. coli* fed with glycerol and volatile fatty acids from waste from the biofuel industry [127]. Succinate is used to produce many chemicals exploited in food, pharmaceutical and polymer industries among others [128]. It has been demonstrated that *Pseudomonas aeruginosa* is able to accumulate PHAs while producing rhamnolipids using cooking oil waste as a carbon source [45]. These molecules are lipids containing a hydrophilic head composed of l-rhamnose and between one and four hydrophobic tails composed of β-hydroxyaliphatic moieties [129]. They are the most typical biosurfactants, which in general are preferred to their chemosynthetic counterparts thanks to their performances and ecological aspects. As surfactants they have many applications, and their production in parallel with PHAs can reduce the price of both.

3.2.6 Improved Extraction Techniques

At the end of PHA production processes, the granules must be extracted from the cells of the biomass and the polymer needs to be purified. These two consecutive steps are termed the "downstream process", which is known to be very expensive since it commonly requires large volumes of chemicals and solvents [130]. Usually, the biomass is lyophilized and then solvents, mostly chlorinated ones, are used to solubilize the polymer, which is then precipitated in cold methanol to purify it from biomass constituents such as proteins and lipids [52]. Soxhlet extraction can be also applied [131, 132]. Another common method of extraction is digestion with hypochlorite of the biomass surrounding PHA granules [133–135]. However, these typical "lab-scale" extraction methods have some drawbacks for the upscaling to the industrial level. Chlorinated solvents and chemicals are harmful both for the environment and human health, and they are very expensive, especially chlorinated solvents [132].

The use of milder digestion agents like sodium hydroxide and potassium hydroxide has been proposed to be more ecofriendly and still efficient compared to other stronger chemicals [121, 136].

An efficient and industrializable green method is the so called "Zeneca method", which consists of the enzymatic digestion of non-PHA biomass [137]. However, it is expensive, and the purity of the extracted PHAs is low [132, 138]. PolyFerm Canada, for example, uses a combination of chemical/enzymatic digestion and solvent extraction to obtain a highly pure polymer [84], and are trying to develop a process needing less solvent. Tianan claims to use a water-based process to extract PHA, but the purity is quite low, and their customers often need to purify it with solvents.

Indeed, the challenge is to find a method which is green, cheap, efficient, and industrializable [138].

Mechanical disintegration of cells containing PHA granules [132] are promising methods suitable for the industrial scale. Among them we can cite high-pressure homogenization (HPH) [132, 138], ultrasound [139], and bead mills [140, 141]. Recently, other methods studied are two-phase water extraction, air extraction, and supercritical CO_2 extraction [132, 138].

There is not just one solution to reduce the cost of downstream processing, because it must be adapted to the type of bacteria, the amount of PHA accumulated in the cells, and the type of PHA. All these parameters influence the efficiency of a certain method. For example, mixed microbial cultures are known to be more resistant compared to pure bacterial cultures, in which the accumulation of PHA granules makes the cell walls more fragile. Interestingly, mutant bacteria have been discovered that are able to produce extracellular PHAs that would significantly reduce the cost of their recovery [24].

3.3 Improving Production Reproducibility of PHAs

Reproducibility is among the challenges to make PHA as a viable alternative to non-biodegradable commodity plastics. Bacteria are very sensitive to external conditions like pH, temperature, oxygen or other nutrients like phosphorus, carbon and nitrogen, but also to the presence of impurities. Variations in just one of these parameters can influence their metabolism, and so their growth or (in the case of PHA-accumulating bacteria) the amount of PHAs accumulated and their chemical composition [87]. Haba et al. demonstrated that an increase in temperature of few degrees during the phase of PHA accumulation could reduce from 30 to 20 wt% the PHA content, and that certain co-monomers were present only at a specific temperature [45].

Even on an industrial level, reproducibility is a challenge. For example, the commercial grade of Tianan's PHBV Y1000 was claimed to contain 8 mol% valerate. However, different papers claimed different percentages after conducting full characterization, indicating that there is a variability from batch to batch.

Moreover, the use of cheap carbon feedstocks could bring a certain variability in term of PHA production. For example, carbon sources coming from waste such as water, food or agriculture contain impurities and contaminants that can impact the efficiency of bacterial growth and PHA expression [87]. Another factor to consider is the variability in composition of a given renewable feedstock, which will probably influence the final composition of the final polyester, causing uncertainty in the final industrial product. Therefore, much effort remains to be made to achieve reproducible production of PHAs, especially if renewable feedstocks are to be used more frequently.

3.4 Improving Performance of PHAs

Once the limitations related to the production of PHAs have been solved, the next significant challenge is to tackle the problems linked to their performance, starting with their poor processing properties and low thermal stability all the way to their crystallization kinetics and so their mechanical properties. The following sections will discuss these three main areas of improvement needed for the future widespread adoption of PHAs in the market. The common strategy encompasses the implementation of additives such as nucleating and reinforcing agents and plasticizers, or the design of blends with other polymers. Fortunately, in some cases a single additive can have an impact on the crystallization *and* mechanical properties, due to their mutual dependence and/or enhanced thermal stability. These different effects will be discussed separately when possible. Circularity is a final aspect of PHA performance that we will consider, with regard to the use of renewable and biodegradable additives.

3.4.1 Controlling Crystallization

In general, PHAs crystallize slowly due to the low density of heterogeneous nuclei that causes the formation of large spherulites [48]. The reason for this is related to the purity of the polyesters. Usually, polymers obtained through synthetic pathways (such as petrochemical-derived ones) contain impurities, inclusions, catalysts, and chemically different end-chains that act as heterogeneous nuclei, decreasing the activation energy required to form a nucleus and so boosting the crystallization process [53]. Since PHAs are produced by bacteria, they do not contain such impurities, so the crystallization mechanism is mainly attributed to a homogeneous nucleation process, in which it is the polymer itself that creates the nucleus from which it starts to crystallize. Homogeneous nucleation requires a high energy of activation, which results in very low density of nuclei. The negative consequences of this slow crystallization are three-fold: poor processing of the polymer, the re-crystallization of the polyester over time, and the already-mentioned brittleness of PHB and PHBV.

Regarding the processing, slow crystallization makes techniques like injection-molding and film-blowing difficult to be applied, since a fast crystallization is required [142, 143]. Moreover, films of PHBV and other copolymers remain sticky for a while after their processing.

PHAs are also known for their "aging" effect due to secondary crystallization events after processing, particularly favored by the chain mobility, since the glass transitions are typically below room temperature [93]. This causes an increase of the crystallinity over time, which is associated with the rise of the Young' modulus, but also a decrease of the elongation at break and of the toughness. This results in an embrittlement effect on the polymeric system [93]. This effect is even more evident for copolymers in which crystallization of the PHB fraction is hindered by the co-monomer during the processing. The brittleness is of course related to the low density of nuclei that results in micrometric spherulites with cracks and imperfections.

To control the crystallization of PHAs, nucleating agents have been employed as additives, and some patents exist [143–145]. Indeed, they increase the nuclei density and boost the crystallization kinetics, resulting in much smaller spherulites which should decrease the brittleness and so avoid embrittlement due to recrystallization during storage. Many nucleating agents have been tested, including boron nitride (which is known to be very efficient), saccharin, cyanuric acid, talc, among others [52, 142, 146, 147]. Clays such as montmorillonite were found to have an effect on the crystallization kinetics of PHAs [148]. Herein, to keep in mind the ultimate goal of a circular economy approach, only the studies using biodegradable additives obtained from renewable resources have been considered. In this context, orotic acid [149], cyclodextrins [150–152] (derived from starch), or uracil and thymine [53] (derived from RNA and DNA), have been successfully employed as nucleating agents for PHB, and for PHBHHx with different percentages of HHx. In general, the increase of heterogeneous nuclei boosted the crystallization rate and resulted in the formation of small spherulites with benefits for the mechanical properties too. Other renewable nucleating agents already employed are lignin [153], polysaccharide nanocrystals (principally derived from cellulose) [154–157], and (in recent years) chitin [158]. As demonstrated in the cited works, nanocrystals are interesting for multiple reasons; even at a very low content, they act as a reinforcement, improving the mechanical performance of the composites, and they also hamper the path of gases, improving the barrier properties.

Although chitin nanocrystals (ChNCs) have been largely used to tune polylactide crystallization [159–161], only a few papers show their effect on PHA crystallization, which varies substantially according to the topochemistry of the nanocrystals. Indeed, Zaccone et al. employed unmodified ChNCs showing a nucleating effect, but coupled with a slower crystallization rate [162]. This effect was confirmed by Wang et al. [163, 164]. In contrast, Li et al. obtained faster crystallization [158], suggesting an influence of both the ChNC extraction method and the processing. Interestingly, the grafting of ChNCs also gave different results according to the type of functions [158, 163, 164]. For example, sulfhydryl silane-coupled ChNCs acted as a nucleating agent, boosting the overall crystallization kinetics of PHBV, although slowing the growth rate compared to pristine ChNCs [158]. In contrast, PHBV-grafted chitin nanocrystals slowed down the crystallization of PHBV [164]. According to the authors, the high density of hydrogen bonds, despite the grafting, reduced the chain mobility of the matrix, hampering the crystallization. However, the same strategy with other matrices boosted crystallization [165].

3.4.2 Improving Mechanical Properties

The importance of choosing biodegradable additives relies on preserving the main characteristics of PHAs, i. e., their capacity to degrade in many environments, giving a significant advantage towards fossil-derived plastics and other bio-sourced polymers like PLA. Non-biodegradable additives could affect the rate of biodegradation. Sridewi et al. demonstrated that TiO_2 nanoparticles visibly slowed down the degradation of PHB in the marine environment [66].

The most environment-friendly additives are natural fiber [166–170] and polysaccharide nanocrystals [156, 171], which thanks to their outstanding mechanical properties, act as reinforcing agents, also enhancing the barrier properties of PHAs. For example, Malmir et al. showed that extruded bionanocomposites of PHBV with cellulose nanocrystals (CNCs) had better mechanical properties compared to neat PHBV [172]. The same group stated out that the addition of 4 wt% CNCs not only increased the density of PHBV spherulites but also quadrupled the resistance to water vapor and oxygen transmission [154]. Yu et al. found that 12 wt% CNCs improved the Young's modulus and tensile strength of PHBV by 300% and 175%, respectively [156].

However, the hydrogen bond formed between the nanocrystals upon drying reduces their easy incorporation into hydrophobic polymer. For example, this results in the formation of aggregates during the melt-processing of PHA with polysaccharide nanocrystals [173, 174], reducing the nano-effect and so the mechanical performance. Thus, the nanocomposites are usually prepared by solvent-casting to avoid the aggregation of nanocrystals [155]. While in some cases a compatibilizer, like polyethylene glycol, has been used to improve the interfacial adhesion [175] the functionalization of the nanocrystals is the most widely adopted strategy to improve the adhesion at the interface [171, 176]. This is due to the hydroxyl groups on the cellulose surface that provide a large spectrum of functionalization opportunities [177].

Yu et al. investigated the hydrogen bonds between PHBV and CNCs, comparing neat CNCs to formic and citrate-modified CNCs. They showed that citrate-modified nanocrystals led to the strongest interface and the best mechanical properties, with an increase in the Young's modulus from ca. 60 to 250 MPa, and an increase in the tensile strength from 14 to 31 MPa, compared to neat PHBV [171].

Chitin nanocrystals were incorporated by Li et al. in PHBV HV20% to produce a scaffold for tissue engineering, improving the compression modulus from 5 to 7 MPa and also promoting cell adhesion [178]. Improved mechanical properties were also obtained by Wang et al. [163], and Zaccone et al. [162] even observed the presence of ChNC aggregates. Like for cellulose nanocrystals, surface modification of chitin was applied to tune the dispersity and the interface interactions of the bionanocomposites [158]. Wang et al. demonstrated that acetylated chitin nanocrystals gave better mechanical properties of PHBV, thanks to better dispersion in the matrix [163].

Carbon nanotubes are another additive used as a reinforcement for PHBV, as studied by Yu et al. [179]. Although at the moment carbon nanotubes cannot be considered renewable, promising efforts are being made to produce them from renewable resources [180, 181] and research their biodegradability [182].

As the rigidity of most PHAs is one of the main drawbacks, plasticizers are preferred as the most common additives. Indeed, plasticizers influence the thermal properties of the polymers, reducing the glass transition and the crystallinity, with an improvement of the ductility as a consequence. As an example, tributyl citrate was added to PHBV, resulting in biocomposites with improved elongation at break. With the same trend, El-Hadi et al. added different plasticizers coupled with saccharin as a nucleating agent, and found improved mechanical properties such as elongation at break and impact strength [46].

The blending of PHAs with other polymers is a strategy largely adopted to tackle the mechanical performance of PHAs, and often it can also help with processing. PHB and PHBV are often blended with less crystalline and softer biodegradable biopolymers such as PBS, PCL, or PLA to improve their ductility, flexibility, and/or impact strength [183–186]. Interestingly, Wang et al. found that PHB mechanical properties such as elongation at break were strongly ameliorated by blending it with amorphous PHAs like PHBHHx or P3/4HB, especially with the synergistic effect of the plasticizer acetyl tributyl citrate (ATBC) [187]. The blending of PHB with PHBHHx has been further studied for tissue engineering, with success.188

Taking advantage of the possibility to tune the structure of PHAs, Srienc et al. discovered that, interestingly, PHA block copolymers withstood aging better than random copolymers [189]. According to this study, in the case of PHBV block copolymer, the elongation at break remained higher than 90% after three months, although recrystallization still occurred. By contrast, the random PHBV become stiff and brittle swiftly. Zhao et al. obtained a terpolymer composed of PHB copolymerized with HV and HHx that showed impressive mechanical properties, with elongation at break reaching 480%, and higher thermal stability [190].

3.4.3 Improving Thermal Stability and Processability

As mentioned above, PHAs suffer from low thermal stability, since they start to degrade around 200 °C through random chain-scission. Together with the high melting temperature (especially for PHB and PHBV (co)polymers with a low content of HV), this results a very narrow processability window. As mentioned before, plasticizers have a positive effect on the thermal behavior, so many studies have been conducted in order to decrease the melting point and the crystallinity [191]. Among them, acetyl tributyl citrate (ABTC) has been often used [148, 192], and in particular Correa et al. showed that ABTC was able to decrease the thermomechanical degradation of PHBV during extrusion and that it had a beneficial effect in mechanical properties in term of impact strength [192]. Other efficient biodegradable plasticizers were dibutyl phthalate (DBP) and triethyl citrate (TEC) used by Choi et al. [193], or vegetable oils like sunflower oil [194] or epoxidized soybean [193] or broccoli oil [195]. However, plasticizers increase the free volume and the diffusion [196], so they can deteriorate the barrier properties of the polyesters [197].

Auriemma et al. employed tannic acid, a natural polyphenolic acid, as a thermal stabilizer for PHB, that increased the maximum temperature of degradation, probably thanks to the crosslinking with PHB chains [198].

The capacity of polysaccharide nanocrystals to improve the thermal stability of PHA is controversial. Some papers stated out that cellulose nanocrystals improved it [156, 171, 176], while others claimed it reduced it [175]. Concerning ChNCs and even their known high thermal stability [199], only one paper studied their influence on PHA thermal degradation, finding it unchanged [162]. However, chitin nanofibrils increased the thermal decomposition temperature of PHB by 24 °C [170].

Looking at PHAs end-of-life, although biodegradability could reduce the impact of plastic waste in the oceans, the natural degradation of biodegradable polymers should not be considered as the unique solution for plastic waste, but it should be seen as the ultimate scenario in the case where they are accidently disposed of in the environment. A more environmentally friendly approach is to take advantage of its biodegradability for organic recycling, i.e., the production of nutrient-rich soils thanks to PHA degradation by microorganisms that can be used for plant growth.

Other types of recycling approaches exist that could contribute to the building of a circular economy, and although they are currently mostly used for non-biodegradable plastics, they have the potential to be applied also for biodegradable bioplastics. The most well-known one is mechanical recycling, which involves the melt-processing of the plastics into raw materials as ready-to-use products without modifying the chemical structure of the polymer. This method is mainly used for fossil-based plastics like PET and HDPE. However, mechanical recycling is often paired with the concept of downcycling, meaning that the recycled plastics have lower performances compared to the virgin ones. Downcycled recycled plastics often come from the recycling of mixed-plastics waste streams, when it is not possible to sort them, or from single-plastic waste stream when additives are present [25].

Chemical recycling is applicable to multi-material waste or added plastics that would otherwise have been incinerated or landfilled with the loss of valuable plastics products [25]. In this sense, chemical recycling supports mechanical recycling when the latter is not applicable. Chemical recycling refers to a group of technologies like pyrolysis, solvolysis, and hydro-cracking [26], which involve chemically modifying the polymer through chemical, thermal or catalytic processes to obtain monomers or other suitable molecules [18]. The recovery of starting monomers or polymerizable entities enables plastic materials to be obtained with the same properties as virgin plastics, and which can be used for the same applications [20, 27]. This is very interesting in the case of food packaging, in which the plastic materials will be free from any contaminations, in contrast to mechanically recycled plastics. In this case, we talk about "upcycled" plastics [19]. Although mechanical recycling can be considered as environmentally better, the products obtained by chemical recycling are of higher quality, together with a broader range of applications.

Bioplastics can be chemically recycled, although this strategy is not yet widely exploited, due to the small amount of bioplastics produced per year. Interestingly, some of them can provide monomers through much easier reactions like the hydrolysis of polyester matrices such as PHAs and PLA [20]. PHAs in this sense have many advantages. They are more hydrolyzable than PLA, so their end-of-life treatment is more compatible with existing composters [24], and they are compostable even at low temperatures, which make them suitable for home composting. In contrast, PLA is compostable only when the temperature is around its glass transition [28, 29]. Moreover, their versatile biodegradability makes them very promising for biological recycling [24]. This innovative technique couples polymer degradation by microorganisms with the concept of chemical recycling: instead of biodegrading polymers until biomineralization, microorganisms and their enzymes can be used to depolymerize the polymers to obtain monomers that are then be collected and revalorized for polymerization again. Although this technology is still in development, PHAs offer a multitude of options compared to PLA and other biodegradable polymers.

Life cycle analysis demonstrates that each parameter can significantly reduce the environmental impact of certain plastics, with a view of moving towards a circular economy. The inherently eco-friendly properties of PHAs, together with their versatility in terms of production and end-of-life scenarios, make them a powerful strategy to accomplish the goal of the circular economy.

3.5 Conclusions

PHAs are promising renewable polyesters produced by bacteria and other microorganisms that have significant potential to replace fossil-derived plastics for the building of a more circular economy. Indeed, they have the advantage of being biodegradable in most natural environments, and they can be readily recycled. However, they suffer from two main drawbacks: their high price, which limits their widespread use, and their poor mechanical performances, which are mainly linked to specific feature of their crystallization process. To make them more appealing, industrial and academic researchers are looking for ways of overcoming these drawbacks.

From a production viewpoint, cheap carbon sources like industries and household waste have been employed to produce PHAs. In the same way, different types of bacteria and mixed cultures have been used to increase production, and advanced purification protocols have been designed. However, controlling the chemical structure of the PHAs produced is still a challenge, and the different characteristics of the bacterial strain and the final polyesters makes it difficult to design adaptable protocols of production and recovery.

Finally, the incorporation of additives, or other polymers, are the most promising strategies to tackle the poor mechanical performance of PHAs, mainly through control of the crystallization kinetics. To improve compatibility with PHAs, additives such as polysaccharide nanocrystals have been functionalized. However, solvent casting, which is not eco-friendly, is the most common method used to produce nanocomposites, and the results obtained are sometimes controversial. Moreover, the potential of some additives like chitin nanocrystals have been poorly investigated.

References for Chapter 3

[1] Hahladakis, J. N.; Iacovidou, E.; Gerassimidou, S. Plastic waste in a circular economy. In *Plastic Waste and Recycling*; Elsevier, 2020; pp 481–512

[2] Shen, M.; Song, B.; Zeng, G.; Zhang, Y.; Huang, W.; Wen, X.; Tang, W. Are biodegradable plastics a promising solution to solve the global plastic pollution? *Environ. Pollut.* 2020, *263*, 114469

[3] European Bioplastics, *https://www.european-bioplastics.org* (accessed December 22, 2021)

[4] Shogren, R.; Wood, D.; Orts, W.; Glenn, G. Plant-based materials and transitioning to a circular economy. *Sustain. Prod. Consum.* 2019, *19*, 194–215

[5] Geyer, R.; Jambeck, J. R.; Law, K. L. Production, use, and fate of all plastics ever made. *Sci. Adv.* 2017, July, 25–29

[6] Karan, H.; Funk, C.; Grabert, M.; Oey, M.; Hankamer, B. Green bioplastics as part of a circular bioeconomy. *Trends Plant Sci.* 2019, *24*, 237–249

[7] Plastic pollution facts: The causes and it's solution – Wikye, *https://www.wikye.com/plastic-pollution-facts-the-causes-and-its-solution/* (accessed December 22, 2021)

[8] Where mismanaged plastic waste is generated and possible paths of change, *https://theoceancleanup.com/updates/where-mismanaged-plastic-waste-is-generated-and-possible-paths-of-change/* (accessed December 22, 2021)

[9] Plastic pollution | Shutterstock *https://www.shutterstock.com/it/search/plastic+pollution?kw=shutterstock&c3apidt=p18709897701&gclid=CjwKCAiAtouOBhA6EiwA2nLKH-hQveq9ChTx6bdnYtCvUT2__tBXG1kji3npshKcX9VKJnCetx6aCRoCn14QAvD_BwE&gclsrc=aw.ds* (accessed December 22, 2021)

[10] Ries, F. European Parliament votes for single-use plastics ban | Environment for Europeans *https://ec.europa.eu/environment/efe/news/european-parliament-votes-single-use-plastics-ban-2019-01-18_en* (accessed December 11, 2019).

[11] Kasulaitis, B. V.; Babbitt, C. W.; Krock, A. K. Rochester Institute of Technology *https://www.rit.edu/sustainabilityinstitute/blog/what-circular-economy* (accessed December 22, 2021)

[12] Popovic, S. Z.; Lazic, V. L.; Hromis, N. M.; Suput, D. Z.; Bulut, S. N. Chapter 8: Biopolymer packaging materials for food shelf-life prolongation. In *Biopolymers for Food Design*; Grumezescu, A. M., Holban, A. M., Eds.; Elsevier: London, 2018; pp 223–278

[13] Luyt, A. S.; Malik, S. S. Can Biodegradable plastics solve plastic solid waste accumulation? In *Plastics to Energy: Fuel, Chemicals, and Sustainability Implications*; William Andrew Publishing, 2019; pp 403–423

[14] Haider, T.P.; Völker, C.; Kramm, J.; Landfester, K.; Wurm, F.R. Plastics of the future? The Impact of biodegradable polymers on the environment and on society. *Angew. Chemie Int. Ed.* 2019, *58*, 50–6

[15] Ali, S.S.; Elsamahy, T.; Al-Tohamy, R.; Zhu, D.; Mahmoud, Y.A.G.; Koutra, E.; Metwally, M.A.; Kornaros, M.; Sun, J. Plastic wastes biodegradation: Mechanisms, challenges and future prospects. *Sci. Total Environ.* 2021, *780*, 146590.

[16] Avérous, L.; Pollet, E. Biodegradable polymers. In *Environmental Silicate Nano-Biocomposites*; Avérous, L., Pollet, E., Eds.; Springer, London, 2012; pp 13–39

[17] Emadian, S.M.; Onay, T.T.; Demirel, B. Biodegradation of bioplastics in natural environments. *Waste Manag.* 2017, *59*, 526–536

[18] Plastics Europe, Enabling a sustainable future *https://plasticseurope.org/* (accessed December 9, 2021)

[19] Zhao, X.; Korey, M.; Li, K.; Copenhaver, K.; Tekinalp, H.; Celik, S.; Kalaitzidou, K.; Ruan, R.; Ragauskas, A.J.; Ozcan, S. Plastic waste upcycling toward a circular economy. *Chem. Eng. J.* 2022, *428*, 131928

[20] Shogren, R.; Wood, D.; Orts, W.; Glenn, G. Plant-based materials and transitioning to a circular economy. *Sustain. Prod. Consum.* 2019, *19*, 194–215

[21] Carus, M.; vom Berg, C. Comprehensive concept of a circular economy *https://renewable-carbon.eu/publications/product/comprehensive-concept-of-a-circular-economy-png/* (accessed December 9, 2021)

[22] Dietrich, K.; Dumont, M.J.; Del Rio, L.F.; Orsat, V. Producing PHAs in the bioeconomy – Towards a sustainable bioplastic. *Sustain. Prod. Consum.* 2017, *9, 58–70*[23] Nitkiewicz, T.; Wojnarowska, M.; Sołtysik, M.; Kaczmarski, A.; Witko, T.; Ingrao, C.; Guzik, M. How sustainable are biopolymers? Findings from a life cycle assessment of polyhydroxyalkanoate production from rapeseed-oil derivatives. *Sci. Total Environ.* 2020, *749*, 141279

[24] Rosenboom, J.-G.; Langer, R.; Traverso, G. Bioplastics for a circular economy. *Nat. Rev. Mater.* 2022

[25] Davidson, M.G.; Furlong, R.A.; McManus, M.C. Developments in the life cycle assessment of chemical recycling of plastic waste – A review. *J. Clean. Prod.* 2021, *293*, 126163

[26] Vollmer, I.; Jenks, M.J.F.; Roelands, M.C.P.; White, R.J.; van Harmelen, T.; de Wild, P.; van der Laan, G.P.; Meirer, F.; Keurentjes, J.T.F.; Weckhuysen, B.M. Beyond mechanical recycling: Giving new life to plastic waste. *Angew. Chemie Int. Ed.* 2020, *59*, 15402–15423

[27] Niaounakis, M. Physical recycling. In *Biopolymers Reuse, Recycling, and Disposal*; William Andrew Publishing, 2013; pp 151–166

[28] Narancic, T.; Verstichel, S.; Reddy Chaganti, S.; Morales-Gamez, L.; Kenny, S.T.; De Wilde, B.; Babu Padamati, R.; O'Connor, K.E. Biodegradable plastic blends create new possibilities for end-of-life management of plastics but they are not a panacea for plastic pollution. *Environ. Sci. Technol.* 2018, *52*, 10441–10452

[29] Delacuvellerie, A.; Benali, S.; Cyriaque, V.; Moins, S.; Raquez, J.M.; Gobert, S.; Wattiez, R. Microbial biofilm composition and polymer degradation of compostable and non-compostable plastics immersed in the marine environment. *J. Hazard. Mater.* 2021, *419*, 126526

[30] Obruca, S.; Sedlacek, P.; Slaninova, E.; Fritz, I.; Daffert, C.; Meixner, K.; Sedrlova, Z.; Koller, M. Novel Unexpected functions of PHA granules. *Appl. Microbiol. Biotechnol.* 2020, *104*, 4795–4810

[31] Bugnicourt, E.; Cinelli, P.; Lazzeri, A.; Alvarez, V. Polyhydroxyalkanoate (PHA): Review of synthesis, characteristics, processing and potential applications in packaging. *Express Polym. Lett.* 2014, *8*, 791–808

[32] Gradíssimo, D.G.; Xavier, L.P.; Santos, A.V. Cyanobacterial polyhydroxyalkanoates: A sustainable alternative in circular economy. *Molecules* 2020, *25*, 4331

[33] Lemoigne, M. Produits de deshydration et de polymerisation de l'acide β-oxybutyrique. *Bull. Soc. Chim. Biol.* 1926, *8*, 770–782

[34] Lu, J.; Tappel, R. C.; Nomura, C. T. Mini-Review: Biosynthesis of poly(hydroxyalkanoates). *Polym. Rev.* 2009, *49*, 226–248

[35] Pollet, E.; Avérous, L. Production, chemistry and properties of polyhydroxyalkanoates. In *Biopolymers – New Materials for Sustainable Films and Coatings*; Plackett, D., Ed.; John Wiley and Sons, 2011; pp 65–86

[36] Choi, J.; Lee, S. Y. Factors affecting the economics of polyhydroxyalkanoate production by bacterial fermentation. *Appl. Microbiol. Biotechnol.* 1999, *51*, 13–21

[37] Zinn, M.; Witholt, B.; Egli, T. Occurrence, synthesis and medical application of bacterial polyhydroxyalkanoate. *Adv. Drug Deliv. Rev.* 2001, *53*, 5–21

[38] Bresan, S.; Sznajder, A.; Hauf, W.; Forchhammer, K.; Pfeiffer, D.; Jendrossek, D. Polyhydroxyalkanoate (PHA) granules have no phospholipids. *Sci. Rep.* 2016, *6*, 1–13

[39] Obruca, S.; Sedlacek, P.; Slaninova, E.; Fritz, I.; Daffert, C.; Meixner, K.; Sedrlova, Z.; Koller, M. Novel unexpected functions of PHA granules. *Appl. Microbiol. Biotechnol. 2020 10411* 2020, *104*, 4795–4810

[40] Maestro, B.; Sanz, J. M. Polyhydroxyalkanoate-associated phasins as phylogenetically heterogeneous, multipurpose proteins. *Microb. Biotechnol.* 2017, *10*, 1323–1337

[41] Meng, D. C.; Shen, R.; Yao, H.; Chen, J. C.; Wu, Q.; Chen, G. Q. Engineering the diversity of polyesters. *Curr. Opin. Biotechnol.* 2014, *29*, 24–33

[42] Yang, S. T. Bioprocessing for value-added products from renewable resources. In *Bioprocessing for Value-Added Products from Renewable Resources*; Shang-Tian, Y., Ed.; Elsevier, 2007; Vol. pg. 591

[43] Reddy, C. S. K.; Ghai, R.; Rashmi; Kalia, V. C. Polyhydroxyalkanoates: An overview. *Bioresour. Technol.* 2003, *87*, 137–146

[44] Policastro, G.; Fabbricino, A.; Panico, M. Improving biological production of poly(3-hydroxybutyrate-co-3-hydroxyvalerate) (PHBV) co-polymer: A critical review. *Rev. Environ. Sci. Bio/Technology 2021 202* 2021, *20*, 479–513

[45] Haba, E.; Vidal-Mas, J.; Bassas, M.; Espuny, M. J.; Llorens, J.; Manresa, A. Poly 3-(hydroxyalkanoates) produced from oily substrates by *Pseudomonas aeruginosa* 47T2 (NCBIM 40044): Effect of nutrients and incubation temperature on polymer composition. *Biochem. Eng. J.* 2007, *35*, 99–106

[46] El-Hadi, A.; Schnabel, R.; Straube, E.; Müller, G.; Henning, S. Correlation between degree of crystallinity, morphology, glass temperature, mechanical properties and biodegradation of poly (3-hydroxyalkanoate) PHAs and their blends. *Polym. Test.* 2002, *21*, 665–674

[47] Mohamed El-Hadi, A. Investigation of the effect of nano-clay type on the non-isothermal crystallization kinetics and morphology of poly(3(R)-hydroxybutyrate) PHB/clay nanocomposites. *Polym. Bull.* 2014, *71*, 1449–1470

[48] Zhu, C.; Chiu, S.; Nakas, J. P.; Nomura, C. T. Bioplastics from waste glycerol derived from biodiesel industry. *J. Appl. Polym. Sci.* 2013, *130*, 1–13

[49] Kai, W.; He, Y.; Inoue, Y. Fast Crystallization of poly(3-hydroxybutyrate) and poly(3-hydroxybutyrate-co-3-hydroxyvalerate) with talc and boron nitride as nucleating agents. *Polym. Int.* 2005, *54*, 780–789

[50] Kamiya, N.; Yamamoto, Y.; Inoue, Y.; Chujo, R.; Doi, Y. Microstructure of bacterially synthesized poly(3-hydroxybutyrate-co-3-hydroxyvalerate). *Macromolecules* 1989, *22*, 1676–1682

[51] Kunioka, M.; Tamaki, A.; Doi, Y. Crystalline and thermal properties of bacterial copolyesters: Poly(3-hydroxybutyrate-co-3-hydroxyvalerate) and poly(3-hydroxybutyrate-co-4-hydroxybutyrate). *Macromolecules* 1989, *22*, 694–697

[52] Zhu, C.; Nomura, C. T.; Perrotta, J. A.; Stipanovic, A. J.; Nakas, J. P. The effect of nucleating agents on physical properties of poly-3-hydroxybutyrate (PHB) and poly-3-hydroxybutyrate-co-3-hydroxyvalerate (PHB-Co-HV) produced by *Burkholderia cepacia* ATCC 17759. *Polym. Test.* 2012, *31*, 579–585

[53] Pan, P.; Liang, Z.; Nakamura, N.; Miyagawa, T.; Inoue, Y. Uracil as nucleating agent for bacterial poly[(3-hydroxybutyrate)-co-(3-hydroxyhexanoate)] copolymers. *Macromol. Biosci.* 2009, *9*, 585–595

[54] Doi, Y.; Kitamura, S.; Abe, H. Microbial synthesis and characterization of poly(3-hydroxybutyrate-co-3-hydroxyhexanoate). *Macromolecules* 1995, *28*, 4822–4828

[55] Pan, P.; Shan, G.; Bao, Y.; Weng, Z. Crystallization kinetics of bacterial poly(3-hydroxylbutyrate) copolyesters with cyanuric acid as a nucleating agent. *J. Appl. Polym. Sci.* 2013, *129*, 1374–1382

[56] Spyros, A.; Marchessault, R.H. Segmental dynamics in poly(3-hydroxybutyrate-co-4-hydroxybutyrate)s above the glass transition temperature: ^{13}C nuclear magnetic relaxation in the amorphous phase. *Macromolecules* 1996, *29*, 2479–2486

[57] Tsang, Y.F.; Kumar, V.; Samadar, P.; Yang, Y.; Lee, J.; Ok, Y.S.; Song, H.; Kim, K.H.; Kwon, E.E.; Jeon, Y.J. Production of bioplastic through food waste valorization. *Environ. Int.* 2019, *127*, 625–644

[58] Wu, F.; Misra, M.; Mohanty, A.K. Challenges and new opportunities on barrier performance of biodegradable polymers for sustainable packaging. *Prog. Polym. Sci.* 2021, *117*, 101395

[59] Modi, S.; Koelling, K.; Vodovotz, Y. Assessment of PHB with varying hydroxyvalerate content for potential packaging applications. *Eur. Polym. J.* 2011, *47*, 179–186

[60] Shogren, R. Water vapor permeability of biodegradable polymers. *J. Environ. Polym. Degrad.* 1997, *5*, 91–95

[61] Corre, Y.M.; Bruzaud, S.; Audic, J.L.; Grohens, Y. Morphology and functional properties of commercial polyhydroxyalkanoates: A comprehensive and comparative study. *Polym. Test.* 2012, *31*, 226–235

[62] Zhou, J.; Ma, X.; Li, J.; Zhu, L. Preparation and characterization of a bionanocomposite from poly (3-hydroxybutyrate-co-3-hydroxyhexanoate) and cellulose nanocrystals. *Cellulose* 2019, *26*, 979–990

[63] Lopez-Llorca, L.V.; Colom Valiente, M.F.; Gascon, A. A study of biodegradation of poly-β-hydroxyalkanoate (PHA) films in soil using scanning electron microscopy. *Micron* 1993, *24*, 23–29

[64] Fernandes, M.; Salvador, A.; Alves, M.M.; Vicente, A.A. Factors affecting polyhydroxyalkanoates biodegradation in soil. *Polym. Degrad. Stab.* 2020, *182*, 109408

[65] Mergaert, J.; Webb, A.; Anderson, C.; Wouters, A.; Swings, J. Microbial degradation of poly(3-hydroxybutyrate) and poly(3-hydroxybutyrate-co-3-hydroxyvalerate) in soils. *Appl. Environ. Microbiol.* 1993, *59*, 3233–3238

[66] Sridewi, N.; Bhubalan, K.; Sudesh, K. Degradation of commercially important polyhydroxyalkanoates in tropical mangrove ecosystem. *Polym. Degrad. Stab.* 2006, *91*, 2931–2940

[67] Volova, T.; Kiselev, E.; Nemtsev, I.; Lukyanenko; Sukovatyi, A.; Kuzmin, A.; Ryltseva, G.; Shishatskaya, E. Properties of degradable polyhydroxyalkanoates with different monomer compositions. *Int. J. Biol. Macromol.* 2021, *182*, 98–114

[68] Wang, Y.W.; Mo, W.; Yao, H.; Wu, Q.; Chen, J.; Chen, G.Q. Biodegradation studies of poly(3-hydroxybutyrate-co-3-hydroxyhexanoate). *Polym. Degrad. Stab.* 2004, *85*, 815–821.

[69] Reis, M.A.M.; Serafim, L.S.; Lemos, P.C.; Ramos, A.M.; Aguiar, F.R.; Van Loosdrecht, M.C.M. Production of polyhydroxyalkanoates by mixed microbial cultures. *Bioprocess Biosyst. Eng.* 2003, *25*, 377–385

[70] Lettner, M.; Schöggl, J.P.; Stern, T. Factors influencing the market diffusion of bio-based plastics: results of four comparative scenario analyses. *J. Clean. Prod.* 2017, *157*, 289–298

[71] Singh, A.K.; Srivastava, J.K.; Chandel, A.K.; Sharma, L.; Mallick, N.; Singh, S.P. Biomedical applications of microbially engineered polyhydroxyalkanoates: An insight into recent advances, bottlenecks, and solutions. *Appl. Microbiol. Biotechnol.* 2019, *103*, 2007–2032

[72] Bio-on | Turn Off Pollution *http://www.bio-on.it/* (accessed December 28, 2021)

[73] 100% Biopolymer Rum Bottle | BioCycle *https://www.biocycle.net/100-biopolymer-rum-bottle/* (accessed December 28, 2021)

[74] Luzier, W. D. Materials Derived from biomass/biodegradable materials. *Proc. Natl. Acad. Sci.* 1992, *89*, 839–842

[75] TerraVerdae Bioworks, *https://terraverdae.com/about/* (accessed December 25, 2021)

[76] Prajapati, K.; Nayak, R.; Shukla, A.; Parmar, P.; Goswami, D.; Saraf, M. Polyhydroxyalkanoates: An exotic gleam in the gloomy tale of plastics. *J. Polym. Environ.* 2021, *29*, 2013–2032

[77] van der Walle, G. A.; de Koning, G. J.; Weusthuis, R. A.; Eggink, G. Properties, modifications and applications of biopolyesters. In *Biopolyesters – Advances in Biochemical Engineering/Biotechnology*; Springer, Berlin, Heidelberg, 2001; Vol. 71, pp 263–291

[78] Noda, I. *Films Comprising Biodegradable PHA Copolymers*, 1998

[79] Elmowafy, E.; Abdal-Hay, A.; Skouras, A.; Tiboni, M.; Casettari, L.; Guarino, V. Polyhydroxyalkanoate (PHA): Applications in drug delivery and tissue engineering. *Expert Rev. Med. Devices* 2019, *16*, 467–482

[80] P4HB made TephaFLEX is our breakthrough Technology, *https://www.tepha.com/technology/overview/* (accessed December 29, 2021)

[81] Sudesh, K.; Loo, C. Y.; Goh, L. K.; Iwata, T.; Maeda, M. The oil-absorbing property of polyhydroxyalkanoate films and its practical application: A refreshing new outlook for an old degrading material. *Macromol. Biosci.* 2007, *7*, 1199–1205

[82] Amelia, T. S. M.; Govindasamy, S; Tamothran, A M; Vigneswari, S.; Bhubalan, K.; Suet, T.; Amelia, M.; Govindasamy, S.; Tamothran, A. M.; Vigneswari, S.; et al. Applications of PHA in agriculture. In *Biotechnological Applications of Polyhydroxyalkanoates*; Springer, Singapore, 2019; pp 347–361

[83] Lu, H.; Madbouly, S. A.; Schrader, J. A.; Kessler, M. R.; Grewell, D.; Graves, W. R. Novel bio-based composites of polyhydroxyalkanoate (PHA)/distillers dried grains with solubles (DDGS). *RSC Adv.* 2014, *4*, 39802–39808.

[84] PolyFerm Canada *http://www.polyfermcanada.com/versamer_tech.html* (accessed December 25, 2021)

[85] Xiang, H.; Wen, X.; Miu, X.; Li, Y.; Zhou, Z.; Zhu, M. Thermal depolymerization mechanisms of poly(3-hydroxybutyrate-co-3-hydroxyvalerate). *Prog. Nat. Sci. Mater. Int.* 2016, *26*, 58–64

[86] Yu, J. Microbial production of bioplastics from renewable resources. In *Bioprocessing for Value-Added Products from Renewable Resources*; Yang, S.-T., Ed.; Elsevier Science, 2007; pp 585–610

[87] Raza, Z. A.; Abid, S.; Banat, I. M. Polyhydroxyalkanoates: Characteristics, production, recent developments and applications. *Int. Biodeterior. Biodegradation* 2018, *126*, 45–56

[88] El-malek, F. A.; Khairy, H.; Farag, A.; Omar, S. The sustainability of microbial bioplastics, production and applications. *Int. J. Biol. Macromol.* 2020, *157*, 319–328

[89] Du, C.; Sabirova, J.; Soetaert, W.; Ki Carol Lin, S. Polyhydroxyalkanoates production from low-cost sustainable raw materials. *Curr. Chem. Biol.* 2012, *6*, 14–25

[90] Page, W. J.; Bhanthumnavin, N.; Manchak, J.; Ruman, M. Production of poly(β-hydroxybutyrate-β-hydroxyvalerate) copolymer from sugars by *Azotobacter salinestris*. *Appl. Microbiol. Biotechnol. 1997 481* 1997, *48*, 88–93

[91] Chen, G. Q.; Page, W. J. The effect of substrate on the molecular weight of poly-β-hydroxybutyrate produced by *Azotobacter vinelandii* UWD. *Biotechnol. Lett.* 1994, *16*, 155–160

[92] eira, L.; Varila, P. Production of polyhydroxyalkanoates from agricultural waste and surplus materials. *Biomacromolecules* 2005, *6*, 561–565

[93] Ashby, R. D.; Solaiman, D. K. Y.; Strahan, G. D.; Zhu, C.; Tappel, R. C.; Nomura, C. T. Glycerine and levulinic acid: Renewable co-substrates for the fermentative synthesis of short-chain poly(hydroxyalkanoate) biopolymers. *Bioresour. Technol.* 2012, *118*, 272–280

[94] Rao, U.; Sridhar, R.; Sehgal, P. K. Biosynthesis and biocompatibility of poly(3-hydroxybutyrate-co-4-hydroxybutyrate) produced by *Cupriavidus necator* from spent palm oil. *Biochem. Eng. J.* 2010, *49*, 13–20

[95] Koller, M.; Bona, R.; Chiellini, E.; Fernandes, E. G.; Horvat, P.; Kutschera, C.; Hesse, P.; Braunegg, G. Polyhydroxyalkanoate production from whey by *Pseudomonas hydrogenovora. Bioresour. Technol.* 2008, *99*, 4854–4863

[96] Ryu, H. W.; Cho, K. S.; Goodrich, P. R.; Park, C. H. Production of polyhydroxyalkanoates by *Azotobacter vinelandii* UWD using swine wastewater: Effect of supplementing glucose, yeast extract, and inorganic salts. *Biotechnol. Bioprocess Eng.* 2009, *13*, 651–658

[97] Pozo, C.; Martínez-Toledo, M. V.; Rodelas, B.; González-López, J. Effects of culture conditions on the production of polyhydroxyalkanoates by *Azotobacter chroococcum* H23 in media containing a high concentration of alpechín (wastewater from olive oil mills) as primary carbon source. *J. Biotechnol.* 2002, *97*, 125–131

[98] Ceyhan, N.; Ozdemir, G. Poly-hydroxybutyrate (PHB) production from domestic wastewater using *Enterobacter aerogenes* 12Bi strain. *African J. Microbiol. Res.* 2011, *5*, 690–702

[99] *https://fullcyclebioplastics.com* (accessed December 24, 2021)

[100] Bluepha – a synthetic biology company from China, with leading technologies for low-cost production of PHAs, *http://en.bluepha.com/pha-bioplastic* (accessed December 25, 2021)

[101] Segura, P. C.; De Meur, Q.; Tanghe, A.; Onderwater, R.; Dewasme, L.; Wattiez, R.; Leroy, B. Effects of mixing volatile fatty acids as carbon sources on *Rhodospirillum rubrum* carbon metabolism and redox balance mechanisms. *Microorganisms* 2021, *9*, 1996

[102] Duque, A. F.; Oliveira, C. S. S.; Carmo, I. T. D.; Gouveia, A. R.; Pardelha, F.; Ramos, A. M.; Reis, M. A. M. Response of a three-stage process for PHA production by mixed microbial cultures to feedstock shift: Impact on polymer composition. *N. Biotechnol.* 2014, *31*, 276–288

[103] Capson-Tojo, G.; Batstone, D. J.; Grassino, M.; Vlaeminck, S. E.; Puyol, D.; Verstraete, W.; Kleerebezem, R.; Oehmen, A.; Ghimire, A.; Pikaar, I.; et al. Purple phototrophic bacteria for resource recovery: Challenges and opportunities. *Biotechnol. Adv.* 2020, *43*, 107567

[104] Higuchi-Takeuchi, M.; Morisaki, K.; Toyooka, K.; Numata, K. Synthesis of high-molecular-weight polyhydroxyalkanoates by marine photosynthetic purple bacteria. *PLoS One* 2016, *11*, e0160981

[105] Belay, A.; Kato, T.; Ota, Y. *Spirulina* (*Arthrospira*): Potential application as an animal feed supplement. *J. Appl. Phycol.* 1996, *8*, 303–311

[106] Chumpol, S.; Kantachote, D.; Nitoda, T.; Kanzaki, H. Administration of purple nonsulfur bacteria as single cell protein by mixing with shrimp feed to enhance growth, immune response and survival in white shrimp (*Litopenaeus vannamei*) cultivation. *Aquaculture* 2018, *489*, 85–95

[107] Numan, M.; Bashir, S.; Mumtaz, R.; Tayyab, S.; Rehman, N. U.; Khan, A. L.; Shinwari, Z. K.; Al-Harrasi, A. Therapeutic applications of bacterial pigments: A review of current status and future opportunities. *3 Biotech* 2018, *8*

[108] Meixner, K.; Kovalcik, A.; Sykacek, E.; Gruber-Brunhumer, M.; Zeilinger, W.; Markl, K.; Haas, C.; Fritz, I.; Mundigler, N.; Stelzer, F.; et al. Cyanobacteria biorefinery – Production of poly(3-hydroxybutyrate) with *Synechocystis salina* and utilisation of residual biomass. *J. Biotechnol.* 2018, *265*, 46–53

[109] De Meur, Q.; Deutschbauer, A.; Koch, M.; Bayon-Vicente, G.; Cabecas Segura, P.; Wattiez, R.; Leroy, B. New perspectives on butyrate assimilation in *Rhodospirillum rubrum* S1H under photoheterotrophic conditions. *BMC Microbiol.* 2020, *20*, 1–20

[110] Fradinho, J. C.; Oehmen, A.; Reis, M. A. M. Photosynthetic mixed culture polyhydroxyalkanoate (PHA) production from individual and mixed volatile fatty acids (VFAs): Substrate preferences and co-substrate uptake. *J. Biotechnol.* 2014, *185*, 19–27

[111] Serafim, L. S.; Lemos, P. C.; Albuquerque, M. G. E.; Reis, M. A. M. Strategies for PHA production by mixed cultures and renewable waste materials. *Appl. Microbiol. Biotechnol.* 2008, *81*, 615–628

[112] Reis, M.; Albuquerque, M.; Villano, M.; Majone, M. Mixed culture processes for polyhydroxyalkanoate production from agro-industrial surplus/wastes as feedstocks. In *Comprehensive Biotechnology, Second Edition*; Academic Press, 2011; Vol. 6, pp 669–683

[113] Fernández-Dacosta, C.; Posada, J.A.; Kleerebezem, R.; Cuellar, M.C.; Ramirez, A. Microbial community-based polyhydroxyalkanoates (PHAs) production from wastewater: Techno-economic analysis and ex-ante environmental assessment. *Bioresour. Technol.* 2015, *185*, 368–377

[114] Jiang, Y.; Hebly, M.; Kleerebezem, R.; Muyzer, G.; van Loosdrecht, M.C.M. Metabolic Modeling of mixed substrate uptake for polyhydroxyalkanoate (PHA) production. *Water Res.* 2011, *45*, 1309–1321

[115] Rosso, D.; Larson, L.E.; Stenstrom, M.K. Aeration of large-scale municipal wastewater treatment plants: State of the art. *Water Sci. Technol.* 2008, *57*, 973–978

[116] Fradinho, J.C.; Domingos, J.M.B.; Carvalho, G.; Oehmen, A.; Reis, M.A.M. Polyhydroxyalkanoates production by a mixed photosynthetic consortium of bacteria and algae. *Bioresour. Technol.* 2013, *132*, 146–153

[117] Fradinho, J.C.; Oehmen, A.; Reis, M.A.M. Effect of dark/light periods on the polyhydroxyalkanoate production of a photosynthetic mixed culture. *Bioresour. Technol.* 2013, *148*, 474–479

[118] Bengtsson, S.; Werker, A.; Christensson, M.; Welander, T. Production of polyhydroxyalkanoates by activated sludge treating a paper mill wastewater. *Bioresour. Technol.* 2008, *99*, 509–516

[119] Anterrieu, S.; Quadri, L.; Geurkink, B.; Dinkla, I.; Bengtsson, S.; Arcos-Hernandez, M.; Alexandersson, T.; Morgan-Sagastume, F.; Karlsson, A.; Hjort, M.; et al. Integration of biopolymer production with process water treatment at a sugar factory. *N. Biotechnol.* 2014, *31*, 308–323

[120] Lee, S.Y.; Choi, J.I. Production of microbial polyester by fermentation of recombinant microorganisms. In *Biopolyesters in "Advances in biochemical engineering/biotechnology"*; Springer, Berlin, Heidelberg, 2001; Vol. 71, pp 183–207

[121] Mohammadi, M.; Hassan, M.A.; Phang, L.Y.; Shirai, Y.; Che Man, H.; Ariffin, H.; Amirul, A.A.; Syairah, S.N. Efficient polyhydroxyalkanoate recovery from recombinant *Cupriavidus necator* by using low concentration of NaOH. *Environ. Eng. Sci.* 2012, *29*, 783–789

[122] Kahar, P.; Tsuge, T.; Taguchi, K.; Doi, Y. High yield production of polyhydroxyalkanoates from soybean oil by *Ralstonia eutropha* and its recombinant strain. *Polym. Degrad. Stab.* 2004, *83*, 79–86

[123] Lee, S.; Choi, J. *Metabolic Engineering*; Lee, S., Papoutsakis, E., Eds.; Marcel Dekker, 1999.

[124] Choi, J.-I.; Lee, S.Y. Efficient and economical recovery of poly(3-hydroxybutyrate) from recombinant *Escherichia coli* by simple digestion with chemicals. *Biotechnol. Bioeng.* 1999, *62*

[125] Yee Chee, J.; Lau, N.-S.; M Abed, R.M.; Sudesh, K.; Chee, J.-Y.; Yoga, S.-S.; Ling, S.-C. Bacterially produced polyhydroxyalkanoate (PHA): Converting renewable resources into bioplastics. 2010

[126] Cohen-Bazire, G.; Sistrom, W.R.; Stanier, R.Y. Kinetic studies of pigment synthesis by non-sulfur purple bacteria. *J. Cell. Comp. Physiol.* 1957, *49*, 25–68

[127] Kang, Z.; Du, L.; Kang, J.; Wang, Y.; Wang, Q.; Liang, Q.; Qi, Q. Production of succinate and polyhydroxyalkanoate from substrate mixture by metabolically engineered *Escherichia coli*. *Bioresour. Technol.* 2011, *102*, 6600–6604

[128] Sauer, M.; Porro, D.; Mattanovich, D.; Branduardi, P. Microbial production of organic acids: Expanding the markets. *Trends Biotechnol.* 2008, *26*, 100–108

[129] Lan, G.; Fan, Q.; Liu, Y.; Chen, C.; Li, G.; Liu, Y.; Yin, X. Rhamnolipid production from waste cooking oil using *Pseudomonas* SWP-4. *Biochem. Eng. J.* 2015, *101*, 44–54

[130] Samorì, C.; Abbondanzi, F.; Galletti, P.; Giorgini, L.; Mazzocchetti, L.; Torri, C.; Tagliavini, E. Extraction of polyhydroxyalkanoates from mixed microbial cultures: Impact on polymer quality and recovery. *Bioresour. Technol.* 2015, *189*, 195–202

[131] Pederson, E.N.; McChalicher, C.W.J.; Srienc, F. Bacterial synthesis of PHA block copolymers. *Biomacromolecules* 2006, *7*, 1904–1911

[132] Koller, M.; Niebelschütz, H.; Braunegg, G. Strategies for recovery and purification of poly[(R)-3-hydroxyalkanoates] (PHA) biopolyesters from surrounding biomass. *Eng. Life Sci.* 2013, *13*, 549–562

[133] Hahn, S.K.; Chang, Y.K.; Kim, B.S.; Chang, H.N. Optimization of microbial poly(3-hydroxybutyrate) recover using dispersions of sodium hypochlorite solution and chloroform. *Biotechnology and Bioengineering.* 1994

[134] Kachrimanidou, V.; Kopsahelis, N.; Papanikolaou, S.; Kookos, I.K.; De Bruyn, M.; Clark, J.H.; Koutinas, A.A. Sunflower-based biorefinery: Poly(3-hydroxybutyrate) and poly(3-hydroxybutyrate-co-3-hydroxyvalerate) production from crude glycerol, sunflower meal and levulinic acid. *Bioresour. Technol.* 2014, *172*, 121–130

[135] García, I.L.; López, J.A.; Dorado, M.P.; Kopsahelis, N.; Alexandri, M.; Papanikolaou, S.; Villar, M.A.; Koutinas, A.A. Evaluation of by-products from the biodiesel industry as fermentation feedstock for poly(3-hydroxybutyrate-co-3-hydroxyvalerate) production by *Cupriavidus necator. Bioresour. Technol.* 2013

[136] Mohammadi, M.; Hassan, M.A.; Shirai, Y.; Man, H.C.; Ariffin, H.; Yee, L.N.; Mumtaz, T.; Chong, M.L.; Phang, L.Y. Separation and purification of polyhydroxyalkanoates from newly isolated *Comamonas* sp. EB172 by simple digestion with sodium hydroxide. *Sep. Sci. Technol.* 2012, *47*, 534–541

[137] Holmes, P.A.; Lim, G.B. *Separation Process*, June 30, 1989

[138] Pérez-Rivero, C.; López-Gómezb, J.P.; Roy, I. A sustainable approach for the downstream processing of bacterial polyhydroxyalkanoates: State-of-the-art and latest developments. *Biochem. Eng. J.* 2019

[139] Hwang, K.-J.; You, S.-F.; Don, T.-M. Disruption kinetics of bacterial cells during purification of poly-β-hydroxyalkanoate using ultrasonication. *J. Chin. Inst. Chem. Engrs.* 2006, *37*, 1–8

[140] Jahanshahi, M.; Najafpour, G. Advanced downstream processing in biotechnology. In *Biochemical Engineering and Biotechnology*; Elsevier, 2007; pp 390–415

[141] Melih Tamer, I.; Moo-Young, M.; Chisti, Y. Disruption of *Alcaligenes latus* for recovery of poly(β-hydroxybutyric acid): Comparison of high-pressure homogenization, bead milling, and chemically induced lysis. *Ind. Eng. Chem. Res.* 1998, *37*, 1807–1814

[142] Pan, P.; Shan, G.; Bao, Y.; Weng, Z. Crystallization kinetics of bacterial poly(3-hydroxylbutyrate) copolyesters with cyanuric acid as a nucleating agent. *J. Appl. Polym. Sci.* 2013, *129*, 1374–1382

[143] Nucleating agents for polyhydroxyalkanoates Patent (Patent # 7,301,000, issued November 27, 2007), *https://patents.justia.com/patent/7301000* (accessed March 10, 2022)

[144] Whitehouse, R.S.; Padwa, A.R. Nucleating agents for polyhydroxyalkanoates. US847023B2, July 16, 2013

[145] Rastogi, S.; Deshmukh, Y.S.; Ma, P.; Wilsens, C.H.R.M. Nucleating agents for biopolymers. WO2013120793A1, 2013

[146] Withey, R.E.; Hay, J.N. The effect of seeding on the crystallisation of poly(hydroxybutyrate), and co-poly(hydroxybutyrate-co-valerate). *Polymer (Guildford).* 1999, *40*, 5147–5152

[147] Liu, W.J.; Yang, H.L.; Wang, Z.; Dong, L.S.; Liu, J.J. Effect of nucleating agents on the crystallization of poly(3-hydroxybutyrate-co-3-hydroxyvalerate). *J. Appl. Polym. Sci.* 2002, *86*, 2145–2152

[148] Branciforti, M.C.; Corrêa, M.C.S.; Pollet, E.; Agnelli, J.A.M.; Nascente, P.A.D.P.; Avérous, L. Crystallinity study of nano-biocomposites based on plasticized poly(hydroxybutyrate-co-hydroxyvalerate) with organo-modified montmorillonite. *Polym. Test.* 2013, *32*, 1253–1260

[149] Jacquel, N.; Tajima, K.; Nakamura, N.; Miyagawa, T.; Pan, P.; Inoue, Y. Effect of orotic acid as a nucleating agent on the crystallization of bacterial poly(3-hydroxybutyrate-co-3-hydroxyhexanoate) copolymers. *J. Appl. Polym. Sci.* 2009, *114*, 1287–1294

[150] He, Y.; Inoue, Y. α-Cyclodextrin-enhanced crystallization of poly(3-hydroxybutyrate). *Biomacromolecules* 2003, *4*, 1865–1867

[151] He, Y.; Inoue, Y. Effect of α-Cyclodextrin on the crystallization of poly(3-hydroxybutyrate). *J. Polym. Sci. Part B Polym. Phys.* 2004, *42*, 3461–3469

[152] Dong, T.; Mori, T.; Aoyama, T.; Inoue, Y. Rapid crystallization of poly(3-hydroxybutyrate-co-3-hydroxyhexanoate) copolymer accelerated by cyclodextrin-complex as nucleating agent. *Carbohydr. Polym.* 2010, *80*, 387–393

[153] Weihua, K.; He, Y.; Asakawa, N.; Inoue, Y. Effect of lignin particles as a nucleating agent on crystallization of poly(3-hydroxybutyrate). *J. Appl. Polym. Sci.* 2004, *94*, 2466–2474

[154] Malmir, S.; Montero, B.; Rico, M.; Barral, L.; Bouza, R. Morphology, thermal and barrier properties of biodegradable films of poly (3-hydroxybutyrate-co-3-hydroxyvalerate) containing cellulose nanocrystals. *Compos. Part A Appl. Sci. Manuf.* 2017, *93*, 41–48

[155] Martínez-Sanz, M.; Villano, M.; Oliveira, C.; Albuquerque, M.G.E.; Majone, M.; Reis, M.; Lopez-Rubio, A.; Lagaron, J.M. Characterization of polyhydroxyalkanoates synthesized from microbial mixed cultures and of their nanobiocomposites with bacterial cellulose nanowhiskers. *N. Biotechnol.* 2014, *31*, 364–376

[156] Yu, H.Y.; Qin, Z.Y.; Liu, L.; Yang, X.G.; Zhou, Y.; Yao, J.M. Comparison of the reinforcing effects for cellulose nanocrystals obtained by sulfuric and hydrochloric acid hydrolysis on the mechanical and thermal properties of bacterial polyester. *Compos. Sci. Technol.* 2013, *87*, 22–28

[157] Ten, E.; Jiang, L.; Wolcott, M.P. Crystallization kinetics of poly(3-hydroxybutyrate-co-3-hydroxyvalerate)/cellulose nanowhiskers composites. *Carbohydr. Polym.* 2012, *90*, 541–550

[158] Li, J.; Wang, Y.; Wang, Z.; Wang, J.; Wu, D. Surface Chain engineering of chitin nanocrystals towards tailoring the nucleating capacities for poly(β-hydroxybutyrate). *Int. J. Biol. Macromol.* 2021, *166*, 967–976

[159] Li, C.; Liu, H.; Luo, B.; Wen, W.; He, L.; Liu, M.; Zhou, C. Nanocomposites of poly(l-lactide) and surface-modified chitin whiskers with improved mechanical properties and cytocompatibility. *Eur. Polym. J.* 2016, *81*, 266–283

[160] Singh, S.; Maspoch, M.L.; Oksman, K. Crystallization of triethyl-citrate-plasticized poly(lactic acid) induced by chitin nanocrystals. *J. Appl. Polym. Sci.* 2019, *136*, 47936

[161] Singh, S.; Patel, M.; Schwendemann, D.; Zaccone, M.; Geng, S.; Maspoch, M.L.; Oksman, K. Effect of chitin nanocrystals on crystallization and properties of poly(lactic acid)-based nanocomposites. *Polymers (Basel).* 2020, *12*, 726

[162] Zaccone, M.; Patel, M.K.; De Brauwer, L.; Nair, R.; Montalbano, M.L.; Monti, M.; Oksman, K. Influence of chitin nanocrystals on the crystallinity and mechanical properties of poly(hydroxybutyrate) biopolymer. *Polymers (Basel).* 2022, *14*, 562

[163] Wang, B.; Li, J.; Zhang, J.; Li, H.; Chen, P.; Gu, Q.; Wang, Z. Thermo-mechanical properties of the composite made of poly (3-hydroxybutyrate-co-3-hydroxyvalerate) and acetylated chitin nanocrystals. *Carbohydr. Polym.* 2013, *95*, 100–106

[164] Wang, J.; Wang, Z.; Li, J.; Wang, B.; Liu, J.; Chen, P.; Miao, M.; Gu, Q. Chitin nanocrystals grafted with poly(3-hydroxybutyrate-co-3- hydroxyvalerate) and their effects on thermal behavior of PHBV. *Carbohydr. Polym.* 2012, *87*, 784–789

[165] Li, C.; Liu, H.; Luo, B.; Wen, W.; He, L.; Liu, M.; Zhou, C. Nanocomposites of poly(l-lactide) and surface-modified chitin whiskers with improved mechanical properties and cytocompatibility. *Eur. Polym. J.* 2016, *81*

[166] Ten, E.; Jiang, L.; Zhang, J.; Wolcott, M.P. Mechanical performance of polyhydroxyalkanoate (PHA)-based biocomposites. In *Biocomposites: Design and Mechanical Performance*; Manjusri Misra, J.K.P. and A.K.M., Ed.; Elsevier Inc., 2015; pp 39–52

[167] Srithep, Y.; Ellingham, T.; Peng, J.; Sabo, R.; Clemons, C.; Turng, L.S.; Pilla, S. Melt Compounding of poly(3-hydroxybutyrate-co-3-hydroxyvalerate)/nanofibrillated cellulose nanocomposites. *Polym. Degrad. Stab.* 2013, *98*, 1439–1449

[168] Jun, D.; Guomin, Z.; Mingzhu, P.; Leilei, Z.; Dagang, L.; Rui, Z. Crystallization and mechanical properties of reinforced PHBV composites using melt compounding: Effect of CNCs and CNFs. *Carbohydr. Polym.* 2017, *168*, 255–262

[169] Torres-Tello, E.V.; Robledo-Ortíz, J.R.; González-García, Y.; Pérez-Fonseca, A.A.; Jasso-Gastinel, C.F.; Mendizábal, E. Effect of agave fiber content in the thermal and mechanical properties of green composites based on polyhydroxybutyrate or poly(hydroxybutyrate-co-hydroxyvalerate). *Ind. Crops Prod.* 2017, *99*, 117–125

[170] Baraki, S.Y.; Zhang, Y.; Li, X.; Ding, L.; Debeli, D.K.; Macharia, D.K.; Wang, B.; Feng, X.; Mao, Z.; Sui, X. Regenerated chitin reinforced polyhydroxybutyrate composites via Pickering emulsion template with improved rheological, thermal, and mechanical properties. *Compos. Commun.* 2021, *25*, 100655

[171] Yu, H.Y.; Yao, J.M. Reinforcing properties of bacterial polyester with different cellulose nanocrystals via modulating hydrogen bonds. *Compos. Sci. Technol.* 2016, *136*, 53–60

[172] Malmir, S.; Montero, B.; Rico, M.; Barral, L.; Bouza, R.; Farrag, Y. PHBV/CNC bionanocomposites processed by extrusion: Structural characterization and properties. *Polym. Compos.* 2017, *40*, E275–E284

[173] Jiang, L.; Morelius, E.; Zhang, J.; Wolcott, M.; Holbery, J. Study of the poly(3-hydroxybutyrate-co-3-hydroxyvalerate)/cellulose nanowhisker composites prepared by solution casting and melt processing. *J. Compos. Mater.* 2008, *42*, 2629–2645

[174] Malmir, S.; Montero, B.; Rico, M.; Barral, L.; Bouza, R.; Farrag, Y. PHBV/CNC bionanocomposites processed by extrusion: Structural characterization and properties. *Polym. Compos.* 2017, *40*, E275–E284

[175] Ten, E.; Turtle, J.; Bahr, D.; Jiang, L.; Wolcott, M. Thermal and mechanical properties of poly(3-hydroxybutyrate-co-3-hydroxyvalerate)/cellulose nanowhiskers composites. *Polymer (Guildford).* 2010, *51*, 2652–2660

[176] Yu, H.; Yan, C.; Yao, J. Fully Biodegradable food packaging materials based on functionalized cellulose nanocrystals/poly(3-hydroxybutyrate-co-3-hydroxyvalerate) nanocomposites. *RSC Adv.* 2014, *4*, 59792–59802

[177] Habibi, Y. Key Advances in the chemical modification of nanocelluloses. *Chem. Soc. Rev.* 2014, *43*, 1519–1542

[178] Li, H.Y.; Li, H.; Wang, B.J.; Gu, Q.; Jiang, Z.Q.; Wu, X.D. Synthesis and properties of poly(3-hydroxybutyrate-co-3-hydroxyvalerate)/chitin nanocrystals composite scaffolds for tissue engineering. *Chinese Chem. Lett.* 2014, *25*, 1635–1638

[179] Yu, H.Y.; Qin, Z.Y.; Sun, B.; Yang, X.G.; Yao, J.M. Reinforcement of transparent poly(3-hydroxybutyrate-co-3-hydroxyvalerate) by incorporation of functionalized carbon nanotubes as a novel bionanocomposite for food packaging. *Compos. Sci. Technol.* 2014, *94*, 96–104

[180] Osman, A.I.; Blewitt, J.; Abu-Dahrieh, J.K.; Farrell, C.; Al-Muhtaseb, A.H.; Harrison, J.; Rooney, D.W. Production and characterisation of activated carbon and carbon nanotubes from potato peel waste and their application in heavy metal removal. *Environ. Sci. Pollut. Res.* 2019, *26*, 37228–37241

[181] Osman, A.I.; O'Connor, E.; McSpadden, G.; Abu-Dahrieh, J.K.; Farrell, C.; Al-Muhtaseb, A.H.; Harrison, J.; Rooney, D.W. Upcycling brewer's spent grain waste into activated carbon and carbon nanotubes for energy and other applications via two-stage activation. *J. Chem. Technol. Biotechnol.* 2020, *95*, 183–195

[182] Chen, M.; Qin, X.; Zeng, G. Biodegradation of carbon nanotubes, graphene, and their derivatives. *Trends Biotechnol.* 2017, *35*, 836–846

[183] Zhang, K.; Mohanty, A.K.; Misra, M. Fully biodegradable and biorenewable ternary blends from polylactide, poly(3-hydroxybutyrate-co-hydroxyvalerate) and poly(butylene succinate) with balanced properties. *ACS Appl. Mater. Interfaces* 2012, *4*, 3091–3101

[184] Ma, P.; Hristova-Bogaerds, D.G.; Lemstra, P.J.; Zhang, Y.; Wang, S. Toughening of PHBV/PBS and PHB/PBS blends via in situ compatibilization using dicumyl peroxide as a free-radical grafting initiator. *Macromol. Mater. Eng.* 2012, *297*, 402–410

[185] Garcia-Garcia, D.; Ferri, J.M.; Boronat, T.; Lopez-Martinez, J.; Balart, R. Processing and characterization of binary poly(hydroxybutyrate) (PHB) and poly(caprolactone) (PCL) blends with improved impact properties. *Polym. Bull.* 2016, *73*, 3333–3350

[186] Nanda, M.R.; Misra, M.; Mohanty, A.K. The effects of process engineering on the performance of PLA and PHBV blends. *Macromol. Mater. Eng.* 2011, *296*, 719–728

[187] Wang, L.; Zhu, W.; Wang, X.; Chen, X.; Chen, G. Q.; Xu, K. Processability modifications of poly(3-hydroxybutyrate) by plasticizing, blending, and stabilizing. *J. Appl. Polym. Sci.* 2008, *107*, 166–173

[188] Zhao, K.; Deng, Y.; Chen, J. C.; Chen, G. Q. Polyhydroxyalkanoate (PHA) scaffolds with good mechanical properties and biocompatibility. *Biomaterials* 2003, *24*, 1041–1045

[189] McChalicher, C. W. J.; Srienc, F. Investigating the structure–property relationship of bacterial PHA block copolymers. *J. Biotechnol.* 2007, *132*, 296–302

[190] Zhao, W.; Chen, G. Q. Production and characterization of terpolyester poly(3-hydroxybutyrate-co-3-hydroxyvalerate-co-3-hydroxyhexanoate) by recombinant *Aeromonas hydrophila* 4AK4 harboring genes *PhaAB*. *Process Biochem.* 2007, *42*, 1342–1347

[191] Mekonnen, T.; Mussone, P.; Khalil, H.; Bressler, D. Progress in bio-based plastics and plasticizing modifications. *J. Mater. Chem. A* 2013, *1*, 13379–13398

[192] Corrêa, M. C. S.; Branciforti, M. C.; Pollet, E.; Agnelli, J. A. M.; Nascente, P. A. P.; Avérous, L. Elaboration and characterization of nano-biocomposites based on plasticized poly(hydroxybutyrate-co-hydroxyvalerate) with organo-modified montmorillonite. *J. Polym. Environ.* 2012, *20*, 283–290

[193] Choi, J. S.; Park, W. H. Effect of biodegradable plasticizers on thermal and mechanical properties of poly(3-hydroxybutyrate). *Polym. Test.* 2004, *23*, 455–460

[194] Mangeon, C.; Modjinou, T.; Rios De Anda, A.; Thevenieau, F.; Renard, E.; Langlois, V. Renewable semi-interpenetrating polymer networks based on vegetable oils used as plasticized systems of poly(3-hydroxyalkanoate)s. *ACS Sustain. Chem. Eng.* 2018, *6*, 5034–5042

[195] Audic, J. L.; Lemiègre, L.; Corre, Y. M. Thermal and mechanical properties of a polyhydroxyalkanoate plasticized with biobased epoxidized broccoli oil. *J. Appl. Polym. Sci.* 2014, *131*, 39983

[196] Kovalcik, A.; Machovsky, M.; Kozakova, Z.; Koller, M. Designing packaging materials with viscoelastic and gas barrier properties by optimized processing of poly(3-hydroxybutyrate-co-3-hydroxyvalerate) with lignin. *React. Funct. Polym.* 2015, *94*, 25–34

[197] Jost, V.; Langowski, H. C. Effect of different plasticisers on the mechanical and barrier properties of extruded cast PHBV films. *Eur. Polym. J.* 2015, *68*, 302–312

[198] Auriemma, M.; Piscitelli, A.; Pasquino, R.; Cerruti, P.; Malinconico, M.; Grizzuti, N. Blending poly(3-hydroxybutyrate) with tannic acid: Influence of a polyphenolic natural additive on the rheological and thermal behavior. *Eur. Polym. J.* 2015, *63*, 123–131

[199] Herrera, N.; Salaberria, A. M.; Mathew, A. P.; Oksman, K. Plasticized polylactic acid nanocomposite films with cellulose and chitin nanocrystals prepared using extrusion and compression molding with two cooling rates: Effects on mechanical, thermal and optical properties. *Compos. Part A – Appl. S.* 2016, *83*, 89–97

4 Manufacture of Biodegradable Polymers Using Reactive Extrusion Processing

Neha Mulchandani and Ramani Narayan

This chapter underlines the commercial importance of reactive extrusion for manufacturing biodegradable polymers and advances in the design of biodegradable polymers by reactive extrusion processing including ring opening polymerization, stereo-complexation, chemical modification of biodegradable polymers, and development of environmentally friendly polymer blends and composites. The merits and challenges associated with reactive extrusion processing of biodegradable polymers are also presented. Furthermore, comments are made on the commercial viability of reactive extrusion processing for designing biodegradable polymers with tailored properties for various applications.

4.1 Introduction

The use of extruders as continuous-flow reactors has witnessed significant attention for the formation and modification of polymers for various applications [1]. Reactive extrusion occurs when a chemical reaction takes place between two or more components that are added to an extruder. The reactions occurring in an extruder include grafting, polymerization, crosslinking, functionalization, compatibilization, as well as controlled depolymerization. Reactive extrusion has been used extensively for producing new polymers from commercially available resins, such as compatibilizing two or more non-compatible polymers using a third material to produce a product with unique properties. The reactive extrusion process has a number of advantages, such as ease of running the reactions at high temperature, increased reaction speed controlled by the reaction temperature, efficient mixing at proper location, sequential addition of reactants to achieve the desired control, ability to achieve reaction in a high-viscosity medium, and so on. However, one of the drawbacks of the reactive extrusion process is the requirement for a shorter reaction time and faster kinetics in order to match the relatively short time in the extruder and obtain a reasonable throughput rate.

The chemical reactions in an extruder can be controlled by various parameters, such as using the correct stoichiometry for addition reactions that are dependent on ingredient feed stream rates, and controlling the polycondensation reactions by water removal. In addition, devolatilization and venting are important to control the reaction sequence. Twin-screw extruders are more effective than single-screw extruders in removing high concentrations of volatiles.

Reactive extrusion can be seen as a combination of several small processes, each dependent on the previous process to obtain the desired final product. The reactive components are combined in one extruder section where volatiles are removed in one section, sequential addition requires another section, mixing to ensure complete reaction occurs in a different section, and a final product is homogenized and extruded through a die. Further, compatibilizers and catalysts may be critical components of a reactive extrusion processes (polymerization or chemical modification of polymers), and their addition in correct sequence is also essential for the kinetics of reaction. Twin-screw extruders also have a greater capability and flexibility for achieving a given objective than single-screw extruders, which justifies their significantly higher cost. A typical reactive extrusion process is shown in Figure 4.1.

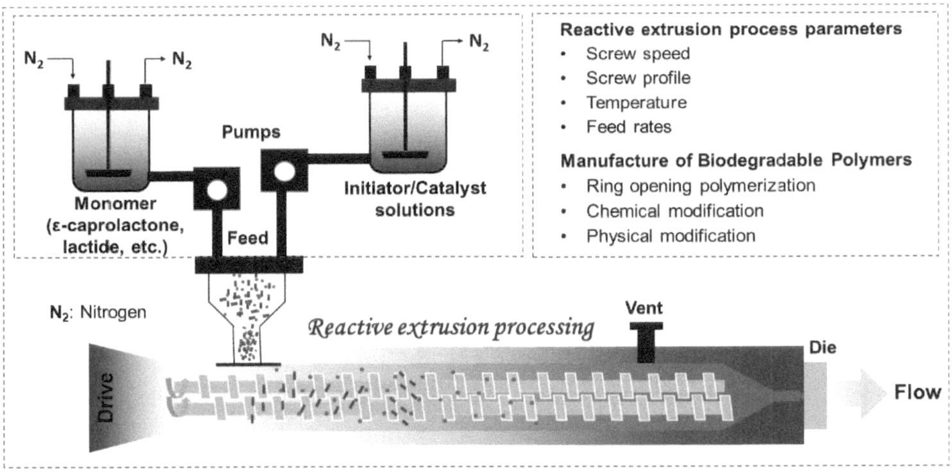

Figure 4.1 Schematic representation of a reactive extrusion process

Reactive extrusion processing serves as an effective tool for manufacturing biodegradable-compostable polymers, which are being explored for their commercial applications, owing to their extraordinary properties and compostable character. Biodegradable-compostable polymers are currently replacing carbon–carbon backbone polymers in single-use applications as a means to reducing the accumulation of plastic waste in the environment. However, the cost-competitiveness of these polymers is restricting their widespread acceptance in the commercial market. In this regard, "just-in-time" manufacturing may serve as a cost-effective strategy in developing biodegradable-compostable polymers, allowing their commercial viability to be increased.

This chapter therefore underlines the commercial importance of reactive extrusion for manufacturing biodegradable polymers and advances in the design of biodegradable polymers by reactive extrusion processing, including ring-opening polymerization, stereocomplexation, chemical modification of biodegradable polymers, and development of environmentally friendly polymer blends and composites. The merits and challenges associated with reactive extrusion processing of biodegradable polymers are also presented. Furthermore, comments have been made on the commercial viability of reactive extrusion processing for designing biodegradable polymers with tailored properties for various applications.

4.2 Biodegradable Polymers for Managed End-of-Life

Carbon-carbon backbone polymers, which are non-biodegradable persistent plastics, have been applied in various fields ranging from commodity and medical applications to the aerospace industry [2]. Such polymers degrade and fragment into microplastics, but do not biodegrade at an acceptable rate, thereby leading to their continual build up in the natural environment [3, 4]. The serious detrimental impacts of carbon-carbon backbone polymers on the environment and human health has been extensively reported in the literature [5, 6]. The widespread usage of carbon-carbon backbone polymers has led to their increased accumulation on land and oceans after their end-of-life [7, 8]. Carbon-carbon backbone polymers comprise nearly 77% of plastics produced globally. Also, the largest demand for plastic by application is for packaging, with 90% of packaging plastics (153.8 million metric tonnes) consisting of non-biodegradable carbon-carbon backbone polymers, which have a very low recycling rate (0–4%). An estimated 12 billion tonnes of plastic waste will have accumulated on earth by 2050, if plastic production, utilization and its waste management follows the current trajectory. In order to address the global environmental and health impacts, carbon-carbon backbone polymers are being replaced by sustainable end-of-life solutions. This can be achieved by incorporating an ester backbone into plastics, which introduces the property of biodegradability-compostability into plastics. These may then be diverted to managed industrial composting systems after their end-of-life [9].

Biodegradation is an end-of-life process, and therefore biodegradable plastics are referred to in accordance with their end-of-life characteristics in a specific environment at a specific time, such as home composting or industrial composting. A biodegradable plastic may be derived from fossil-based resources or renewable (biobased) resources, but it is nevertheless biodegradable in a specific environment after its end-of-life [10]. Government legislation, environmental regulations and social concerns have led to an increased awareness regarding biodegradable plastics and their role in reducing the accumulation of plastic waste on land and oceans [11]. Biodegradable-compostable plastic can replace carbon–carbon backbone polymers in single-use applications such as packaging (trash-bags, food containers, laminated papers), disposable non-woven fabrics, agricultural mulch films, and so on. The utilization of biodegradable polymers as bulk commodity materials requires cost-effective strategies for their production due to the cost-competition with carbon–carbon backbone polymers. In addition, biodegradable polymers are limited by their thermo-mechanical stability compared to carbon–carbon backbone polymers [12].

Reactive extrusion offers a facile, solvent-free and cost-effective route for processing of biodegradable polymers, which increases their commercial viability and cost-competitiveness [13]. Various processes including melt-blending, as well as chemical reactions such as polymerization, grafting, branching, and functionalization, can be effectively carried out using reactive extrusion in order to improve the overall properties of biodegradable polymers, including their thermo-mechanical stability [14].

4.3 Manufacturing Biodegradable Polymers by Reactive Extrusion

Reactive extrusion represents a unique tool for manufacturing biodegradable polymers in a cost-effective way, because it offers different types of reactive modification, which enhances the commercial ability of biodegradable polymers [15]. The reactive agents can be introduced at optimum points in a sequence during the reactive extrusion process. The process also involves homogenizing the reactants, and allowing necessary time for the completion of the reactions. The reactants, typically, are fed through a hopper, however, various liquids and gases may be introduced at specific points in the sequence by injection, and the reaction mixture is conveyed through the extruder and continued until the desired degree of completion is attained. After the completion of the reaction and removal of volatile by-products, the polymeric melt is collected through a die, which is followed by quenching, solidifying and pelletizing processes. The reactive extrusion process thus integrates the production and processing of polymers into a one-stage process, providing an inexpensive route for the production of polymers with desired properties [16].

The reactive extrusion process uses both single- and twin-screw extruder configurations, however, the twin-screw configuration offers superior heat and mass transfer along with improved mixing and residence time distribution [17]. The single-screw extruder is often utilized for melting, plasticizing and production of profiles, pipes, etc. The conveying mechanism for twin-screw extruders is largely dependent on the geometrical configuration of the screw, whereas in the case of single-screw extruders, it is dependent on the frictional forces in the conveying zone and viscous forces in the pumping zone.

There are different types of twin-screw extruders commercially available which may be chosen for a particular purpose depending on the application. The extruders usually have two parallel shafts which either rotate in the same direction or opposite directions. The screws are said to be intermeshing if the centerline distance between the shafts is less than the screw diameter, whereas those with the distance equal to the screw diameter are said to be non-intermeshing [18], as shown in Figure 4.2. Further, twin-screw extruders can be divided into "co-rotating" extruders, in which the screws rotate in the same direction, or "counter-rotating" extruders, in which the screws rotate in opposite directions [19]. For co-rotating machines, depending on the direction of screw rotation, both the screws have either left-handed or right-handed thread. On the other hand, one screw in the counter-rotating extruders has a left-handed thread and the other has a right-handed thread. The co-rotating and counter-rotating screws (viewed from the die end) are shown in Figure 4.3. The co-rotating screws have a radial shearing and plasticizing effect, whereas the counter-rotating screws have an axial shearing and plasticizing effect. For continuous reactive extrusion processes, co-rotating intermeshing twin-screw extruders have been found to be suitable due to their unique design, high speeds, high shearing force, self-wiping feature, and high output, as well as the constant surface renewal, which is favorable for degassing conditions [20].

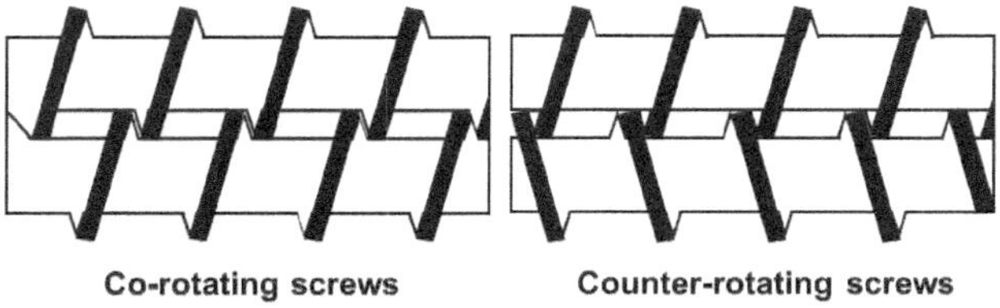

Figure 4.2 Co-rotating and counter-rotating fully intermeshing screws

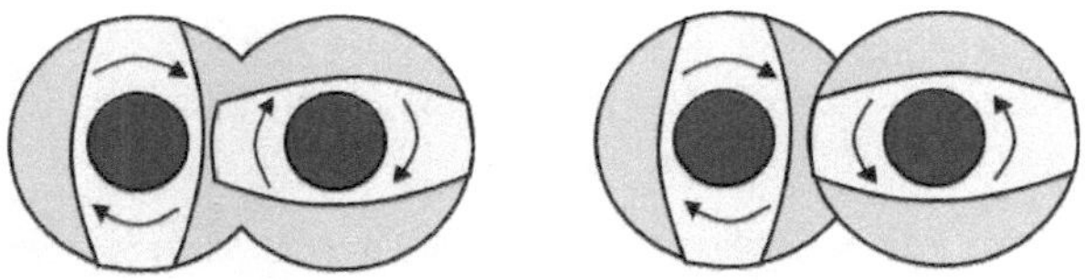

Figure 4.3 Screw rotation in co-rotating and counter-rotating extruders

In addition, the sequence of screw elements is responsible for controlling the filling ratio inside the extruder. These comprise kneading blocks (used for the dispersion of polymer particles and additives), mixing gears (for homogenization of feed streams), and conveying screw elements (for moving the material through the extruder) for co-rotating intermeshing twin-screw extruders. The arrangement of the screw elements determines the residence time distribution in twin-screw extruders. A typical co-rotating intermeshing twin-screw extruder (length/diameter (*L*/*D*) 45:1) has a residence time of about 7 min at low screw speeds and short-pitch elements. However, screw designs with longer residence times (10–75 min) are also available and may be used for specific applications.

The advantages of reactive extrusion over conventional polymer processing are as follows:

- Reactive extrusion is a continuous process with a short reaction time, thereby preventing prolonged exposure to elevated operating temperatures and in turn undesired thermal degradation of polymers.
- It is a safe and inexpensive route, as it eliminates the costs of solvent inlet and recycling streams, as well as harmful solvents.
- The screw elements of the reactive extruder provide high shear forces that create thin layers of material in the area between the screws, thereby providing a high surface area for the reactions. In addition, it ensures efficient heat and mass transfer during polymerization reactions by minimizing the temperature gradients that may arise due to the increase in viscosity.
- The gaseous by-products generated during reactive extrusion processing are removed by downstream devolatilization, which results in higher efficiency.
- It allows for incorporation of multiple input streams into the system as well as control over several processing parameters, which renders it ideal for a diverse range of applications.

4.4 Ring-Opening Polymerization by Reactive Extrusion

Aliphatic polyesters such as poly(ε-caprolactone) (PCL), poly(lactic acid) (PLA) and poly(glycolic acid) (PGA) have had significant attention in various applications, ranging from packaging to the biomedical sector [21]. Aliphatic polyesters are most commonly prepared by ring-opening polymerization (ROP) of cyclic lactone monomers, i. e., ε-caprolactone, lactide, and glycolide. The polymerization of cyclic lactones is usually performed on laboratory scale by melt polymerization, or industrially using a batch process. Reactive extrusion offers a continuous process for the polymerization of cyclic lactones, which is being investigated for its commercial applicability [22]. Some important studies on ROP of cyclic lactones by reactive extrusion processing are explained below.

4.4.1 Ring-Opening Polymerization of ε-Caprolactone

The ROP of ε-caprolactone results in the synthesis of PCL, which is a hydrophobic, semi-crystalline polymer widely investigated for various applications [23] due to its miscibility with commercial polymers such as polyvinyl chloride and polycarbonate. The melting and glass transition temperatures of PCL are 60 °C and -60 °C, respectively. PCL is marketed under the trade name CAPA® by Ingevity, and is prepared by the ROP of ε-caprolactone in the presence of stannous octoate as a catalyst and an alcohol initiator with a maximum molecular weight (M_n) of 80 kg/mol. PCL has high flexibility and toughness, although its lower melt strength and tear strength restricts its usage in some applications. The drawbacks of PCL are overcome by blending or chemically reacting it with other polymers such as poly(L-lactic acid) (PLLA) or poly(ethylene glycol) (PEG).

The ROP of ε-caprolactone has been reported by our group using reactive extrusion with a residence time of less than 5 minutes, with aluminum tri-*sec*-butoxide (ATSB) as an initiator [24]. The screw configuration of the twin-screw extruder entirely used conveying elements in order to prevent the undesired scission of PCL chains. The screw diameter was 30 mm and the *L/D* ratio was 30. The purified monomer was fed to an extruder using a peristaltic pump followed by injecting the initiator solution (5 wt% solution in anhydrous toluene) by maintaining a constant feed ratio of monomer to initiator. The use of ATSB initiator resulted in the formation of a three-armed PCL structure as shown in Figure 4.4. The molecular weight of each arm was calculated from the feed ratio of monomer to initiator. When the extrusion melt temperature was 220 °C and the screw speed was 30 rpm, the monomer conversion was 92% and the molecular weight of PCL was >100,000 g/mol.

However, when the extrusion melt temperature was reduced to 150 °C (with the same screw speed), the conversion dropped to 62% and the molecular weight fell to ~74,000 g/mol. This suggests the significance of polymerization temperature on the molecular weight and conversion of monomer. The polymerization was inferred to propagate through a co-ordination insertion mechanism initiated by the aluminum alkoxide.

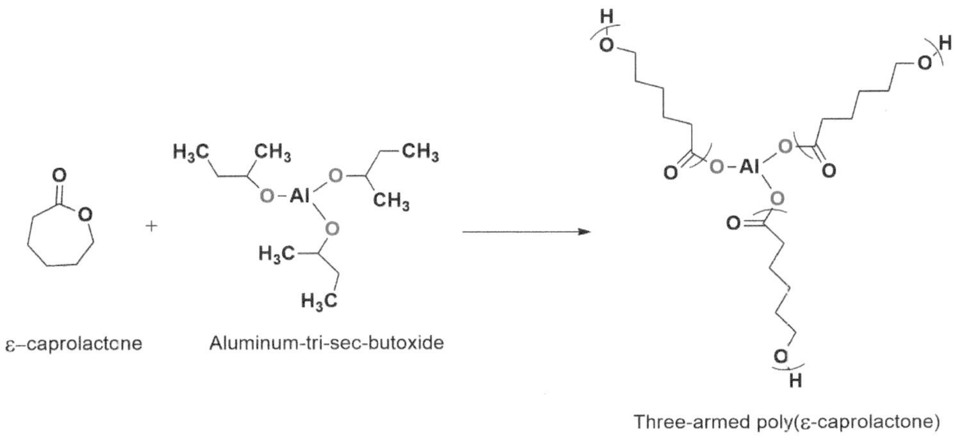

Figure 4.4 Synthesis of three-armed PCL from ε-caprolactone via ROP

The ROP of ε-caprolactone in the presence of chemically modified titanium propoxide was reported by Xia et al., using a glass reactor as well as an extruder. The ROP in a glass reactor was carried out at 125 °C with a constant ratio of monomer to initiator. For the reaction in an extruder, an intermeshing co-rotating twin-screw extruder system was used to prepare functionalized PCL in a continuous manner [25]. The screw diameter was 30 mm, and the *L/D* ratio was 36. The feeding rate of monomer and initiator was 1 kg/h with a screw speed of 100 rpm, and a barrel temperature of 180 °C. The kneading zones allowed for a specific residence time and thorough mixing, resulting in successful synthesis of functional polymers. The feeding ratio of monomer to initiator was varied to give polymers with different molecular weights, which were characterized by size exclusion chromatography (SEC) and nuclear magnetic resonance (NMR). The number average molecular weight (M_n) of PCL was close to the theoretical molecular weight, whereas the polydispersity index increased with the conversion, which was attributed to the transesterification reactions. The polymerization was quasi-living, and it was inferred that the active chain end of PCL could undergo reaction with different end groups and unsaturated monomers to allow preparation of customized graft polymers.

4.4.2 Ring-Opening Polymerization of Lactide

The homopolymerization of L-lactide and D-lactide monomers results in their homopolymers [26], poly(L-lactic acid) (PLLA) and poly(D-lactic acid) (PDLA), respectively, which have a melting temperature of ~175 °C, and a glass transition temperature of ~60 °C. PLLA and PDLA can also be prepared by dehydrative condensation of L-lactic acid and D-lactic acid, respectively. However, the industrially preferred route of manufacturing PLA is using lactide monomer, which is a cyclic dimer, formed by a "backbiting" reaction of oligomeric PLA (MW ~1000–5000 Da), which in turn is obtained by condensation of lactic acid (Figure 4.5). Further, PLLA and PDLA are known to have high ultimate tensile strength and lower elongation at break, which render them suitable for load-bearing applications [27]. The biocompatibility of PLLA has led to its acceptance in biomedical applications [28, 29], such as sutures and orthopedic fixation devices [30, 31]. PLA has gained significant attention as a biobased and compostable material [32–34]. The strength and modulus of PLA are similar to some of the conventional plastics, which renders it suitable for several commodity applications [35–37]. However, certain inherent properties such as low impact toughness, poor heat and hydrolytic stability, and weak melt strength [38, 39] have restricted its widespread use in applications requiring higher toughness, elongation and heat stability [40, 41] such as flexible packaging. In this regard, tailoring the backbone architecture of PLA may lead to the production of PLA-based materials with desired properties [42, 43].

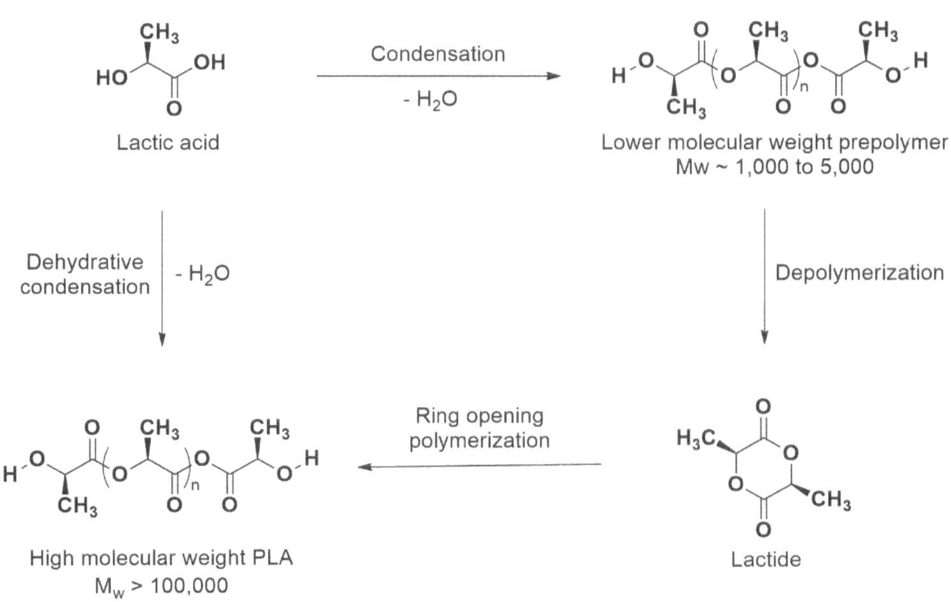

Figure 4.5 Synthesis of PLA from lactic acid and lactide precursors

A mathematical model for the ring-opening polymerization of L-lactide using a twin-screw extruder was reported by Banu et al., who carried out experimental studies to investigate the process kinetics and performance of PLA synthesis [44]. The ROP of L-lactide was investigated in a batch process using a Haake Rheocord Mixer. A kinetic model was developed based on the experimental data, considering a co-ordination insertion mechanism. The reactive extrusion experiments were carried for the same polymerization process in a co-rotating twin-screw extruder having a barrel length of 1.2 m, a screw external diameter of 34 mm, and an *L/D* ratio of 35. An equimolar mixture of stannous octoate and triphenylphosphine were mixed in anhydrous toluene, the resulting solution was manually mixed with the monomer, and the mixture was fed to an extruder using a gravimetric feeder. The reactive extrusion experiments were performed at a temperature of 200 °C, with a flow rate of 0.75–1.5 kg/h, a screw speed of 50–300 rpm, and a residence time of 10 min. The screw speed of 50 rpm was used for the kinetic experiments. Monomer-to-initiator ratios of 2250 and 4500 were used to perform the experiments. The simulation package Ludovic® was used to determine the flow and mixing characteristics of the reaction mixture, which were described by two flow models, i. e., the axial dispersion model and a compartment model. The extrusion processes were built by combining the flow models with the polymerization kinetic model, which were used to predict the average molecular weight of the polymer and conversion of the monomer. The experimental values were found to be in agreement with the calculated values, which confirmed the validity of the modelling approach. Further, the axial dispersion model was found to be a valuable tool for modelling the reactive extrusion process.

4.5 Chemical Modification by Grafting

Natural polymers such as starch, cellulose, and chitosan have been explored for their use in several applications due to their non-toxic, biocompatible, and biodegradable nature [45, 46]. The widespread use of chitosan is restricted owing to its poor mechanical and thermal properties [47]. In order to improve its commercial applicability, chitosan is physically or chemically bonded with other polymers [48, 49]. The graft copolymerization of chitosan with several biodegradable and non-biodegradable polymers has been reported [50, 51], with the aim of developing graft copolymers with improved properties.

4.5.1 Poly(ε-caprolactone)-Grafted Chitosan

Vaidya et al. reported a scalable route for grafting PCL onto chitosan by lipase-catalyzed reactive extrusion [52]. The graft copolymers were prepared by a co-rotating twin-screw extruder with an *L/D* ratio of 40:1, and a screw speed of 300 rpm. The pellets of PCL and chitosan powder in the ratio of 90:10 along with lipase (obtained from CALB and NovoCor) were fed to the extruder at a rate of 3 g/min, with the barrel and die temperatures set above 60 °C (above the T_m of PCL). The extruded samples were allowed to cool to room temperature, washed with water to remove the unreacted lipase, followed by vacuum drying and Soxhlet extraction using chloroform to remove any unreacted PCL, dried, and then characterized. The grafting yield using NovoCor enzyme was higher (96.3%) than with CALB (5.9%), which was attributed to the higher thermal stability of NovoCor. However, the grating ratio of PCL:chitosan was low for both enzymes, and no grafting of PCL occurred on chitosan in the absence of enzyme, thereby confirming enzyme-catalyzed grafting of PCL on chitosan via reactive extrusion processing.

4.5.2 Poly(lactic acid)-Grafted Maleic Anhydride

PLA, which is gaining commercial acceptance in single-use packaging, is often combined with natural and synthetic biodegradable polymers in order to reduce costs and improve its properties. However, simply blending PLA with other polymers or incorporating fillers doesn't necessarily improve the properties of the resulting blend or composite, which may be due to the lack of interfacial adhesion [53]. On the other hand, chemically modifying PLA by introducing functional groups onto the backbone of PLA serves as an effective strategy to prepare composites, blends/alloys, laminates, and coated items with improved properties, and this can be performed by reactive extrusion processing [54].

An investigation into grafting of maleic anhydride onto a PLA backbone in the presence of 2,5-di-(*tert*-butylperoxy)hexane (Lupersol 101) was reported by our group using reactive extrusion processing [55]. A physically mixed blend of PLA, maleic anhydride and Lupersol 101 was fed into a co-rotating intermeshing twin-screw extruder, where the diameter of each screw was 3 cm and the length was 42 cm. Each screw consisted of two sets of mixing paddles, whereas the die had two 3-mm diameter holes. The screw speed was set to 100 rpm, and a concentration of 2 wt% maleic anhydride was used in the presence of Lupersol 101, as a free-radical initiator. The reaction was performed at 180 °C and 200 °C with a peroxide initiator concentration of 0–0.5 wt%, in order to functionalize PLA chains with maleic anhydride, as shown in Table 4.1. This resulted in 0.066–0.672 wt% grafting of maleic anhydride onto the PLA chains. The maleated PLA polymers were characterized by triple-detector SEC, melt flow index analysis (MFI) and thermogravimetric analysis (TGA). An increase in the initiator concentration resulted in an increase in grafting of maleic anhydride, with a reduction in the molecular weight of the polymer. In order to determine the impact of anhydride-grafted PLA onto the interfacial adhesion in PLA blends, several experiments were performed to compare the interfacial adhesion properties between starch and PLA blends obtained by melt-blending. The scanning electron microscopic (SEM) analysis of a 40:60 starch/PLA blend showed poor interfacial adhesion, whereas a substantial improvement in the interfacial adhesion was inferred for the 30:70 starch/maleated PLA blend, where the maleated PLA consisted of PLA, 2% maleic anhydride and 0.5% Lupersol 101. The results suggested that incorporation of even small amounts of maleic anhydride on the PLA backbone can significantly improve the interfacial adhesion of PLA-based blends and composites.

Table 4.1 Temperature Parameters for Maleation of PLA in a Twin-Screw Extruder

Temperature (°C)	Temperature type	Temperature (°C)			
		Zone 1	Zone 2	Zone 3	Die
180	Set	170	170	190	160
	Melt	148	184	180	162
200	Set	170	190	210	155
	Melt	146	201	199	175

4.5.3 Functionalization of Poly(lactic acid)

We have also reported the functionalization of PLA by reaction with a multifunctional epoxy polymer (Joncryl®) using reactive extrusion processing. A model compound study using stearic acid and polyethylene glycol, each in the weight ratio of 1:1 with Joncryl, was performed using differential scanning calorimetry (DSC), which showed that the epoxy functional groups on Joncryl react selectively with the carboxylic acid chain-ends of PEG [56]. This was determined by the presence of an exothermic peak in the DSC scan of the stearic acid/Joncryl mixture, which was not observed with PEG/Joncryl mixture.

The process was as follows. The dried PLA pellets along with Joncryl (1, 5, and 10 wt%) were fed to a co-rotating twin-screw extruder with an *L/D* ratio of 30, using gravimetric feeders. The samples, after being compounded at a throughput of 4.8 kg/h, were cooled in a water bath and then pelletized. A series of epoxy-functionalized PLA formulations (1–10% Joncryl) were prepared with no gel formation. The epoxy-functionalized PLA with 5 wt% Joncryl, after drying at 70 °C, was further compounded with neat PLA and extruded to obtain specimens with 0.25, 0.5, and 1 wt% Joncryl. The screw speed was 125 rpm, and the temperature profile for compounding in the six zones of the equipment was 140/165/170/175/175/170 °C, where the three major kneading zones were used to impart additional mechanical energy into the extrusion system driving the chemistry. The reaction mechanism between PLA and Joncryl revealed that only the carboxyl end groups of PLA reacted with the epoxy groups of Joncryl below 200 °C (Figure 4.6). The epoxy-functionalized PLA, as a rheology modifier, was synthesized by reactive extrusion processing,

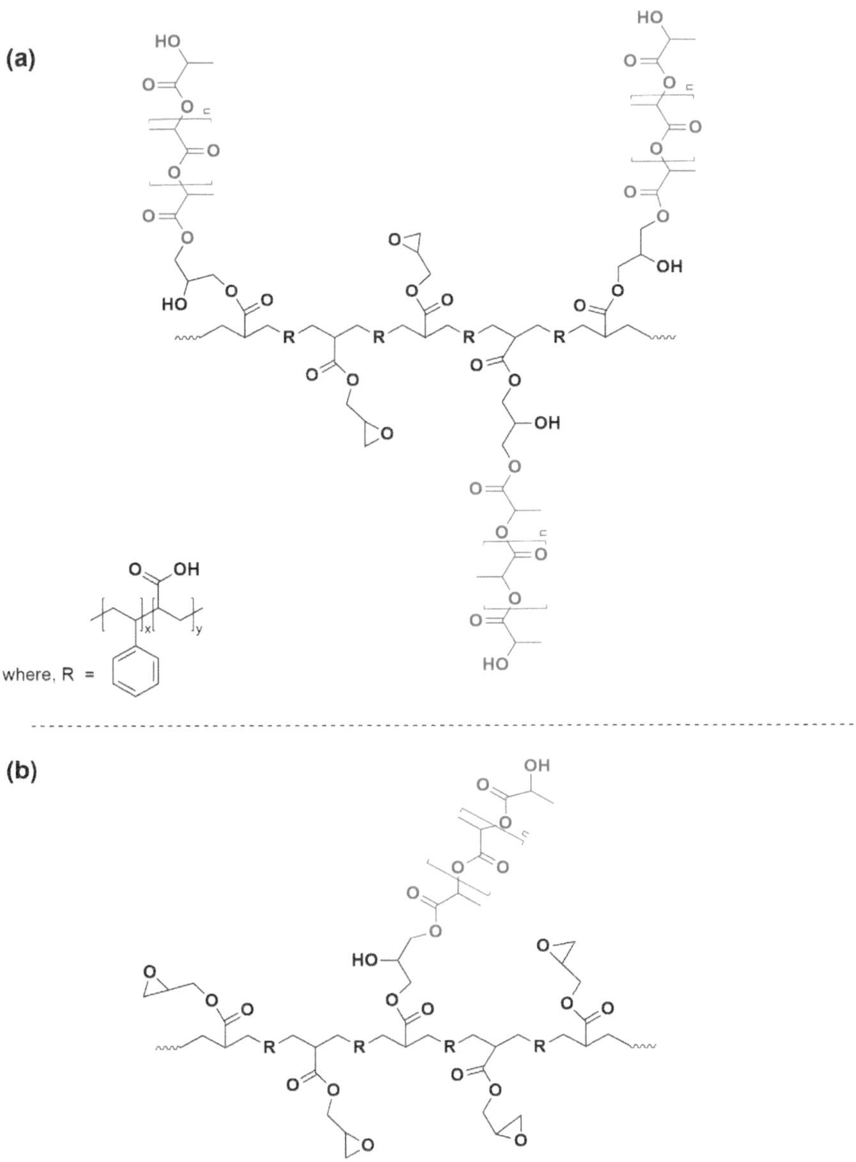

Figure 4.6 Molecular structure of epoxy-modified PLA at (a) low and (b) high concentration of Joncryl

and its reactive blending with PLA led to significant improvements in the rheological properties of PLA (increased viscosity and strain hardening) that are essential for blown film and foam applications. The improved rheological properties were attributed to the distribution of Joncryl through the material. The materials showed significant strain-hardening rheological behavior, and the films of the material were blown which were capable of blow-up ratios much higher than those for PLA. A statistical model was also developed to simulate the molecular weight characteristics and reactivity of the epoxy-functionalized PLA and further provide a fundamental understanding of the reaction.

4.5.4 Poly(lactic acid)-Grafted Vinyltrimethoxysilane

Our group has reported the free-radical grafting of vinyltrimethoxysilane (VTMOS) and silanol-terminated poly(dimethyl siloxane) (OH-PDMS) onto the backbone of PLA using reactive extrusion [57]. PLA resin was pre-dried to remove the moisture in order to avoid the hydrolytic degradation of PLA by "back-biting" and intermolecular transesterification reactions. The physically mixed PLA, VTMOS, OH-PDMS, and Luperox were added to the hopper of a co-rotating twin-screw extruder (*L/D* ratio of 30) using a gravimetric feeder. The samples were compounded at a throughput of 6 kg/h and a screw speed of 125 rpm, with a temperature profile in the six zones of the equipment of 120/140/160/170/170/165 °C. The temperature of the first heating zone was set to 120 °C in order to prevent the flash-evaporation of VTMOS in the hopper. The grafted material was moisture-cured to form a siloxane-crosslinked PLA material (Figure 4.7). The methoxy groups of the silane provided the modified PLA sites for crosslinking through a moisture-induced pathway. A grafting efficiency of VTMOS of up to 90% was achieved, as can be seen in Table 4.2. The addition of OH-PDMS resulted in improved tensile toughness by lengthening the siloxane crosslinks. An improvement in modulus, strength and impact properties was achieved with a reduction in ductility, for samples crosslinked with only VTMOS. However, incorporation of OH-PDMS into the crosslinking reaction resulted in an increased ductility and tensile toughness (due to the longer siloxane linkages) in addition to improved modulus and impact toughness (due to crosslinking). The presence of stretched siloxane crosslinks for the impact samples was observed from SEM fractured surfaces, and a siloxane pullout was observed for the tensile samples. This resulted in an increased ability of the material to deform and absorb energy, which explains the increased toughness of the sample.

Table 4.2 Percent Grafting of VTMOS After Reactive Extrusion Processing and Crosslinking, Along with Tensile Properties of the Siloxane-Crosslinked PLA Samples

VTMOS (%)	OH-PDMS (%)	Luperox (%)	Grafting (initial) (%)	Grafting (2 weeks) (%)	Tensile modulus (GPa)	Yield stress (MPa)	Elongation at break (%)
0.5	1	0.10	78.2	70.5	3.40	54.68	53.07
0.5	1	0.25	66.1	75.1	3.25	55.02	5.94
0.5	3	0.10	63.5	50.3	3.67	44.95	9.97
0.5	3	0.25	63.7	73.1	3.07	45.51	4.45
1.0	1	0.10	89.5	83.6	3.43	55.36	3.59
1.0	1	0.25	86.4	86.5	3.33	53.16	4.05
1.0	3	0.10	79.3	77.8	3.71	45.44	3.24
1.0	3	0.25	76.4	79.8	3.48	45.3	4.49
1.0	0	0.10	82.9	79.9	3.73	58.67	2.40
1.0	0	0.25	87.4	86.8	3.46	55.92	1.92
0	0	0	–	–	2.62	58.67	2.93

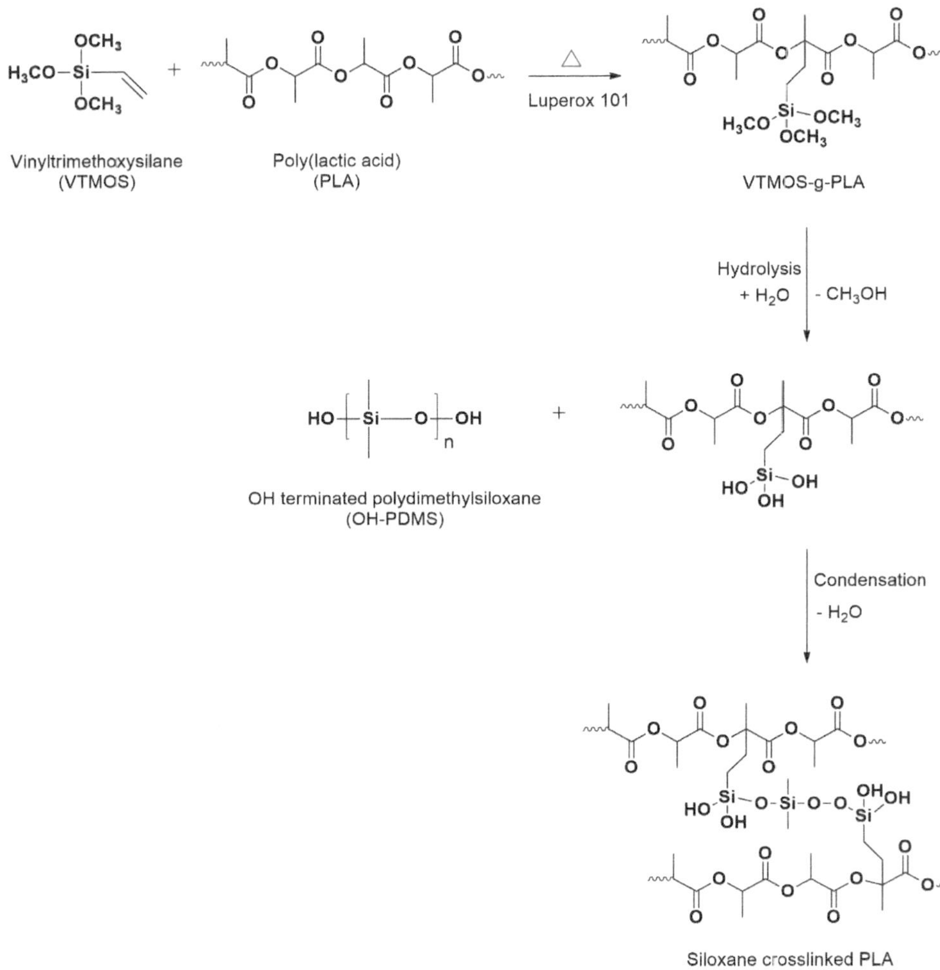

Figure 4.7 Reaction scheme for the free-radical-initiated grafting of VTMOS onto the backbone of PLA, followed by hydrolysis of the methoxy groups and condensation resulting in the crosslinked PLA

4.5.5 Chemical Modification of Starch

Starch is an inexpensive, naturally occurring anhydroglucose polymer, which is also biodegradable. Starch comprises two isomers – amylopectin and amylose – in varying ratios depending on its source, which leads to different physical and mechanical properties. Amylopectin exists as a branched structure, where D-glucose units are linked to each other through α(1→4) linear linkages and α(1→6) linkages at the branching nodes. Amylose, on the other hand, comprises α(1→4) linkages, and exists as a linear chain structure (Figure 4.8). Despite several advantages of starch, its use in commercial applications as a substitute for conventional plastics is limited by its hydrophilicity and poor melt-processability.

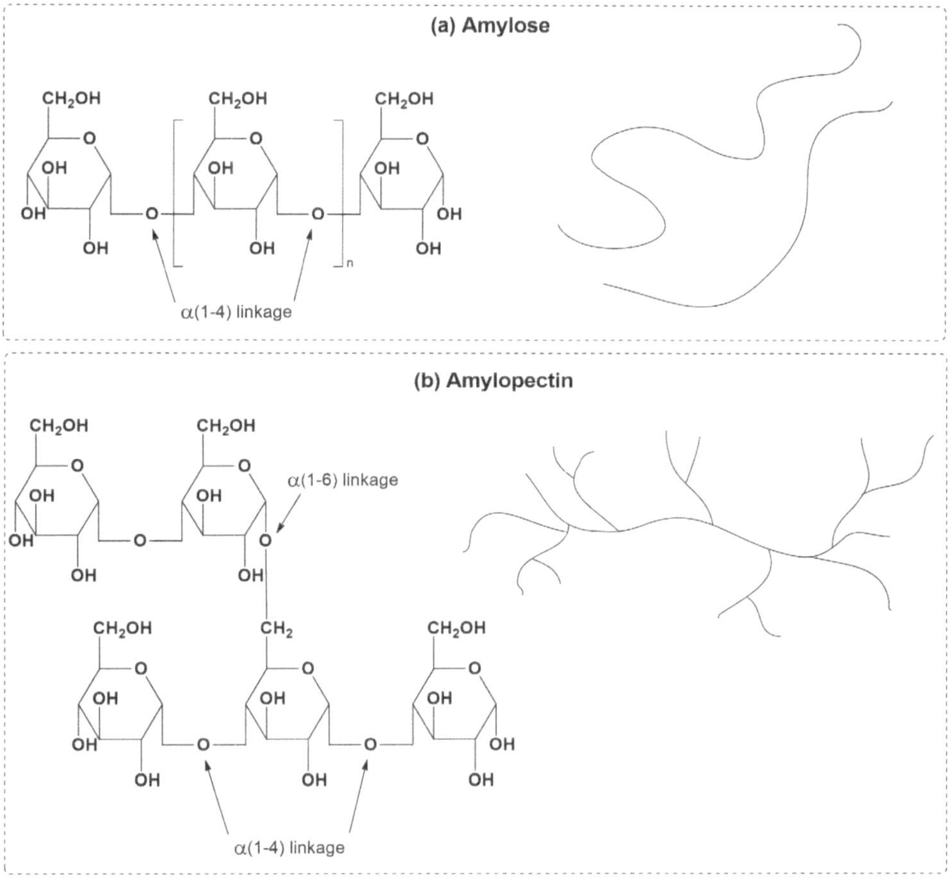

Figure 4.8 Amylose and amylopectin: Isomers of starch

The presence of multiple hydroxyl groups on the backbone of starch is responsible for its hydrophilic nature [58, 59], which can be reduced by melt-blending it with more hydrophobic polymers, or reacting it with an acid anhydride or dibasic acid in the presence of a plasticizer. The poor melt-processability is attributed to the fact that it does not melt, as it is not a thermoplastic material. The addition of a plasticizer such as water or glycerol breaks down the semi-crystalline structure of starch at elevated temperature in an extruder, allowing it to be melt-processed to produce injection-molded articles or blown films.

The physical and chemical modification of starch has been carried out by reactive extrusion in order to expand its commercial applicability. An example of such a modification is the graft copolymerization approach developed by Siyamak et al., which uses a method based on reactive extrusion to produce starch-g-polyacrylamide using four different types of starches (native wheat starch, soluble wheat starch, waxy maize starch, and tapioca starch) [60]. The screw profile consisted of six forward-conveying screw elements, six mixing elements oriented at 90°, and three reverse conveying elements oriented at 60°. The screw speed was set to 90 rpm while the die temperature was 80 °C. The reactive extrusion conditions were optimized for grafting acrylamide monomer onto various types of starches. The feed rates were adjusted so as to give a substrate/monomer ratio of 3:1, and 1 wt% of initiator (ammonium persulfate) relative to the starch feed. This was followed by the saponification of starch-g-copolymers. An average monomer conversion of 80% and a grafting efficiency of ~74% was achieved for all the samples, as confirmed from elemental analysis. Further, the mechanism of grafting reaction in the extruder was influenced by the type of starch, which was also responsible for the swelling and thermal behavior of graft copolymers. The water absorption capacity of the graft copolymers was improved by saponification. The starch copolymers prepared using a cationic substrate also possessed pH sensitivity. This reactive extrusion process thus allows the properties of starch copolymers to be controlled, and this approach can be further applied for the large-scale manufacture of commodity products for producing complex systems based on starch-g-copolymers.

Further, we performed a detailed study on the preparation of thermoplastic starch and maleated thermoplastic starch by reactive extrusion, followed by their reactive extrusion with poly(butylene adipate-co-terephthalate) (PBAT) with the aim of enhancing the structure of the starch by reactive extrusion [56]. Glycerol was used as a plasticizer at a constant concentration (20 wt%) with respect to starch. Thermoplastic starch was prepared by reactive extrusion, where corn starch was fed to a co-rotating twin-screw extruder which had a screw diameter of 30 mm and an *L/D* ratio of 42. Glycerol was pumped to the extruder by a peristaltic pump such that the starch-to-glycerol mass composition was 80:20. Further, powdered maleic anhydride was mixed with starch using a blender, and a separate mixture of starch with Luperox 101 was prepared by blending. The starch/maleic anhydride mixture was then blended with the starch/Luperox mixture for 15 min, before being introduced into the extruder via a two-screw feeder. For the preparation of thermoplastic starch and maleated thermoplastic starch, excess moisture and unreacted maleic anhydride were removed by applying a vacuum at the vent-port. The temperature profile (in the six zones from the feed throat to the die) was 70/140/155/155/155/155 °C, and the screw speed was 125 rpm. The obtained thermoplastic starch and maleated thermoplastic starch were pelletized and dried in an oven at 70 °C overnight, prior to preparing the starch graft polyesters. The blends of starches with PBAT were prepared by feeding individual starches with PBAT using an external feeder, and adjusting the feed rate such that the mass ratio of two components was 60:40. The temperature profile for the reactive extrusion process in the six zones was 90/145/160/165/160/150 °C and the screw speed was 150 rpm. The resulting blends were pelletized and dried in an oven prior to analysis.

The grafting of glycerol on the starch backbone was determined by Soxhlet extraction in acetone, whereas dichloromethane was the solvent for determining the grafting of PBAT on the prepared starches. The extent of glycerol grafting onto the starch backbone was determined by gas chromatography (GC) and Fourier transform infrared spectroscopy (FTIR), with values found to be 83% for maleated thermoplastic starch and 26% for thermoplastic starch, when the amount of maleic anhydride used was 2 wt%. The Soxhlet extractions in dichloromethane resulted in 40% grafting of maleated thermoplastic starch with PBAT and 14% grafting of thermoplastic starch with PBAT. It was inferred that the presence of maleic acid in the thermoplastic starch enhanced the transesterification reaction between grafted glycerol and PBAT. The higher grafting yield of PBAT on maleated thermoplastic starch was also attributed to the presence of starch as a dispersed phase in the continuous PBAT matrix, as observed using transmission electron microscopy (TEM).

4.6 Stereocomplexation in Poly(lactic acid)

It is known that the properties of PLA are governed by its stoichiometry, where a reduction in melting point and crystallinity are observed upon increasing the content of D-lactide or meso-lactide when copolymerizing with L-lactide. However, physically blending PLLA and PDLA may not lead to the same behavior as that obtained by copolymerizing L-lactide and D-lactide. A stereoselective interaction is said to occur when the interaction between polymers of varying configuration dominates over the one between polymers with similar configurations, which is known as "stereocomplexation".

Stereocomplexation in PLA can occur in the solution or melt state as well as during polymerization. It is observed that the PLLA and PDLA chains have better stability in the solution state than in the melt state, which leads to the formation of stereocomplex crystals due to the self-assembly of chains [28]. However, stereocomplexation in PLA is affected by the optical purity and molecular weight of the homopolymers in addition to the tacticity, mixing ratio, and steric hindrance of the pendant chain. Stereocomplex crystals (with a melting point ~50 °C higher than that of

PLA) with increased crystallinity are formed when PLLA and PDLA are blended in a 1:1 ratio above their melting point. However, the formation of stereocomplex PLA crystals is often accompanied by homocrystals with an increased molecular weight of PLLA and PDLA. The formation of exclusive stereocomplex PLA crystals improves the heat stability of the material, which is the desired property for many end-use applications. In addition, solution blending is not a commercially viable process, and therefore, designing a process, which is continuous, may improve the acceptability of stereocomplex PLA in the commercial market.

In this regard, our group has recently reported a scalable, solvent-free, and additive-free method for the continuous manufacturing of stereocomplex PLA via reactive extrusion. The homopolymers i.e. PLLA and PDLA (M_n > 90 kDa) were blended in 1:1 ratio in a co-rotating twin-screw extruder having an *L/D* ratio of 40 and screw diameter of 27 mm, to achieve exclusive formation of stereocomplex PLA crystals. The processing conditions for stereocomplexation were determined using DSC in order to identify the optimum temperature profile and time for processing of stereocomplex PLA. It was observed that a specific temperature window resulted in high stereocomplex formation in a twin-screw extruder due to the alignment of chains during extrusion. High molecular weight PLLA and PDLA (1:1 ratio) were fed to the feeding zone of a co-rotating twin-screw extruder which comprised of a heating system divided into 10 zones. The temperature profile for processing stereocomplex PLA (zone 1 to die) was 160/170/180/180/180/210/220/220/220/230 °C. The temperature profile was so chosen as to heat the beginning of the barrel up to 180 °C, which is sufficient to allow the PLLA/PDLA blend to flow, but low enough to prevent excessive stereocomplex PLA formation at an early stage. The end of the barrel was heated to a higher temperature (210–220 °C) in order to accelerate the formation of stereocomplex crystals within the last minute of the process, which in turn prevents the solidification and thermal dissociation of the final product, thereby allowing continuous manufacturing of stereocomplex PLA. For continuous manufacturing, the feed rate was 3 kg/h and screw speed was set to 40 rpm. The screw configuration was constructed as shown in Figure 4.9 (kneading elements placed between conveying elements), in order to allow for mixing in the first and last minute of processing.

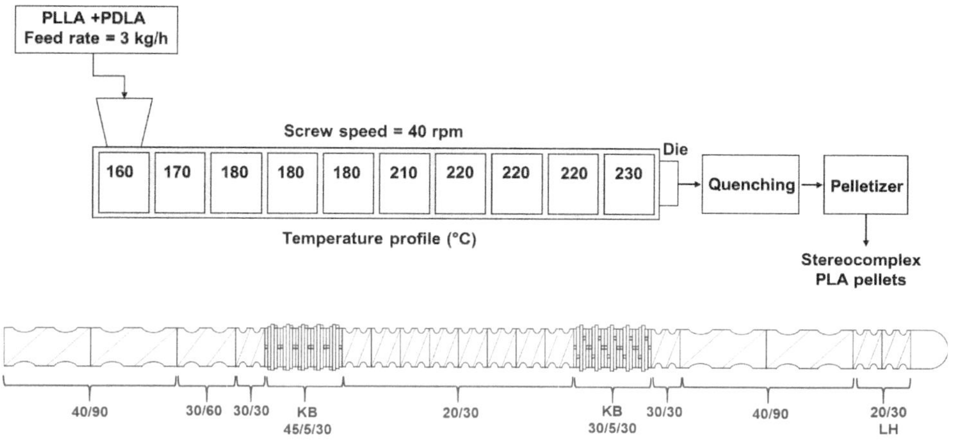

Figure 4.9 Processing conditions for preparation of stereocomplex PLA by reactive extrusion process

After a residence time of 2.5 min, the product was extruded out of a 5 mm diameter strand die followed by quenching in a water bath and pelletizing. The samples were collected at intervals of 5, 10, 20, and 40 min to identify the dispersity of the stereocomplex, followed by drying in an oven at 55 °C for 24 h. The extruded samples and PLLA were injection-molded using injection and mold temperatures of 220 and 55 °C, with a dwell time of 10 s. The samples were then annealed at 100 °C for 2 h. The wide-angle X-ray diffraction studies indicated formation of stereocomplex crystals corresponding to a triclinic unit cell, unlike the pseudo-orthorhombic crystal structure of PLLA and PDLA. A melting point of 240 °C was obtained for the sample with stereocomplex crystals, which had a crystallinity of 56–64% (Table 4.3). The intermolecular interactions between PLLA and PDLA in an extruder also led to a significant improvement in the tensile properties such as tensile strength, yield strength, modulus, resilience, and toughness.

Table 4.3 Thermal Properties and Crystallinities of Stereocomplex PLA, Obtained From DSC, at Different Collection Times

Sample collection time (min)	Fraction of stereo-complex crystallites (f_{sc}) (%)	Crystallinity (X_c) (%)	$T_{m,sc}$ (°C)	$T_{m,hc}$ (°C)	T_g (°C)
5	85	64	241	178	63
10	92	61	240	175	64
20	93	62	240	175	65
40	95	56	240	175	61

sc = stereocomplex crystallites; hc = homocrystals

4.7 Copolymerization by Reactive Extrusion

Poly(3-hydroxybutyrate) (PHB), which belongs to the class of poly(hydroxyalkanoate)s (PHAs) [61], is obtained from renewable sources [62], and is well-known for its biocompatibility [63, 64] as well as being a promising substitute for non-biodegradable polymers in commercial applications [65]. However, the poor processability, high brittleness [66], and high stiffness - due to its high degree of crystallinity - limits its practical implementation in various applications [67, 68].

In this regard, the transesterification of PHB and PCL was carried out by Sakaguti et al. to synthesize poly(3-hydroxybutyrate-b-ε-caprolactone) (PHB-b-PCL) by reactive extrusion in the presence of catalytic stannous octanoate, as shown in Figure 4.10 [69]. The polymers PHB and PCL (1:1 molar ratio) were dried followed by adding stannous octoate catalyst (5 wt%). The components were mixed on a heating plate until a homogenous molten mixture was formed, which was cooled and reduced to small pieces. The small pieces were fed to the hopper of an extruder for reactive extrusion processing, where the processing temperature (mixing chamber and die) was 165 °C. The obtained copolymer, after purification, was characterized using various analytical techniques, which confirmed the modification of PHB, the predominance of PCL segments, and the reduced crystallinity in the copolymer. The obtained copolymer was further subjected to electrospinning to prepare fibrous mats, where the average fiber diameter was 900–1200 nm, with voids (a few microns to few tens of microns) suitable for cell diffusion in biomedical applications.

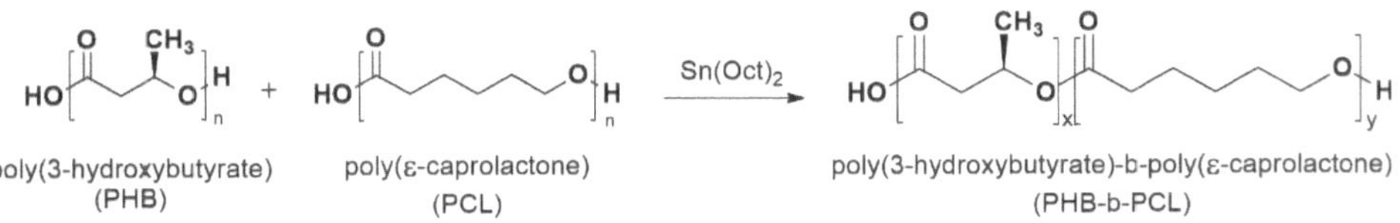

Figure 4.10 Synthesis of poly(3-hydroxybutyrate)-b-poly(ε-caprolactone) (PHB-b-PCL) by reactive extrusion

Tang et al. have reported the synthesis of poly(ethylene terephthalate)-b-poly(ε-caprolactone) (PET-b-PCL) by reactive extrusion processing [70]. The pellets of PET were fed into a counter-rotating twin-screw extruder at a rate of 5.8 kg/h, followed by heating and melting in the first two barrel sections at 290 °C. The molten PET was forwarded to an injection piston and mixed with the monomer and catalyst (ε-caprolactone and stannous octoate), which were injected using a piston pump at a rate of 1 kg/h. A uniform mixture of molten PET and ε-caprolactone was achieved in 40 s, which was then forwarded to a reaction zone set at 260 °C to initiate the ring-opening polymerization of ε-caprolactone by the hydroxyl end groups of molten PET. The reaction scheme for the synthesis of PET-b-PCL is shown in

Figure 4.11. After the residence time of 4 min, unreacted monomer was removed in the devolatilization section by vacuum, and a block copolymer with minimum transesterification was obtained. The block copolymer was characterized by intrinsic viscosity, gel permeation chromatography (GPC), ^{1}H-NMR, and DSC. The obtained copolymer was melt-spun at 275 °C where the speed of the godet was 280 m/min. The as-spun fibers were drawn with a ratio of 6.6:1 followed by annealing at 175 °C to obtain fibers with high tenacity, which were characterized by small-angle X-ray scattering (SAXS) and wide-angle X-ray diffraction (WAXD). The fiber morphology was so tailored as to obtain high degree of crystalline orientation and a large fraction of the unoriented amorphous phase, which may be suitable for applications such as automotive seat belts.

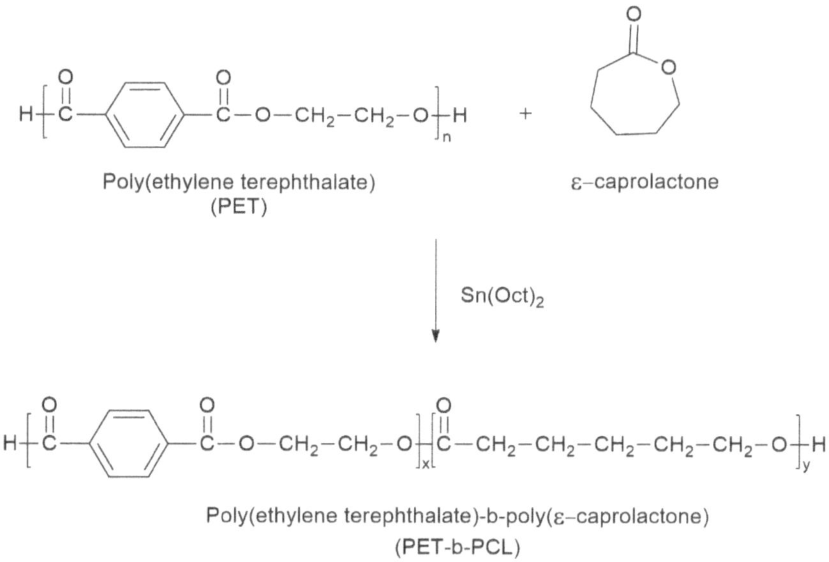

Figure 4.11 Synthesis of poly(ethylene terephthalate)-b-poly(ε-caprolactone) (PET-b-PCL) by reactive extrusion

■ 4.8 Preparation of Composites by Reactive Extrusion

With the aim of obtaining fully biodegradable polymer composites with improved mechanical performance and reduced costs, Nitz et al. grafted maleic anhydride (MA, 10 wt%) onto the chains of PCL to obtain PCL-g-maleic anhydride, which was incorporated with lignin and wood flour components to form composites by reactive extrusion [71]. PCL-g-MA acted as a compatibilizer to improve adhesion between reinforcing agents and polymer matrix. The twin-screw co-rotating intermeshing extruder system (*L/D* 42:1) was used for the preparation of compounds as well as compatibilizers, in different steps. However, the screw configuration was changed for the preparation of compounds, whereby the PCL-g-MA compatibilizers and PCL were mixed in a tumbling mixer. For preparing the compounds, all the components were fed separately using a gravimetric feed system with a processing temperature of 160 °C, a screw speed of 200 rpm, a mass flow rate of 5 kg/h, and a residence time of ~1 min. This was followed by the preparation of injection-molded test specimens at a melt temperature of 200 °C, which were used for characterization. The reactive extrusion processing parameters for the preparation of biocomposites are shown in Table 4.4. The mechanical properties (Young's modulus, yield stress, and impact strength) of the wood flour composites were found to be better than lignin-based composites. However, the dispersion of lignin in the PCL matrix was found to be very effective as observed using TEM. Further, the composites containing high content of lignin (40%) resulted in very high elongation at break (greater than 500%). In addition, it was inferred that the addition of lignin enhanced the biostability of PCL-based composites, as revealed from biodegradation studies, which suggests the potential of lignin as a non-toxic biostabilizer in enhancing the product lifetime of PCL composites in outdoor applications.

Table 4.4 Biodegradable Composites Prepared by Reactive Extrusion Processing

	Polymer matrix/ Compatibilizer	Filler	Initiator/ Catalyst	Type	*L/D* ratio	Screw speed (rpm)	Processing temp. (°C)
Nitz et al. [71]	PCL-g-MA	Lignin/ Wood flour	-	Co-rotating, intermeshing	42	200	160
Gupta et al. [72]	PBAT-g-MA	Hemp	Dicumyl peroxide	Co-rotating, intermeshing	40	60	130
	PBAT/ PBAT-g-MA	Hemp	-	Co-rotating, intermeshing	40	100	180

Furthermore, Gupta et al. developed flexible and heat-stable hemp powder based PBAT composites by reactive extrusion processing (Table 4.4). In the first step, PBAT pellets were mixed with maleic anhydride (5 wt%) and kept at 80 °C for 30 min in order to melt maleic anhydride and create a coating on PBAT pellets [72]. The mixture, after cooling, was mixed with dicumyl peroxide (1 wt%) as an initiator, followed by feeding in a twin-screw extruder at a rate of 500 g/h in order to conduct the reactive extrusion modification. The temperature profile of the extruder was 130/135/140/150/150/140/135/130 °C, with a screw speed of 60 rpm and an L/D ratio of 40. The modified PBAT was pelletized, and the unreacted maleic anhydride was removed by drying the sample in a vacuum oven under a reduced pressure. In the next step, dried PBAT and hemp were mixed with maleic anhydride-g-PBAT in different ratios, followed by processing in a twin-screw extruder at a processing temperature of 180 °C and a screw speed of 100 rpm. The obtained biocomposites were pelletized prior to analysis.

A significant improvement in the tensile strength (209%) of the biocomposite containing 50% PBAT, 40% hemp powder, and 10% PBAT-g-maleic anhydride was observed. Additionally, PBAT composites (with 20 to 40% hemp) exhibited substantial improvement in the impact resistance (~90%), toughness (~300%) and elongation at break (63–172%) upon addition of PBAT-g-maleic anhydride as a compatibilizer, which was attributed to the interfacial adhesion developed by the addition of the compatibilizer. Also, the presence of compatibilizer reduced the melt viscosity due to its plasticization effect, which was inferred from the rheological analysis. Additionally, the incorporation of 40% hemp powder into PBAT led to an increased heat deflection temperature (93 °C) when compared to neat PBAT (39 °C), which was considered to be important for applications requiring high dimensional stability at higher temperatures, such as hot beverages. The developed biocomposites were said to have potential in replacing single-use plastics as sustainable alternatives requiring heat stability and low cost along with biodegradability. A similar study was carried out by the same group using poly(3-hydroxybutyrate-co-3-hydroxyvalerate) (PHBV) as a matrix material for rigid packaging applications [73].

4.9 Conclusions and Outlook

This chapter has provided a broad overview on the facile, "just-in-time" manufacture of biodegradable plastics by reactive extrusion processing. The application of reactive extrusion for the chemical and physical modification of polymers, ring-opening polymerization, copolymerization, and preparation of composites has been detailed. Reactive extrusion, being a solvent-free and continuous process, integrates the extrusion streams along with polymerization processes, as well as providing good control over residence time and residence time distribution, enabling the manufacture end products with versatile properties.

Biodegradable polymers could effectively replace carbon–carbon backbone polymers in single-use packaging in order to minimize waste disposal in landfills. However, the use of biodegradable polymers as bulk materials is restricted due to their higher cost of production and poor mechanical properties when compared with commodity polymers such as polyethylene. In this regard, reactive extrusion would serve as a facile and relatively inexpensive route for polymer processing that could improve the commercial viability of biodegradable polymers.

This chapter has also highlighted the research and developments that have been undertaken in the area of biodegradable polymers produced by reactive extrusion processing. Reactive extrusion is particularly suitable for the synthesis of biodegradable polyesters such as PCL and PLA by ROP of their individual monomers in presence of efficient catalyst systems. Furthermore, chemical modification of biodegradable polyesters by grafting with natural polymers or fillers offers an effective way of controlling the properties of the resulting product by varying the processing conditions. Development of stereocomplex PLA blends with an improved fraction of stereocomplex crystallites was also demonstrated by varying the temperature profile and screw configuration during reactive extrusion processing.

This chapter has also highlighted the copolymerization of biodegradable polymers such as PHB and PCL, and the formation of composites (using lignin and wood flour with PCL, or hemp powder with PBAT) by reactive extrusion processing, which has resulted in tailored properties of the end-product due to an improved dispersion and strong interfacial adhesion between the polymer and filler. Reactive extrusion processing could serve as a facile and cost-competitive tool for designing biodegradable polymers with chemically modified structures and improved overall properties that would comply with the end-use-applications, and in turn drive the future growth of biodegradable polymers to provide sustainable solutions.

References for Chapter 4

[1] Tzoganakis, C., *Advances in Polymer Technology: Journal of the Polymer Processing Institute* (1989) 9, pp. 321–330.

[2] Farkas, V., Nagyházi, M., Anastas, P.T., Klankermayer, J., Tuba, R., *ChemSusChem* (2023), p. e202300553.

[3] Yang, X., Man, Y.B., Wong, M.H., Owen, R.B., Chow, K.L., *Science of the Total Environment* (2022) 825, p. 154025.

[4] Narayan, R., in *Soil Degradable Bioplastics for a Sustainable Modern Agriculture*, Malinconico, M. (Ed.) (2017), Springer, pp. 23–34.

[5] Gaur, V.K., Sirohi, R., Bhat, M.I., Gautam, K., Sharma, P., Srivastava, J.K., Pandey, A., *Chemosphere* (2022), p. 136877.

[6] Huo, Y., Dijkstra, F.A., Possell, M., Singh, B., *Environmental Pollution* (2022), p. 119892.

[7] Lebreton, L., *Nature Reviews Earth & Environment* (2022) 3, pp. 730–732.

[8] Wang, T., Zhao, S., Zhu, L., McWilliams, J.C., Galgani, L., Amin, R.M., Nakajima, R., Jiang, W., Chen, M., *Nature Reviews Earth & Environment* (2022) 3, pp. 795–805.

[9] Zumstein, M.T., Narayan, R., Kohler, H.-P.E., McNeill, K., Sander, M. *Environmental Science & Technology* (2019) 53 pp. 9967–9969.

[10] Law, K.L., Narayan, R., *Nature Reviews Materials* (2022) 7, pp. 104–116.

[11] Filiciotto, L., Rothenberg, G., *ChemSusChem* (2021) 14, pp. 56–72.

[12] Mulchandani, N., Narayan, R., *Molecules* (2023) 28, p. 3832.

[13] Hopmann, C., Adamy, M., Cohnen, A., *Reactive Extrusion: Principles and Applications* (2017), pp. 1–10.

[14] Raquez, J.M., Narayan, R., Dubois, P., *Macromolecular Materials and Engineering* (2008) 293, pp. 447–470.

[15] Graiver, D., Waikul, L., Berger, C., Narayan, R., *Journal of Applied Polymer Science* (2004) 92, pp. 3231–3239.

[16] Stagner, J., Dias Alves, V., Narayan, R., Beleia, A., *Journal of Polymers and the Environment* (2011) 19, pp. 589–597.

[17] Andersen, J., Starbuck, H., Current, T., Martin, S., Mack, J., *Green Chemistry* (2021) 23, pp. 8501-8509.

[18] Kimura, Y., Chaudhary, A.K., Spalding, M.A., *Handbook of Industrial Polyethylene and Technology: Definitive Guide to Manufacturing, Properties, Processing, Applications and Markets* (2017), pp. 357–380.

[19] White, J.L., Keum, J., Jung, H., Ban, K., Bumm, S., *Polymer-Plastics Technology and Engineering* (2006) 45, pp. 539–548.

[20] Zhuang, Y., Saadatkhah, N., Morgani, M.S., Xu, T., Martin, C., Patience, G.S., Ajji, A., *The Canadian Journal of Chemical Engineering* (2023) 101, pp. 59–77.

[21] Zhang, S., Yan, D., Zhao, L., Lin, J., *Composites Communications* (2022) 34, p. 101268.

[22] Li, C., Pan, S., Xu, W., Lu, Y., Wang, P., Zhang, F., Gross, R.A., *Green Chemistry* (2020) 22, pp. 662–668.

[23] Mulchandani, N., Masutani, K., Kumar, S., Sakurai, S., Kimura, Y., Katiyar, V., *Materials Today Communications* (2022) 33, p. 104040.

[24] Balakrishnan, S., Krishnan, M., Narayan, R., Dubois, P., *Polymer Engineering & Science* (2006) 46, pp. 235–240.

[25] Xia, Z., Zerroukhi, A., Chalamet, Y., Chen, J., *Journal of Applied Polymer Science* (2008) 109, pp. 1772–1780.

[26] Inkinen, S., Hakkarainen, M., Albertsson, A.-C., Södergård, A., *Biomacromolecules* (2011) 12, pp. 523–532.

[27] Mulchandani, N., Katiyar, V., in *Advances in Sustainable Polymers: Synthesis, Fabrication and Characterization*, Katiyar, V., Kumar, A., Mulchandani, N. (Eds.) (2020), Springer, pp. 1–20.

[28] Mulchandani, N., Prasad, A., Katiyar, V., in *Materials for Biomedical Engineering*, Grumezescu, V., Grumezescu, A.M. (Eds.) (2019), Elsevier, pp. 87–125.

[29] Mulchandani, N., Gupta, A., Katiyar, V., in *Cellulose-Based Superabsorbent Hydrogels*, (2019), Springer, pp. 1537–1559.

[30] Mulchandani, N., Gupta, A., Masutani, K., Kumar, S., Sakurai, S., Kimura, Y., Katiyar, V., *ACS Applied Polymer Materials* (2019) 1, pp. 3354–3365.

[31] Mulchandani, N., Masutani, K., Kumar, S., Yamane, H., Sakurai, S., Kimura, Y., Katiyar, V., *Polymer Chemistry* (2021) 12, pp. 3806–3824.

[32] Oluwabunmi, K., D'Souza, N.A., Zhao, W., Choi, T.-Y., Theyson, T., *Scientific Reports* (2020) 10, p. 17771.

[33] Nagarajan, V., Mohanty, A.K., Misra, M., *ACS Sustainable Chemistry & Engineering* (2016) 4, pp. 2899–2916.

[34] Ghosh, K., Jones, B.H., *ACS Sustainable Chemistry & Engineering* (2021) 9, pp. 6170–6187.

[35] Farah, S., Anderson, D.G., Langer, R., *Advanced Drug Delivery Reviews* (2016) 107, pp. 367–392.

[36] Naser, A.Z., Deiab, I., Darras, B.M., *RSC Advances* (2021) 11, pp. 17151–17196.

[37] Lay, M., Thajudin, N.L.N., Hamid, Z.A.A., Rusli, A., Abdullah, M.K., Shuib, R.K., *Composites Part B: Engineering* (2019) 176, p. 107341.

[38] Oksman, K., Skrifvars, M., Selin, J.F., *Composites Science and Technology* (2003) 63, pp. 1317–1324.

[39] Razavi, M., Wang, S.-Q., *Macromolecules* (2019) 52, pp. 5429–5441.

[40] Zhang, K., Nagarajan, V., Misra, M., Mohanty, A.K., *ACS Applied Materials & Interfaces* (2014) 6, pp. 12436–12448.

[41] Yeo, J.C.C., Muiruri, J.K., Tan, B.H., Thitsartarn, W., Kong, J., Zhang, X., Li, Z., He, C., *ACS Sustainable Chemistry & Engineering* (2018) 6, pp. 15517–15527.

[42] Corneillie, S., Smet, M., *Polymer Chemistry* (2015) 6, pp. 850–867.

[43] Rabanel, J.-M., Faivre, J., F. Tehrani, S., Lalloz, A., Hildgen, P., Banquy, X., *ACS Applied Materials & Interfaces* (2015) 7.

[44] Banu, I., Puaux, J.P., Bozga, G., Nagy, I., Modeling of L-lactide polymerization by reactive extrusion, in *Proceedings of the Macromolecular Symposia* (2010) pp. 108–118.

[45] Kumari, S.V.G., Pakshirajan, K., Pugazhenthi, G., *International Journal of Biological Macromolecules* (2022).

[46] Shokri, Z., Seidi, F., Saeb, M.R., Jin, Y., Li, C., Xiao, H., *Materials Today Chemistry* (2022) 24, p. 100780.

[47] Peter, S., Lyczko, N., Gopakumar, D., Maria, H.J., Nzihou, A., Thomas, S., *Waste and Biomass Valorization* (2021) 12, pp. 4777–4804.

[48] Ghosh, T., Nakano, K., Mulchandani, N., Katiyar, V., *Food Packaging and Shelf Life* (2021) 28, p. 100658.

[49] Mulchandani, N., Shah, N., Mehta, T., *Polymers and Polymer Composites* (2017) 25, pp. 241–246.

[50] Kumar, D., Gihar, S., Shrivash, M.K., Kumar, P., Kundu, P.P., *International Journal of Biological Macromolecules* (2020) 163, pp. 2097–2112.

[51] Sorokin, A.V., Olshannikova, S.S., Lavlinskaya, M.S., Holyavka, M.G., Faizullin, D.A., Zuev, Y.F., Artukhov, V.G., *Polymers (Basel)* (2022) 14, p. 2279.

[52] Vaidya, A.A., Hussain, I., Gaugler, M., Smith, D.A., *Carbohydrate Polymers* (2019) 217, pp. 98–109.

[53] Mulchandani, N., Katiyar, V., *International Journal of Biological Macromolecules* (2019) 141, pp. 831–842.

[54] Gupta, A., Prasad, A., Mulchandani, N., Shah, M., Ravi Sankar, M., Kumar, S., Katiyar, V., *ACS Omega* (2017) 2, pp. 4039–4052.

[55] Carlson, D., Nie, L., Narayan, R., Dubois, P., *Journal of Applied Polymer Science* (1999) 72, pp. 477–485.

[56] Schneider, J., Shi, X., Manjure, S., Gravier, D., Narayan, R., *Journal of Applied Polymer Science* (2015) 132.

[57] Schneider, J., Bourque, K., Narayan, R., *Express Polymer Letters* (2016) 10, p. 799.

[58] Ojogbo, E., Ogunsona, E.O., Mekonnen, T.H., *Materials Today Sustainability* (2020) 7, p. 100028.

[59] Chatterjee, S., Mahmood, S., Hilles, A.R., Thomas, S., Roy, S., Provaznik, V., Romero, E.L., Ghosal, K., *International Journal of Biological Macromolecules* (2023), p. 125757.

[60] Siyamak, S., Laycock, B., Luckman, P., *Carbohydrate Polymers* (2020) 227, p. 115066.

[61] Turco, R., Corrado, I., Zannini, D., Gargiulo, L., Di Serio, M., Pezzella, C., Santagata, G., *Bioresource Technology* (2022) 363, p. 127954.

[62] Zoghbi, L., Argeiti, C., Skliros, D., Flemetakis, E., Koutinas, A., Pateraki, C., Ladakis, D., *Biochemical Engineering Journal* (2023) 191, p. 108808.

[63] Tyubaeva, P.M., Varyan, I.A., Nikolskaya, E.D., Mollaeva, M.R., Yabbarov, N.G., Sokol, M.B., Chirkina, M.V., Popov, A.A., *Nanomaterials (Basel)* (2023) 13, p. 236.

[64] Suo, D., Rao, J., Wang, H., Zhang, Z., Leung, P.H.-M., Zhang, H., Tao, X., Zhao, X., *Biomaterials Science* (2022) 10, pp. 3493–3502.

[65] Popa, M.S., Frone, A.N., Panaitescu, D.M., *International Journal of Biological Macromolecules* (2022) 207, pp. 263–277.

[66] Garcia-Garcia, D., Quiles-Carrillo, L., Balart, R., Torres-Giner, S., Arrieta, P. M., *European Polymer Journal* (2022), p. 111505.

[67] Moreira, J. B., Kuntzler, S. G., da Silva Vaz, B., da Silva, C. K., Costa, J. A. V., de Morais, M. G., in *Biodegradable Polymers, Blends and Composites* (2022), Elsevier, pp. 389–413.

[68] Luoma, E., Rokkonen, T., Tribot, A., Nättinen, K., Lahtinen, J., *Polymers from Renewable Resources* (2022) 13, pp. 83–101.

[69] Sakaguti, K. Y., Wang, S. H., *Journal of the Brazilian Chemical Society* (2021) 32, pp. 355–362.

[70] Tang, W., Murthy, N. S., Mares, F., Mcdonnell, M. E., Curran, S. A., *Journal of Applied Polymer Science* (1999) 74, pp. 1858–1867.

[71] Nitz, H., Semke, H., Landers, R., Mülhaupt, R., *Journal of Applied Polymer Science* (2001) 81, pp. 1972–1984.

[72] Gupta, A., Chudasama, B., Chang, B. P., Mekonnen, T., *Composites Science and Technology* (2021) 215, p. 109014.

[73] Gupta, A., Lolic, L., Mekonnen, T. H., *Composites Part A: Applied Science and Manufacturing* (2022) 156, p. 106885.

5 Thermal Processing of Natural-Based Biodegradable Polymers

Mondli Abednicko Masanabo, Mpho Phillip Motloung, M. Naushad Emmambux, and Suprakas Sinha Ray

Natural-based biodegradable polymers can be suitable for the development of eco-friendly plastic materials. However, due to their complex structures and susceptibility to thermal degradation, it is challenging to process them using continuous thermal processing techniques. Nevertheless, studies focusing on the possibilities of processing natural-based biodegradable polymers in continuous thermal processing techniques have been conducted. Natural-based polymers require chemical modifications before they can be thermally processed, and even then still require plasticizers to assist with processability. In this chapter, thermal processing of selected natural-based biodegradable polymers is discussed. In particular, thermal processing of starch, cellulose, and proteins is covered. The chapter further discusses challenges and recommendations regarding thermal processability of these polymers.

5.1 Introduction

Biodegradable polymers have received substantial attention in recent years [1]. The interest in this type of material is mostly propelled by stringent regulations, which encourage the development and utilization of sustainable, and environmentally friendly polymers [1–3]. Innumerable research studies have been conducted that investigate the synthesis, processability, application, and biodegradation of biodegradable polymers [4, 5].

Biodegradable polymers can be fossil-based or natural-based. There has been a growing interest in utilization of biodegradable polymers in various application fields such as packaging, biomedical, water treatment, agriculture, automotive, and others. However, before using them in these applications, the processability of various biodegradable polymers should be understood. Research studies have demonstrated the processability of various biodegradable polymers through different processing techniques. Most studies conducted are based on the processing of synthetic polymers [6–9].

Natural polymers such as starch, cellulose, and proteins have garnered interest in different applications. However, most of the research on the processing of these natural polymers has focused on solvent-based processes, with little attention given to the thermal processing that is important for large-scale production [9, 10]. In addition, thermal-based processes are efficient, less time consuming, and environmentally friendly. In this chapter, the processing of natural biodegradable polymers via melt processing techniques, such as extrusion, film blowing, and injection molding is outlined.

5.2 Thermal Processability of Natural Polymers

Most natural polymers do not have good thermoplastic properties compared to commercially available synthetic polymers for making plastics due to their chemical nature. This is a limitation during conventional thermal processing. The thermoplasticity of these natural polymers can be improved either by chemical modifications or addition of processing aids, such as plasticizers, with processing under specific conditions. Herein, we briefly discuss how natural polymers, mainly starch, cellulose, and proteins, can be manipulated to behave like thermoplastics and thus become processable by various melt-processing techniques.

5.2.1 Starch

Starch is a storage polysaccharide in higher plants, and it occurs as semi-crystalline granules [11]. Amylose is an essentially linear molecule that is made up of d-glucopyranose monomers linked together by α-1,4-glycosidic bonds. In contrast, amylopectin is a highly branched molecule made up of d-glucopyranose monomers in which the main linear backbone is linked by α-1,4-glycosidic bonds and the branches are linked by α-1,6-glycosidic bonds [11, 12].

The inter- and intramolecular hydrogen bonds in starch and the highly branched nature of starch makes it resistant to processing as a thermoplastic material [13]. Starch has unique and complex thermal processing properties compared to conventional thermoplastic polymers. Various physical and thermal changes occur in starch during thermal processing, such as water diffusion, granular swelling, gelatinization, depolymerization, decomposition, melting, and crystallization.

Gelatinization is the most important transition, and is the basis of the conversion of starch into thermoplastic starch (TPS) [11, 14]. Starch gelatinization usually occurs in the presence of excess water, and it is characterized by the swelling of the starch granules, melting of the native crystalline structure, and molecular solubilization [15]. During thermal processing of starch into a thermoplastic, gelatinization occurs under limited moisture conditions, and it is referred to as "melting" of starch [16]. Plasticizers, such as water and, glycerol are added, which break the starch-to-starch interactions and result in starch–plasticizer interactions, thus increasing the flexibility, and lowering the glass transition temperature (T_g), to allow for native starch to be processable into TPS [17]. Therefore, TPS can be defined as plasticized starch that was thermally processed under low moisture conditions, and under the action of thermal and mechanical forces. TPS can be made from native starch or its polymers, amylose and amylopectin [18]. In addition, normal starch, high amylose starch, waxy starch, and pre-gelatinized starch can be used to prepare TPS with different material properties. The source of starch has significant influence on the resulting material properties of TPS.

5.2.2 Cellulose

Cellulose is also made up of d-glucopyranose monomers that are linked by β-1,4-glycosidic bonds instead of α-1,4-glycosidic bonds as in starch. Like starch, it is also not a conventional thermoplastic, due to its chemical nature. The inter- and intramolecular hydrogen bonds in cellulose makes it not processable by melt-processing techniques [19]. The modification of cellulose, which generally involves the breaking of the cellulose hydrogen bonds, is necessary to induce thermoplasticity. The hydroxyl (-OH) groups of cellulose can be converted into aliphatic chains by esterification, etherification, and/or sialylation to improve the thermoplasticity of cellulose [20, 21].

Some examples of commercially traded cellulose derivatives with short side-chains produced by esterification include cellulose acetate propionate (CAP), cellulose acetate (CA), and cellulose acetate butyrate (CAB). However, these derivatives have a narrow melt-processing window between the decomposition and melt temperature, thus requiring the use of large amounts of low molecular weight plasticizers in order to produce thermoplastic cellulose [19]. The migration of the plasticizers is an inevitable shortcoming that will affect the material properties of these thermoplastic cellulose derivatives. To address the inevitable plasticizer migration, the grafting of long side-chains that can act as internal plasticizers on the cellulose backbone has been explored. Some research has shown that an increase in the length of the side-chains grafted into the cellulose backbone significantly reduces the melting temperature of cellulose, so aiding processability into thermoplastic cellulose [22, 23]. Examples of cellulose esters with long side-chain lengths include cellulose octanoate (C_8) and cellulose palmitate (C_{16}). These were previously processed without the use of plasticizers [24, 25].

5.2.3 Proteins

Proteins have been of interest in the development of biodegradable plastics. Compared to polysaccharides such as starch and cellulose, proteins have various reactive functional groups that can be modified to tailor their properties for various applications [26]. In addition, proteins are heteropolymers, unlike synthetic polymers and other natural polymers, such as starch and cellulose. Proteins, such as soybean, wheat proteins, sunflower proteins and, prolamin proteins (zein and kafirin) have been investigated for their film-forming properties [9, 27, 28]. Of the film-forming proteins, zein, a by-product of maize processing, has stood out because of its unique thermoplasticity and hydrophobicity [29]. Zein is relatively hydrophobic due to its high content of non-polar amino acids [30], giving it better moisture barrier properties compared to other proteins (such as wheat casein) or other polysaccharides (such as starch) [29].

The film-forming properties of proteins are rather complex compared to conventional thermoplastic polymers. The thermoplastic properties of proteins can be improved by the addition of plasticizers. These plasticizers increase the free volume, reduce protein–protein interactions, and result in more protein–plasticizer interactions, so reducing the T_g. When cereal proteins are heated above their T_g, they can change from a glassy to a rubbery state, leading to an increase in molecular chain mobility and a decrease in viscosity [31]. Some authors have reported that when proteins are heated they unfold, thus exposing the hydrophobic groups and resulting in the aggregation of the protein. Taylor et al. [27] reported that the film formation of prolamin proteins is made possible by controlled aggregation. The mechanism by which proteins can form films during thermoprocessing is unclear. However, protein aggregation and heating the protein above its T_g seem to be the driving forces in producing thermoplastic proteins.

5.3 Processing Techniques of Natural Polymers

5.3.1 Extrusion and Batch Processing

Extrusion processing is the most efficient process for large-scale processing of polymers, compared to lab-scale processes such as rheomixers. Generally, extrusion serves as the first processing step, where the materials are mixed to form blends or composites, before they can be processed in any other techniques. However, for other product lines such as in pipe-making, extrusion can be used as a standalone technique. Compared to batch mixers such as rheomixers, the extrusion process is complex, as it comprises screws with varying configurations. Usually, the screw in extruders consists of knitting elements, which provide efficient mixing in the barrel compared to batch mixing. The residence time of the material mostly depend on the screw speeds. However, the material properties such as melt viscosity and melt flow index (MFI) also tend to influence the residence time inside a barrel. In this section, extrusion processing of starch, cellulose, and proteins is discussed.

5.3.1.1 Starch

As explained in the earlier section, TPS can only occur in limited moisture conditions and in the presence of a plasticizer and shear stress conditions. Therefore, rheomixers, single and twin-screw extruders can be adapted to produce TPS. The processing of starch-based polymers using batch mixers usually involves the mixing of starch with plasticizers (for example, water and glycerol) using high-speed mixers. This is followed by thermal processing of the plasticized starch in a rheomixer barrel under defined temperature, screw speed, and time. When processed under limited moisture conditions, and in the presence of the plasticizer, native starch forms a highly viscous polymer melt, posing processing challenges during batch mixing. Therefore, extrusion is a widely preferred thermal processing technique to convert native starch into TPS. This is because it can withstand highly viscous starch polymer melts, offers flexibility in terms of processing, and allows for a wide range of processing conditions compared to batch mixers [32, 33]. High-amylose starch is often difficult to process by thermal techniques compared to normal starch or waxy starch because it forms a polymer melt with very high viscosity and has a higher melting temperature compared to other starches. However, this can be circumvented by raising the moisture levels, raising the barrel temperature, increasing the screw speed, and decreasing the feed rate [34]. Starch undergoes several physiochemical changes during extrusion, and these have been extensively covered elsewhere [14]. These physicochemical changes include the diffusion of water molecules into the starch granules, the swelling of the starch granules, gelatinization, melting, loss of crystallinity, depolymerization, and decomposition [11, 14].

Extrusion can be used as a stand-alone technique to produce thermoplastic starch films/sheets, or it can be followed by another processing technique, such as compression-molding, injection molding, extrusion film-blowing, or extrusion blow-molding. During extrusion, TPS pellets are produced from plasticized native starch to which cellulose or proteins may be added to improve the processability or tailor the material properties of the resulting starch-based blends [35, 36]. Alternatively, a two-stage extrusion process may be followed, in which during the first extrusion, TPS starch pellets are produced, followed by a second extrusion in which other toughening polymers, fillers, proteins, etc. can be added. The resulting starch-based pellets may then be later processed by extrusion blow-molding, cast extrusion, or injection molding.

Galicia-García et al. [37] showed the thermal processability of starch-based composites through the extrusion-calendering technique. Starch obtained from different sources such as maize and potato were used. The maximum barrel temperature was 60–95 °C, while the screw speed was in the range 13–25 rpm. Rodríguez-Castellanos et al. [36] showed extrusion blow-molding of maize starch/gelatin containing cellulose as a reinforcing agent. The authors observed a decrease in the viscosity of starch/gelatin when cellulose was incorporated, resulting in processability at the relatively low temperature of 100 °C. Figure 5.1 depicts the starch/gelatin/cellulose samples obtained after extrusion blow-molding. Suitable conditions for producing the rounded masses (parisons) was 95 °C, according to the authors. The results indicated an increase in stiffness and strength of the prepared materials upon the addition of cellulose.

Figure 5.1 Extrusion blow-molding to produce thermoplastic starch bottles [37]

5.3.1.2 Cellulose

As mentioned earlier, it is difficult to process cellulose as a thermoplastic because of the hydrogen bonds between the chains. Therefore, cellulose, and other natural polymers such as chitin cannot be processed in any continuous melt-processing technique. However, cellulose derivatives can be thermally processed. Cellulose derivatives such as hydroxyethyl cellulose (HEC), hydroxypropyl cellulose (HPC), hydroxypropyl methylcellulose (HPMC), carboxymethyl cellulose (CMC), and cellulose acetate (CA) have the potential of being processed in continuous melt-processing techniques.

Taheri et al. [38] showed the one-step extrusion of HEC and cellulose fibers, followed by compression-molding. The tensile strength of the compression-molded samples increased with inclusion of cellulose fibers. Matsuoka et al. [39] reported a similar study on the thermal extrusion of cellulose using HPMC to develop an all-cellulose thermoplastic. The authors evaluated thermal extrusion using capillary diameter. The cellulose extrusion was successfully obtained by increasing the loading pressure and the amounts of glycerol, which assisted with the flow of HPMC and reduced the extrusion temperatures.

Prodduturi et al. [40] reported the hot-melt extrusion of HPC films, extruded together with the drug clotrimazole (CT). The films were prepared from HPC with different molecular weights. The processing temperatures increased with increasing the molecular weight of HPC, possibly due to restricted flow when molecular weight increases. The maximum temperature reached was 160 °C for HPC with the highest molecular weight. The tensile strength and strain at break of the prepared films increased with the molecular weight. Hartzke et al. [41] also showed the thermal processability of HEC-HPC as a carrier for the drug diclofenac. The materials were prepared by melt extrusion, followed by fused deposition modeling. In their study, the authors demonstrated not only the melt extrusion of HEC, but also showed the 3D printability of HEC. 3D printing is one of the techniques that have garnered immense interest in the development of complex structures and shapes. Vo et al. [42] reported a similar study on melt extrusion and fused deposition modeling of HPC to develop floating tablets of cinnarizine. The samples were prepared in a co-rotating extruder, operated at 130 °C, whereas the printing temperature was 165 °C.

5.3.1.3 Proteins

Batch mixers such as rheomixers were used in the earlier work to process proteins, such as zein and their blends with starch. The processing of zein using a rheomixer involved the mixing of zein powder with a plasticizer such as glycerol, followed by thermal processing using a rheomixer and subsequent compression-molding into films [43, 44]. Corradini et al. [43] plasticized zein with 22, 30, and 40% glycerol, and processed it for 6 min at 160 °C and 50 rpm, which was followed by compression-molding. The tensile properties of the thermoplastic zein films revealed that the increase in the amount of glycerol decreased the ultimate tensile stress and Young's modulus, with the elongation at break remaining constant due to the poor interaction between zein and glycerol. As mentioned, zein is hydrophobic and glycerol is hydrophilic, this might have led to the poor interaction between zein and glycerol.

Masanabo et al. [35] produced glycerol-plasticized thermoplastic zein films and their blends with starch using twin-screw extrusion, with a die temperature of 120 °C. The resulting thermoplastic zein films had a porous cross-sectional structure, possibly due to trapped air and water droplets [45]. Zein has four main subunits, mainly α-, β-, γ-, and δ-zein: commercial zein is mostly made up of α-subunits, whereas "total zein" contains all the subunits [46]. Zhang et al. [47] studied the effect of extrusion temperature and the α-subunit content on the structure of zein extrudates. The authors extruded commercial zein, total zein and α-subunit-rich zein using a twin-screw extruder. Notably, zein was only plasticized with water (22% moisture dry basis) and extruded at different die temperatures of 80, 100, 120, and 140 °C respectively and the extrudates are shown in Figure 5.2. Their results demonstrated that higher extrusion temperatures negatively affected the ability of zein to form a viscoelastic plasticized network, as the bubbling of zein is observed. This was especially seen with commercial zein (mostly α-subunits), which formed fibrous structures when stretched at a temperature above 100 °C. In contrast, total zein and α-rich zein formed a compact structure, as observed from morphological analysis. Most studies on the production of thermoplastic zein by solvent casting, thermal processing use commercial zein (mostly α-subunits). However, laboratory-extracted total zein can form a more compact structure when processed at a temperature below 100 °C.

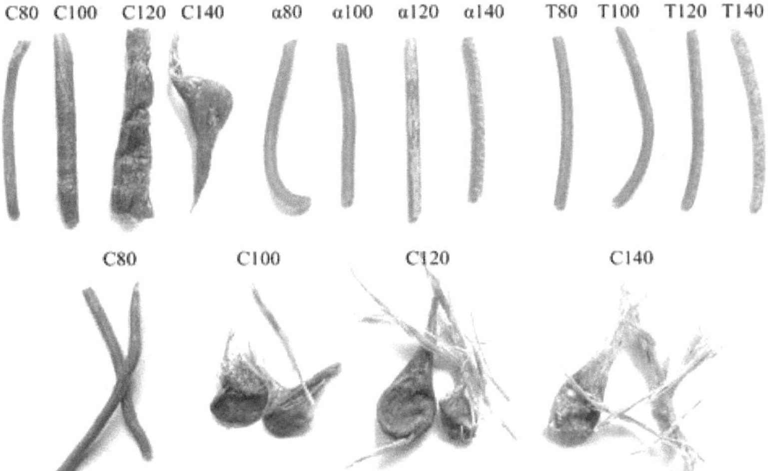

Figure 5.2 Effect of die processing temperature on the properties of commercial zein (C), total zein (T), and α-subunit-rich zein (α) [47]

5.3.1.4 Extrusion of Natural Polymer Blends

As has been described in the earlier section, extrusion can be used to process natural polymers. However, the poor melt-processability of natural polymers and their inferior material properties compared to their synthetic polymer counterparts is a challenge. Therefore, blending of natural polymers by extrusion is one strategy to produce polymer blends with improved material properties compared to the individual polymer counterparts. In natural polymer blends produced by extrusion, starch is a preferred major matrix due to its relative ease of processability, and its ability to form a continuous matrix compared to proteins and cellulose [48, 49]. However, the relatively hydrophilic nature of thermoplastic starch and inferior mechanical properties compared to synthetic polymers remains a challenge.

Since starch and cellulose have chemical similarities, they have the potential to produce compatible thermoplastic blends with improved material properties compared to the individual counterparts. e Moraes et al. [49] produced TPS/plasticized cellulose acetate sheets by an extrusion calendaring process at different temperatures (180 and 200 °C). This involved two-step extrusion, in which the pellets were produced by single-screw extrusion and the sheets were produced by a twin-screw extruder equipped with a flat die at 180 and 200 °C, respectively. This study reported that at 180 °C there was incomplete melting of thermoplastic cellulose acetate and starch, resulting in lower melt viscosity. As a result, the melt did not fill the die sheets, resulting in narrow sheets. However, higher temperatures resulted in complete melting of plasticized cellulose acetate, resulting in better plasticization and compatibility in the blend, especially in the 50:50 and 75:50 blends (TPS–cellulose acetate).

Starch and protein blending is another strategy to produce natural polymer blends with improved material properties. Of the film-forming proteins, zein is attractive due to its unique melt-processability and relative hydrophobicity. However, glycerol-plasticized starch–zein blends produced by batch mixers (rheomixers) were reported to be incompatible, resulting in starch–zein blends with inferior material properties compared to starch and zein alone [50]. Leroy et al. [51] showed that starch–zein blends produced using a conical twin-screw mini-compounder plasticized with an ionic liquid (1-butyl-3-methylimidazolium chloride [BMIM]Cl) result in better plasticization of the blend with better material properties compared to glycerol-plasticized blends. In a later study, Masanabo et al. [35] showed that glycerol-plasticized starch–zein blend films can be produced using twin-screw extrusion followed by compression-molding. In this study, it was demonstrated that the addition of 0.2 M sodium hydroxide during extrusion of the starch–zein blends resulted in better compatibility between starch and zein, and efficient plasticization, with higher elongation at break compared to the starch–zein blend films without NaOH. Therefore, the blending of starch with other natural polymers, such as cellulose and proteins, such as zein by extrusion, has potential to not only improve the processability of natural polymers, but also enhance their material properties compared to the individual counterparts.

5.3.2 Extrusion Film-Blowing

Polymer films are widely used in flexible packaging, mulch films, shrink-wraps, etc. The extrusion film-blowing technique is the most used thermal processing technique to produce films at an industrial scale. One of the critical pre-requisites for film-processing is adequate melt strength, which determines film-blowability, extensibility, and bubble stability during film-processing. Film-blowing of natural biodegradable polymers has been reported in the literature. This section focuses on film-blowing of natural-based biodegradable polymers.

5.3.2.1 Starch

Starch-based biofilms prepared through melt-processing techniques have been reported. Film-blowing of starch alone is a challenging task. Generally, plasticizers are added to starch to attain TPS. The critical factor in such cases is the concentrations of the plasticizers. Plasticizers are usually added to enhance plasticizer-polymer interactions. Moreover, the amylose content also affects the film-blow processability of starch. In essence, the amount of amylose content and thus the source of starch plays a pivotal role in the film-processability of starch.

Starch has poor melt tenacity that limits its film-blowability. In addition, the moisture sensitivity and stickiness during film-blowing restricts film-processing of starch. Thunwall et al. [52] conducted a study on film-blowing of potato starch with amylose content of about 21%. In their study, the authors used native potato starch and oxidized and hydroxypropylated (HONPS) starches. The ratio of modified starch to glycerol was varied, whereas the native starch to glycerol was fixed at ratio of 100:30 by parts. The authors identified possible limiting factors to film-processing of starch. Among factors identified, stickiness, which results in double-walled films, was identified as major challenge in extrusion of TPS, and this was attributed to the presence of glycerol and its concentrations. Also, reducing the hydrophilicity of starch through modification can minimize the issue of stickiness. Dang and Yoksan [53] used plasticized chitosan to assist with film-blowing of TPS. The authors reported an improved extrusion processability, albeit the film extensibility decreased in the presence of chitosan.

Zullo and Iannace [54] evaluated the influence of different sources of starch and plasticizers on film-blowing of TPS. Starch obtained from maize, potato, and wheat was used, while glycerol, urea, and formamide were used as plasticizers. In summary, the combination of urea and formamide enhanced the mechanical properties and restrained retrogradation. Besides, it was found that starch with high amylose content, combined with 30% of urea-formamide plasticizers resulted into homogeneous film. With this formulation, the resulting bubble was not sticky. Also, migration of plasticizers and foaming were not observed. Dang and Yoksan [55] demonstrated the effects of different plasticizers and content on properties of thermoplastic cassava starch. The authors used glycerol, glycerol-xylitol, and glycerol-sorbitol as plasticizers as varying concentrations from 38, 40, and 42 parts per hundred parts of starch. It was noticed that the stickiness and separable double walls can be minimized by using larger-sized plasticizers such as xylitol or sorbitol. The xylitol- and sorbitol-based starch films exhibited enhanced strength, water vapor and oxygen barrier properties.

5.3.2.2 Cellulose

Although cellulose derivatives show thermoplasticity and processability in other melt-processing techniques, film-blowing of cellulose has not been widely reported in the literature. Beckers et al. [56] patented a method of extrusion film-blowing of cellulose. Their invention was based on extruding cellulose solution comprising water and amine oxide through the amine-oxide extrusion process. Fink et al. [57] also reported on the development of cellulose film by blow-extrusion using a cellulose-amine oxide process. Though this technology appeared promising for producing blown cellulose films, research work around this technology did not continue as expected.

5.3.2.3 Proteins

The extrusion film-blowing of proteins such as zein has long been investigated. However, very few studies report on the extrusion film-blowing of proteins. In the earlier-mentioned work, Wang and Padua [45] successfully produced extrusion-blown films from zein. Briefly, zein was mixed with oleic acid and distilled monoglyceride, dissolved in 75% ethanol, and mixed at 60 to 65 °C; the solution was then poured into ice-water to precipitate zein and form a zein resin. The zein resin was then extruded in either single- or twin-screw extruders with a minimum of three cycles to produce a film-blowable resin. Extrusion film-blowing was then used to process the zein resin into films ranging from 0.15 to 0.25 mm in thickness. The effect of barrel and blowing head temperatures and initial moisture content were evaluated. This study reported that higher initial moisture content led to the collapse of the film during blow-molding, while low moisture content resulted in dry and hard resin that lacked plasticity and was not processable by extrusion film-blowing. Higher temperatures were reported to cause excessive loss of moisture and plasticity of the resin. The optimal moisture content before film-blowing of the resin was found to be 14–15%, while the optimal temperature of the extruder barrel was between 20 to 35 °C. Lastly, the temperature of the blowing head was 45 °C.

In another study, Luecha et al. [58] prepared zein-montmorillonite (MMT) nanocomposite films by extrusion film-blowing. However, as in the work by Wang and Padua [45], their extrusion film-blowing method involved multiple and complex processing steps, as shown in Figure 5.3. Firstly, zein-MMT resin was prepared by a similar method described earlier (steps 1-3). The zein resin was then passed into an extruder four times at room temperature to get rid of excess water. Lastly, the zein-MMT resin was processed into a film by extrusion blow-molding (step 5). The effect of MMT loading on the processability and material properties of zein/MMT nanocomposite films produced by extrusion film-blowing was evaluated. These authors reported that MMT could be fabricated into zein by blow-molding to form partially exfoliated nanocomposite structures, but was limited to a low MMT loading (1-3 wt%). In summary, the extrusion film-blowing techniques used in these two studies were time-consuming and involved a complex series of steps that would cause logistical challenges in processing at an industrial scale.

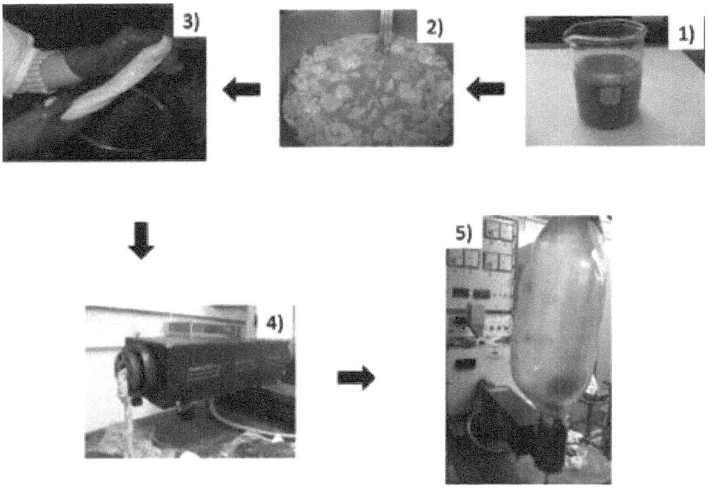

Figure 5.3 Processing steps to produce zein-MMT nanocomposite films by extrusion blow-molding: (1-3) Zein resin fabrication with MMT; (4) Cold extrusion to remove water; (5) Blowing of the resin [58]

To address the expensive and complex multi-step process of producing zein films by extrusion film-blowing reported earlier [45, 58], Oliviero et al. [29] directly plasticized zein powder with 25% poly(ethylene glycol) 400 (PEG400) in a twin-screw extruder to obtain thermoplasticized zein pellets. This was followed by extrusion film-blowing at temperatures between 70 to 100 °C or 85 to 115 °C for different zein batches, as shown in Figure 5.4. As earlier mentioned, the complex structure of zein is important to consider when producing protein films by thermal processing techniques. The higher number of α-helices in the native zein powder seems to have a positive effect on the melt-processability of zein by extrusion film-blowing. The zein films produced from zein powder with higher amounts of α-helices had higher tensile stress and modulus.

Figure 5.4 Extrusion film-blowing of thermoplasticized zein [29]

In this section, it has been demonstrated that natural polymers can be processed by extrusion film-blowing. However, there are still processability challenges. For example, starch is sensitive to moisture, it sticks during the film-blowing process, and has low poor stability during extrusion film-blowing, whereas proteins have poor flexibility during film-blowing. [59, 60]. Therefore, modifying the composition of natural polymers can improve their processability during extrusion film-blowing and improve the material properties of the resulting films.

Huntrakul et al. [60] prepared acetylated edible cassava starch–pea protein films by extrusion blowing process for oil packaging, as shown in Figure 5.5. Firstly, the acetylated starch–pea protein blend plasticized with glycerol and water was compounded using twin-screw extrusion to obtain pellets. The pellets were then processed by extrusion film-blowing to obtain acetylated starch–pea protein blend films. The addition of pea protein was reported to reduce stickiness and stabilized the acetylated starch bubble formation during extrusion film-blowing. In addition, the addition of pea protein enhanced the surface hydrophobicity, oxygen transfer, and vapor transfer.

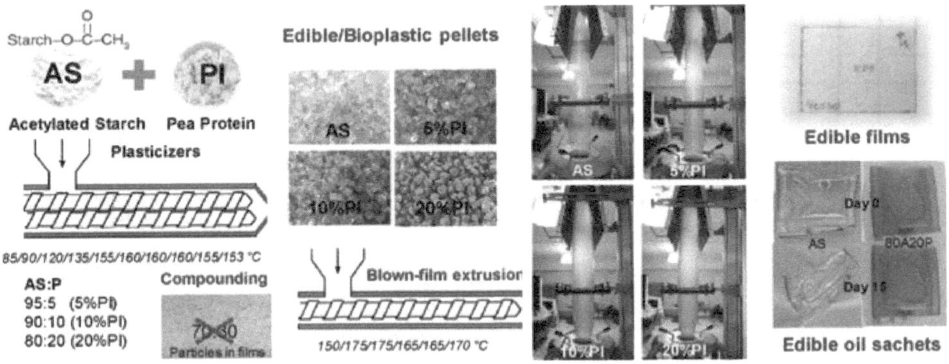

Figure 5.5 Processing steps to produce starch–pea protein edible films by extrusion film-blowing [60]

The use of pea protein in the development of packaging material may not be ideal, since it contains essential amino acids and has nutritional value. Thus, alternative proteins with film-forming properties, such as zein, should be considered. This is because zein has a lower nutritional value, since it lacks essential amino acids (lysine and tryptophan) [61]. Therefore, starch–zein blends should be investigated for their film forming properties by extrusion film-blowing since they have complementary properties. Previous studies have shown that compatible starch–zein blends can be produced by extrusion followed by compression molding [35, 51]. Masanabo et al. [35] demonstrated that compatible starch–zein blend films with 0.2 M NaOH can be produced by twin-screw extrusion. These may be potentially suitable to process by extrusion film-blowing.

5.3.3 Injection Molding

Injection molding is one of the most important techniques for developing rigid, complex plastic items on a commercial scale. Rigid plastic materials are used for various purposes including packaging, cutlery, automotive parts, rigid containers, etc. This section summarizes studies conducted on injection molding of natural-based polymers.

5.3.3.1 Starch

Lim and Jane [62] demonstrated the injection moldability of starch–zein blends. The temperature of the metering-zone was varied from 150 to 165 °C. The molding temperatures were shown to affect the resulting mechanical properties. For instance, the average tensile strength and elongation at break decreased at a temperature of 165 °C, possibly due to the degradation of starch and/or proteins at this temperature. Further, the authors observed low melt-flowability at 150 °C; however, at 160 °C, the processability was much better due to increased viscosity.

Huang et al. [63] have conducted research work on extrusion and injection-molding of starch-protein blends. The injection molding temperatures were varied from 80 to 140 °C. The tensile properties varied with injection molding temperature, and the optimum was 130 °C. At this temperature, the maximum tensile strength and modulus of 3.9 MPa and 46 MPa, respectively, were obtained. Reis et al. [64] investigated the injection molding of starch-polyvinyl alcohol blends. Tábi and Kovács [65] investigated on the effects of injection molding pressure and holding pressure during injection molding of maize starch plasticized with glycerol.

Tanetrungroj and Prachayawarakorn [66] showed the effect of starch type on material properties of TPS processed by injection molding, using glycerol as plasticizer. Cassava, mung bean, and arrowroot starches were used, which differ in amylose and amylopectin content. The study clearly demonstrated that material properties of injection molded TPS depend on the amylose and amylopectin contents. TPS produced from mung bean starch, which contained the highest amylose content (37.9%), showed the highest modulus, hardness, and stress at maximal load, and higher crystallinity than cassva starch, which has lower amylose content.

5.3.3.2 Cellulose

The injection molding of cellulose derivatives has been reported in the literature. As mentioned previously, direct injection molding of cellulose is not possible. Willberg-Keyriläinen et al. [25] investigated on injection molding processability of cellulose esters; namely, cellulose octanate (C_8) and cellulose palmitate (C_{16}). Both cellulose derivatives and their blends were prepared in twin screw micro-compounder at 130 °C. The tensile and impact test specimens were injection molded with a mold temperature of 35 °C. It is important to mention that the plasticizers were not used in their study, which is unusual. The tensile modulus and strength were not significantly different, whereas the elongation at break of C_8 was higher than C_{16}. The Charpy impact strength of C_{16} was lower than C_8; however, during blending, no significant difference was noticed compared to C_8.

Charvet et al. [67] reported on the mechanical behavior of injection molded cellulose acetate in the presence of triacetine and diethyl phthalate plasticizers. It was reported that high impact strength at room temperatures can be obtained at a plasticizer content above 25 wt%. Elsewhere (WO1995001935A1), injection molding of cellulose acetates, starches, and other materials has been patented. The invention discloses injection molding of cellulose acetates and starch in the presence of glycerol and glycerol acetates as plasticizers. The processes begin with extrusion processing to make the pellets, which are then used for injection molding.

5.3.3.3 Proteins

Proteins have been long investigated for their film-forming properties by either solvent casting or thermoprocessing. However, despite all the work done, it seems that the thermal processing of proteins by injection molding has only been recently investigated [68]. Proteins, such as pea proteins, soy isolates, and rice concentrates have been recently processed by injection molding [68–70]. However, in these studies the laboratory-scale mini-piston-jet injection molders were used that can only process a few grams of materials at a time. The injection molding of the aforementioned proteins involved mixing of the protein powders with plasticizers using batch mixers at 25 °C for about an hour to obtain a dough-like mass of protein. This was followed by the injection molding using the mini-piston-jet injection molders as shown in Figure 5.6.

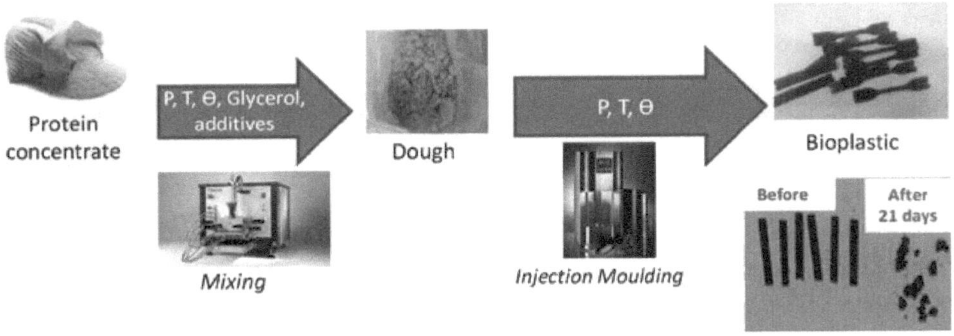

Figure 5.6 Processing of protein-based materials by injection molding [70]

Surprisingly, there are no studies reporting on the injection molding of zein, despite its greater potential in bio-packaging compared to other proteins. Zein and kafirin are prolamin proteins are reported to be more hydrophobic compared to other proteins and polysaccharides [28]. Therefore their injection-moldability should be investigated

Lim and Jane [62] investigated the injection-moldability of starch–zein blends plasticized with glycerol and water. In their study, plasticized starch–zein blends were thermoprocessed at temperatures above 100 °C, and the pellets were processed further by injection molding. This demonstrates that starch–zein blends are processible via injection molding; however, more research is needed to produce complex shapes for various applications where biodegradability and biocompatibility are desired.

5.4 Discussion

In this chapter, the thermal processability of selected natural polymers is discussed. In particular, the thermal processing of starch, cellulose, and proteins is summarized. Thermal processing is mostly preferred for mass-production of polymers. Thermal-based processes such as extrusions, injection molding, etc. are free from solvents and quick. Natural polymers such as starch, cellulose, and proteins are difficult to process via conventional thermal polymer processing techniques compared to commercial synthetic polymers. However, their thermal processability can be achieved through various modification strategies. The summarized studies showed that, indeed, the thermal processability of natural polymers can only be attained after either physical or chemical modifications. Even so, processing aids such as plasticizers are still required to assist with the flow in the heating barrels. In addition, processing temperatures can be reduced when plasticizers are used, which is important because natural polymers are susceptible to thermal degradation.

It is worth mentioning that an enormous amount of work has been carried out on the thermal processability of starch. TPS is usually achieved through the incorporation of plasticizers such as glycerol and water. The film-blowability of starch and zein alone has been demonstrated. However, there are some limitations - for example, the stickiness of starch during film-blowing, and the time-consuming multistep processing needed during the film-blowing of zein. In recent years, starch–protein blending has been shown to improve processability by film-blowing and the material properties of the resulting films. Little work has been carried out on injection molding processing of starch and proteins alone. This necessitates the need for more research into the injection molding of natural polymers, such as starch and proteins.

The thermoplasticity of native cellulose is difficult to attain even in the presence of plasticizers. Only chemical modifications allow breaking of the strong hydrogen bonds between cellulose chains, thus allowing its processability. However, cellulose derivatives are more easily thermally processed than native cellulose. Most recent studies have clearly demonstrated the 3D-printability of cellulose derivatives, and numerous studies have been reported on the additive manufacturing of cellulose derivatives as drug carriers. 3D printing, particularly fused-deposition modeling (FDM), is the most promising for the development of complex structures. Cellulose derivatives also have film-forming properties, but there are few research reports on film-blowing them. The injection molding of cellulose derivatives has been reported in the literature, and the findings demonstrate the possibility of developing different injection molded parts from cellulose derivatives.

Proteins, such as gluten proteins, soy isolates, casein, and kafirin have been long used to make protein films. Of the film-forming proteins, zein has been the most thoroughly studied because of its relative hydrophobicity compared to other proteins and polysaccharides. In addition, zein has lower nutritional value and is often not dedicated for human consumption, unlike other proteins, such as soy proteins. The mechanism by which proteins can form films during thermal processing is not thoroughly understood in the literature due to the complex nature of proteins from different sources. The protein aggregation and reduction of T_g when proteins are processed with a plasticizer are said to be the driving force making it possible for proteins to be processable as thermoplastics. The processability of proteins is often improved by combining them with other natural polymers (such as starch), or synthetic polymers (such as polycaprolactone (PCL)).

In the context of developing environmentally friendly biodegradable plastics, natural polymers can play a huge role in the transition to minimize the use of conventional polymers such as polyolefins. The summarized studies showed the possibility of developing thermoplastics from natural polymers, namely starch, cellulose, and proteins. Furthermore, it is plausible to process these polymers using the continuous melt-processing techniques used industrially, although few studies have been reported. Minimal attention to these materials could be the result of the physical or chemical modifications required prior to thermal processing, which could make the entire process expensive compared to commercial synthetic polymers. In addition, natural polymer-based materials have poor material properties, such as load-bearing and thermal stability, compared to synthetic biodegradable polymers. However, blending natural-based thermoplastics with low-melting polymers such as PCL, poly(butylene succinate), and poly(butylene succinate-co-adipate), which have melt temperatures below 150 °C, is recommended. In addition, composite systems containing native natural fillers such as cellulose and/or other natural fibers can enhance the mechanical, thermal, and other properties of natural-based thermoplastics.

5.5 Conclusions

In this chapter, research work conducted on the thermal processing of three natural polymers – cellulose, starch, and protein – was discussed. Natural polymers can be processed using thermal processing techniques, although these materials require either chemical or physical modifications to induce thermoplasticity and processability. However, more attention has been given to TPS, compared to thermoplastic cellulose and proteins. Therefore, it is imperative to focus on the thermal processing of cellulose and proteins, to have a broad perspective on how thermoplastic natural polymers can be used in many different applications. The focus on natural-based biodegradable polymers would also pave the way to develop eco-friendly materials for edible packaging, biomedical applications such as in tissue engineering, and many other applications.

References for Chapter 5

[1] Mtibe, A., Motloung, M.P., Bandyopadhyay, J., and Ray, S.S., *Macromolecular Rapid Communications* (2021) 42(15) p. 2100130

[2] George, A., Sanjay, M., Srisuk, R., Parameswaranpillai, J., and Siengchin, S., *International Journal of Biological Macromolecules* (2020) 154 p. 329–338

[3] Samir, A., Ashour, F.H., Hakim, A.A., and Bassyouni, M., *NPJ Materials Degradation* (2022) 6(1) p. 68

[4] Das, A., Ringu, T., Ghosh, S., and Pramanik, N., *Polymer Bulletin* (2022) p. 1–66

[5] Díez-Pascual, A.M., *Biopolymer Composites: Synthesis, Properties, and Applications* 2022, MDPI. p. 2257

[6] Ren, J. and Ren, J., *Biodegradable Poly (Lactic Acid): Synthesis, Modification, Processing and Applications* (2010) p. 142–207

[7] Raquez, J.M., Narayan, R., and Dubois, P., *Macromolecular Materials and Engineering* (2008) 293(6) p. 447–470

[8] Ojijo, V. and Ray, S.S., *Progress in Polymer Science* (2013) 38(10–11) p. 1543–1589

[9] Flórez, M., Cazón, P., and Vázquez, M., *Polymers* (2023) 15(3) p. 641

[10] Prasad, K. and Sharma, M., *Current Opinion in Green and Sustainable Chemistry* (2019) 18 p. 72–78

[11] Liu, H., Xie, F., Yu, L., Chen, L., and Li, L., *Progress in Polymer Science* (2009) 34(12) p. 1348–1368

[12] Pérez, S. and Bertoft, E., *Starch-Stärke* (2010) 62(8) p. 389–420

[13] Khan, B., Bilal Khan Niazi, M., Samin, G., and Jahan, Z., *Journal of Food Process Engineering* (2017) 40(3) p. e12447

[14] Lai, L. and Kokini, J., *Biotechnology Progress* (1991) 7(3) p. 251–266

[15] Blazek, J. and Gilbert, E.P., *Carbohydrate Polymers* (2011) 85(2) p. 281–293

[16] Souza, R.C. and Andrade, C.T., *Advances in Polymer Technology: Journal of the Polymer Processing Institute* (2002) 21(1) p. 17–24

[17] Mohammadi Nafchi, A., Moradpour, M., Saeidi, M., and Alias, A.K., *Starch-Stärke* (2013) 65(1–2) p. 61–72

[18] Paes, S.S., Yakimets, I., and Mitchell, J.R., *Food Hydrocolloids* (2008) 22(5) p. 788–797

[19] Luan, Y., Wu, J., Zhan, M., Zhang, J., Zhang, J., and He, J., *Cellulose* (2013) 20 p. 327–337

[20] Kostag, M., Gericke, M., Heinze, T., and El Seoud, O.A., *Cellulose* (2019) 26 p. 139–184

[21] Hou, D.-F., Li, M.-L., Yan, C., Zhou, L., Liu, Z.-Y., Yang, W., and Yang, M.-B., *Green Chemistry* (2021) 23(5) p. 2069–2078

[22] Chen, H., Yang, F., Du, J., Xie, H., Zhang, L., Guo, Y., Xu, Q., Zheng, Q., Li, N., and Liu, Y., *Cellulose* (2018) 25 p. 6935–6945

[23] Onwukamike, K.N., Grelier, S., Grau, E., Cramail, H., and Meier, M.A., *ACS Sustainable Chemistry & Engineering* (2018) 6(7) p. 8826–8835

[24] Willberg-Keyriläinen, P., Rokkonen, T., Malm, T., Harlin, A., and Ropponen, J., *Journal of Applied Polymer Science* (2020) 137(48) p. 49588

[25] Willberg-Keyriläinen, P., Orelma, H., and Ropponen, J., *Materials* (2018) 11(12) p. 2358

[26] Huo, W., Wei, D., Zhu, W., Li, Z., and Jiang, Y., *Journal of Cereal Science* (2018) 79 p. 354–361

[27] Taylor, J., Anyango, J.O., and Taylor, J.R., *Cereal Chemistry* (2013) 90(4) p. 344–357

[28] Taylor, J. R. and Taylor, J., *Cereal Chemistry*

[29] Oliviero, M., Di Maio, E., and Iannace, S., *Journal of Applied Polymer Science* (2010) 115(1) p. 277–287

[30] Wilson, C. M., *Plant Physiology* (1991) 95(3) p. 777–786

[31] Reddy, N. and Yang, Y., *Journal of Applied Polymer Science* (2013) 130(2) p. 729–738

[32] Zhang, Y., Rempel, C., and Liu, Q., *Critical Reviews in Food Science and Nutrition* (2014) 54(10) p. 1353–1370

[33] Surendren, A., Mohanty, A. K., Liu, Q., and Misra, M., *Green Chemistry* (2022)

[34] Li, M., Liu, P., Zou, W., Yu, L., Xie, F., Pu, H., Liu, H., and Chen, L., *Journal of Food Engineering* (2011) 106(1) p. 95–101

[35] Masanabo, M. A., Ray, S. S., and Emmambux, M. N., *International Journal of Biological Macromolecules* (2022) 208 p. 443–452

[36] Rodríguez-Castellanos, W., Martínez-Bustos, F., Rodrigue, D., and Trujillo-Barragán, M., *European Polymer Journal* (2015) 73 p. 335–343

[37] Galicia-García, T., Martínez-Bustos, F., Jiménez-Arévalo, O., Arencón, D., Gámez-Pérez, J., and Martínez, A., *Journal of Applied Polymer Science* (2012) 126(S1) p. E327-E336

[38] Taheri, H., Hietala, M., and Oksman, K., *Cellulose* (2020) 27 p. 8105–8119

[39] Matsuoka, T., Kumagai, K., and Nonaka, H., *Cellulose* (2022) 29(5) p. 2975–2983

[40] Prodduturi, S., Manek, R. V., Kolling, W. M., Stodghill, S. P., and Repka, M. A., *Journal of Pharmaceutical Sciences* (2004) 93(12) p. 3047–3056

[41] Hartzke, D., Pössl, A., Schlupp, P., and Runkel, F. E., *Pharmaceutics* (2022) 14(10) p. 2103

[42] Vo, A. Q., Zhang, J., Nyavanandi, D., Bandari, S., and Repka, M. A., *Carbohydrate Polymers* (2020) 246 p. 116519

[43] Corradini, E., Carvalho, A. J. F. d., Curvelo, A. A. d. S., Agnelli, J. A. M., and Mattoso, L. H. C., *Materials Research* (2007) 10(3) p. 227–231

[44] Corradini, E., Mattoso, L., Guedes, C., and Rosa, D., *Polymers for Advanced Technologies* (2004) 15(6) p. 340–345

[45] Wang, Y. and Padua, G. W., *Macromolecular Materials and Engineering* (2003) 288(11) p. 886–893

[46] Shewry, P. R. and Tatham, A. S., *Biochemical Journal* (1990) 267(1) p. 1

[47] Zhang, Y., Xu, M., Zhang, X., Li, J., Wu, C., Cao, S., Hu, Y., and Luan, G., *Food Research International* (2022) 162 p. 112129

[48] Ghanbarzadeh, B., Almasi, H., and Entezami, A. A., *Innovative Food Science & Emerging Technologies* (2010) 11(4) p. 697–702

[49] e Moraes, A. R. F., Pola, C. C., Bilck, A. P., Yamashita, F., Tronto, J., Medeiros, E. A. A., and Soares, N. d. F. F., *Materials Science and Engineering: C* (2017) 78 p. 932–941

[50] Chanvrier, H., Colonna, P., Della Valle, G., and Lourdin, D., *Carbohydrate Polymers* (2005) 59(1) p. 109–119

[51] Leroy, E., Jacquet, P., Coativy, G., Laure Reguerre, A., and Lourdin, D., *Carbohydrate Polymers* (2012) 89(3) p. 955–963

[52] Thunwall, M., Kuthanova, V., Boldizar, A., and Rigdahl, M., *Carbohydrate Polymers* (2008) 71(4) p. 583–590

[53] Dang, K. M. and Yoksan, R., *Carbohydrate Polymers* (2015) 115 p. 575–581

[54] Zullo, R. and Iannace, S., *Carbohydrate Polymers* (2009) 77(2) p. 376–383

[55] Dang, K. M. and Yoksan, R., *International Journal of Biological Macromolecules* (2021) 188 p. 290–299

[56] Beckers, S. D., Weigel, P., Fink, H.-P., Dosz, M., and Purz, H.-J., *Cellulose Extrusion* 2006, Google Patents

[57] Fink, H.-P., Weigel, P., and Bohn, A., *Journal of Macromolecular Science - Physics* (1999) 38(5-6) p. 603-613.

[58] Luecha, J., Sozer, N., and Kokini, J. L., *Journal of Materials Science* (2010) 45(13) p. 3529-3537

[59] Matzinos, P., Tserki, V., Kontoyiannis, A., and Panayiotou, C., *Polymer Degradation and Stability* (2002) 77(1) p. 17-24

[60] Huntrakul, K., Yoksan, R., Sane, A., and Harnkarnsujarit, N., *Food Packaging and Shelf Life* (2020) 24 p. 100480

[61] Zhang, Y., Cui, L., Che, X., Zhang, H., Shi, N., Li, C., Chen, Y., and Kong, W., *Journal of Controlled Release* (2015) 206 p. 206-219

[62] Lim, S. and Jane, J., *Journal of Environmental Polymer Degradation* (1994) 2 p. 111-120.

[63] Huang, H., Chang, T., and Jane, J., *Journal of the American Oil Chemists' Society* (1999) 76(9) p. 1101-1108

[64] Reis, R. L., Cunha, A. M., Allan, P. S., and Bevis, M. J., *Polymers for Advanced Technologies* (1996) 7(10) p. 784-790

[65] Tábi, T. and Kovács, J., *Express Polymer Letters* (2007) 1(12) p. 804-809

[66] Tanetrungroj, Y. and Prachayawarakorn, J., *Songklanakarin Journal of Science & Technology* (2015) 37(2)

[67] Charvet, A., Vergelati, C., and Long, D. R., *Carbohydrate Polymers* (2019) 204 p. 182-189

[68] Perez, V., Felix, M., Romero, A., and Guerrero, A., *Food and Bioproducts Processing* (2016) 97 p. 100-108

[69] Jiménez-Rosado, M., Rubio-Valle, J. F., Perez-Puyana, V., Guerrero, A., and Romero, A., *Journal of Applied Polymer Science* (2021) 138(20) p. 50412

[70] Félix, M., Lucio-Villegas, A., Romero, A., and Guerrero, A., *Industrial Crops and Products* (2016) 79 p. 152-159

6 Processing of Polysaccharides

Fengwei Xie

This chapter reviews the different aspects of knowledge and progress in the processing of polysaccharides including starch, chitosan, alginate, and cellulose, which show properties and a processability very different from those of traditional synthetic polymers. They are important renewable resources, biodegradable and functional materials to meet specific applications needs and to address sustainability issues. This chapter will first introduce the fundamental characteristics and modification strategies of these different polysaccharides, particularly related to processing. Then, strategies to formulate polysaccharide-based materials and composites will be discussed. Following that, different techniques for the processing of polysaccharides will be highlighted, such as solution-based methods, thermomechanical processing (with limited solvents), electrospinning, and 3D printing. Finally, challenges and future perspectives in this area will be addressed.[1,2]

6.1 Introduction

Biopolymers represent a large family of polymers that are biologically synthesized by plants, animals, and microorganisms (in contrast to polymers that are chemically synthesized). The word "natural" is added to emphasize that they naturally exist in the biosphere, and for this reason, biopolymers are also called "natural polymers". Biopolymers mainly include polysaccharides (for example, starch, cellulose, chitosan, pectin, alginate, agar - a mixture of agarose and agaropectin - and carrageenan) and proteins (for example, wheat gluten, soy protein, collagen, and gelatin-hydrolyzed collagen). Polysaccharides are carbohydrate polymers formed by connecting monosaccharide units via glycosidic linkages. The chemical struc-

[1] This work was supported by the Engineering and Physical Sciences Research Council (EPSRC) [grant number EP/V002236/2].

[2] Data sharing is not applicable to this article as no new data was created or analyzed in this review.

tures of typical polysaccharides are shown in Figure 6.1. Proteins are formed by amino acid residues linked via peptide bonds.

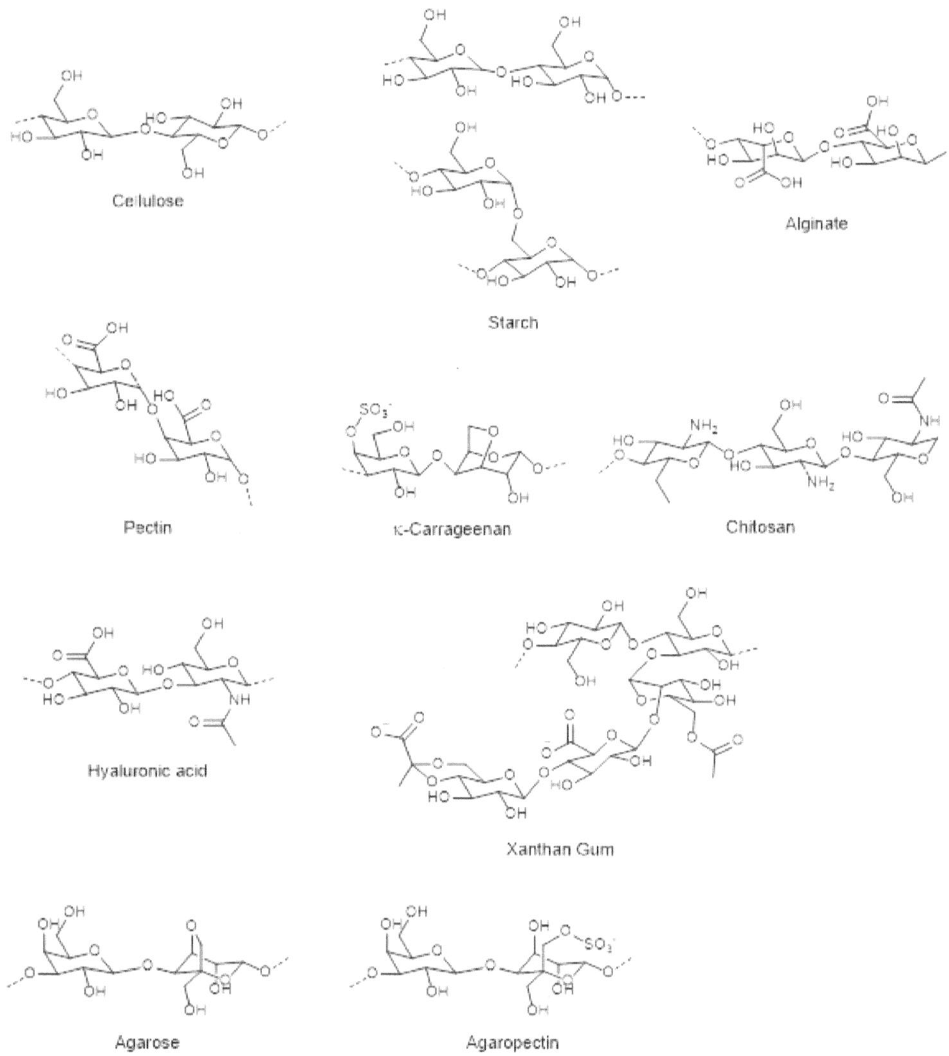

Figure 6.1 Chemical structures of typical polysaccharides

It is important to note that biopolymers are frequently distinct from biobased polymers. The latter category consists of polymers produced from biomass monomers chemically, biologically, or through a mix of both, and examples of them are biobased polyesters (e.g., polyhydroxyalkanoates (PHAs), poly(lactic acid) (PLA), polyethylene terephthalate (PET), and polyethylene furanoate (PEF)), polycarbonates, epoxy resins, polyamides, polyurethanes, polybenzoxazines, polyimines, among others [1]. Therefore, only polymers that are produced biologically (by microbes) from biomass carbon sources (e.g., sugars and lipids) can be referred to as both biopolymers and biobased polymers. Some examples of these are PHAs, bacterial cellulose, gellan gum, xanthan gum, and curdlan.

In the past decades, tremendous research has been devoted to the development of biopolymer-based materials, for which there are several reasons. First, biopolymers have their appealing or unique properties and functionality such as gelation, emulsifying, and film-forming properties [2–4], excellent oxygen and CO_2 barrier properties [5–12], antioxidant and antimicrobial properties [13–15], biocompatibility [16–20], other bioactive functions (e.g., immunomodulatory, cell-proliferative, and angiogenic functions [19], the abilities to stimulate cell adhesion, diffusion, propagation, and diversity [20]), and their tendency to interact with and absorb specific pollutants and heavy metals [21–23], making them important materials for food packaging, food, textile, biomedical, environmental, construction, and other applications. Especially for food, food contact, pharmaceutical, and biomedical applications, the additional nontoxicity and even edibility of biopolymers are big advantages. Additionally, biopolymers are biodegradable in the natural environment [24, 25], which makes them suitable to make materials for applications where biodegradability/compostability is indeed an advantage, such as food packaging (as food-contaminated packages are hard to recycle), agriculture (e.g., controlled release of fertilizers and pesticides), and environmental treatment. Thus, biopolymers can play an important role in sustainability.

This chapter focuses on the processing of the most important polysaccharides, including starch, cellulose, alginate, and pectin. First, their fundamental aspects and modification strategies (e.g., to reduce hydrophilicity and improve processability) are introduced. Then, updated approaches to formulate polysaccharide-based materials are examined. Afterward, the processing techniques for the fabrication of different forms of polysaccharide materials and the required processability are discussed. The challenges and future perspectives of polysaccharide processing are presented at the end. It is hoped that this chapter will provide essential knowledge for the utilization of polysaccharides and guide the future development of polysaccharide-based materials for various applications.

6.2 Fundamentals of Polysaccharides

6.2.1 Starch

Starch is a polysaccharide with D-anhydro-glucopyranose as the repeating unit. There are mainly two types of starch macromolecule, namely, amylose, which has a primarily linear structure joined by an α-1,4-glycosidic linkage with limited branches connected to the main backbone by α-1,6-glycosidic bonds, and amylopectin, which has a branched structure (contains about 5% of branch points); see Figure 6.1. Native starch is in the form of granules, which have a hierarchical structure (from small to large scale, molecular chains, helices, crystalline and amorphous lamellae, growth rings, and the granule) constructed basically by hydrogen bonding between hydroxyl groups. When starch is heated with abundant water, the hydrogen bonds will be disrupted and its granules will be dramatically swollen and then ruptured, with the diffusion of molecules out of the granule and the destruction of its crystalline, lamellar, and granule structures. This irreversible process is called gelatinization. Starch is produced by plants as a major way of energy storage, such as in corn, potato, cassava, wheat, and rice. Different cultivars of starch vary in amylose/amylopectin ratio, molecular mass, granule morphology and structure, gelatinization temperature, and many other properties. For cornstarches with different amylose/amylopectin ratios, the amylose fractions have different weight-average molecular masses ($M_w = 1.03–4.89 \times 10^5$ g/mol) and branching ratios (0.3–1.0); for amylopectin, some fractions have different M_w values ($7.08–9.88 \times 10^7$ g/mol) but statistically similar branching ratios (1.5–4.3) [26].

Besides starch cultivar, the amylose/amylopectin ratio is another important factor influencing the properties of starch. Studies have shown that cornstarch with a higher amylose content has a more compact granule structure [27, 28] and a higher gelatinization temperature [29]. With lower water content (below the abundant water level), starch gelatinization will also occur at a higher temperature [29].

Starch is the most widely studied natural polymer to develop bioplastics. The major reason for this is its low cost, gelatinization behavior, and film-forming ability. Film can be produced from a gelatinized starch paste using solution casting. Starch can also be processed with a low content of water (and other plasticizers) using thermomechanical processes such as extrusion, and in this way, starch "melts" and plasticized starch can be generated [30]. This is also commonly called "thermoplastic starch", although it is more like being "thermomechanoplastic" as shear needs to be applied to make it flowable. Plasticized starch has a shear-thinning behavior and the extent of shear-thinning and the viscosity highly depend on parameters such as starch type, amylose/amylopectin ratio, temperature, plasticizer, and additives [31–33]. For cornstarch with a higher amylose/amylopectin ratio, a higher viscosity and a stronger shear-thinning behavior has been observed, which can be ascribed to the higher tendency of amylose chains to entangle [32]. Along with this, a higher amylose content can increase the difficulty of extrusion processing, although this could be overcome by increasing the processing temperature, moisture content, and moisture equilibration time [34]. High-amylose starches are still interesting as they can result in materials with better mechanical properties [34], presumably due to a greater extent of chain interactions between linear amylose chains.

While water acts as a plasticizer for starch processing, it is volatile and easy to lose; thus, starch-based materials processed with only water usually have unstable properties. To address this issue, additional, nonvolatile plasticizers are commonly used, including polyols (e.g., glycerol, sorbitol, and glycol), urea, amides (e.g., formamide, acetamide, ethylenebisformamide, and ethanolamine), and citric acid [30, 31, 35]. It has been shown that a higher temperature, a higher content of glycerol, and a longer storage duration are all favorable for the interaction of glycerol with starch molecules [36].

A recent study indicated that the extrusion process to plasticize potato starch that included glycerol and urea was more effective than the one that used only glycerol, and the plasticizer combination may have contributed to the stable chemical structure of the plasticized starch [37]. Urea can effectively disrupt the crystalline structure of starch [37], while glycerol can do a better job than urea to increase starch chain mobility [38]. While glycerol endows starch with adequate flexibility and processability, it also compromises the water and gas barrier properties [39].

More plasticizers have been tested recently for starch processing. For example, Alee et al. [40] tested the plasticization efficiency of monosaccharides, disaccharides, and low-molecular-mass polysaccharides for starch and concluded that monosaccharides with linear structures, such as fructose and xylose, are more efficient to destroy the ordered structures and enhance the movement of starch chains. Area et al. [41] found that, compared to that with glycerol, the processing of cornstarch with isosorbide required lower temperatures but greater shear and a longer processing duration, resulting in a more homogeneous material. Additionally, the plasticization with isosorbide led to higher crystallinity without retrogradation, whereas retrogradation was observed for glycerol-plasticized samples [41, 42]. Isosorbide leads to stiffer plasticized cornstarch with greater water-resistance and biodegradability [42].

Ionic liquids (ILs) have also been shown to be effective plasticizers for starch [43–47]. Similarly, polymeric ionic liquid (PIL) was found to be able to effectively destroy the crystalline structure of cornstarch and endow the starch with excellent plasticity and mechanical properties [48]. Zdanowicz et al. [49] reported that for thermocompression-molded potato starch films, a deep eutectic solvent (DES) containing a citrate anion provided both cross-linking and plasticization effects for the starch matrix. A higher choline citrate (CCit) content in the DES led to higher tensile strength, a more amorphous structure, and lower water sorption. Moreover, CCit-based DES-plasticized starch films did not retrograde even after one year of storage [49].

Zhang et al. [50] synthesized hyperbranched polyesters, terminated with either hydroxyl or carboxyl groups, from citric acid and glycerol, and used them as plasticizers for solution-cast starch films. Compared with glycerol-plasticized starch films, poly(citrate glyceride)-plasticized starch films had lower crystallinity, a lower glass transition temperature, and better mechanical and thermal properties [50].

Xie and co-workers [51] studied the thermal transition of cornstarches with different amylose/amylopectin contents in a mixture of water and an ionic liquid (IL) 1-ethyl-3-methylimidazolium acetate ([C_2mim][OAc]), and found that the thermal transition temperature of the starch depends on the water/[C_2mim][OAc] ratio. In the 7.2:1 (mol/mol) water/[C_2mim][OAc] mixture (4:6 (w/w) water/[C_2mim][OAc] mixture), starches (especially waxy and normal cornstarches) undergo the phase transition mostly efficiently, at a significantly reduced temperature [51]. It has also been revealed that, for this water/[C_2mim][OAc] ratio, some starches such as waxy and normal cornstarches can be transformed into gel even at room temperature within about 20 minutes [52]. It has been suggested that such a rapid transition in the water/IL mixture of a certain ratio occurs to starches that have pores on the granule surface, while for starches with a compact structure without pores on the granule surface (e. g., high-amylose starches and potato starch), the transition under these conditions becomes much less effective [53]. It was demonstrated that the use of such aqueous IL allows normal cornstarch to be processed into optically transparent electroconductive films by compression molding at a mild temperature (55 °C or 65 °C), much lower than the temperature generally used for starch thermal processing (over 150 °C) [54].

Xie and co-workers [55, 56] found that aqueous metal chloride salts ($ZnCl_2$, $MgCl_2$, and $CaCl_2$) of a certain concentration can gelate starches, even a high-amylose starch, at a moderate temperature (50 °C, 4 h), and thus are more advantageous than the aqueous ILs mentioned above. Based on this discovery and using aqueous $CaCl_2$, the same group [57, 58] reported the facile preparation (70 °C for 1 h or 80 °C for a few minutes) of a high-amylose cornstarch hydrogel (Gelose 80) with excellent mechanical properties and electrical conductivity and of galvanic cell-type batteries and strain sensors based on this hydrogel, useful for wearable electronics applications.

Starch can be modified via versatile chemistry such as cross-linking, grafting, esterification, and etherification, which have been extensively discussed elsewhere [59–62]. Chemical modification varies the properties of starch, leading to broad applications such as drug delivery, tissue engineering, and wastewater treatment

[62]. Research has shown that acetylated starch had increased hydrophobicity [63, 64] and could be used to produce extrusion-blown films [65]. Hydroxypropylation reduces the gelatinization temperature of starch by improving the interaction between starch chains and water and the tendency of amylose and amylopectin molecules to associate and thus retrograde [66]. A hydroxypropylated starch containing 80% amylose content displayed a lower viscosity and improved processability during extrusion processing compared with its unmodified counterpart [67].

6.2.2 Cellulose

Cellulose is the most abundant natural polymer on Earth. It can be extracted from plants or produced by microorganisms. Cellulose is a polysaccharide with the same repeating unit (D-anhydro-glucopyranose) as starch but joined together by a β-1,4-glycosidic bond (see Figure 6.1). Unlike starch, which can undergo gelatinization with water under heating, the native structure of cellulose is much more recalcitrant to be destructed. Cellulose is insoluble in water and many organic solvents. It has been suggested that cellulose is actually amphiphilic and that the low aqueous solubility should be largely contributed to by hydrophobic interaction [68]. Carbon disulfide (CS_2), which is highly toxic, is a solvent traditionally used to dissolve cellulose (subsequently, the regenerated cellulose is called cellophane). A few less toxic, non-derivatizing solvent systems for dissolving cellulose have been discovered, such as *N*-methylmorpholine-*N*-oxide (NMMO), ILs, LiCl/*N*,*N*-dimethylacetamide (LiCl/DMAc), alkali/urea and NaOH/thiourea aqueous solution, NaOH aqueous solution, tetrabutylammonium fluoride/dimethyl sulfoxide (TBAF/DMSO) system, metal complex solutions (e.g., cuprammonium hydroxide), molten inorganic salt hydrates, and concentrated salt solutions (e.g., zinc chloride, ammonium, calcium and sodium thiocyanate solutions) [69–71]. It has been proposed that the development of new solvents for cellulose should focus on the elimination of not only hydrogen bonding but also hydrophobic interactions [70].

After dissolution, cellulose could then be regenerated from the solutions using coagulants (usually non-solvents) to form films. For example, for the regeneration of cellulose from NMMO solutions, water can be used as a coagulant, leading to a phase-separation process [70]. Alcohol can be used to regenerate cellulose dissolved in ILs [72]. The structure and properties of regenerated cellulose depend on the coagulant, temperature, time, and pathway [69]. For example, Zhang and coworkers found that, for the regeneration of cellulose in NaOH/urea, among all the coagulants (H_2SO_4, CH_3COOH, H_2SO_4/Na_2SO_4, Na_2SO_4, $(NH_4)_2SO_4{\cdot}H2O$, C_2H_5OH, and $(CH_3)_2CO$ aqueous solutions), H_2SO_4/Na_2SO_4 resulted in the highest optical transmittance, the most homogeneous structure, and the best mechanical properties [73, 74].

Among these different solvents, only NMMO is actually used to make cellulose fibers on the industrial level (the Lyocell process), to replace the viscose process; however, this is rather energy-consuming and has some disadvantages, such as the occurrence of oxidative side reactions and some thermal instability [70, 71, 75]. The other solvents listed above are toxic, too expensive, or have low cellulose-dissolution capability [75].

While cellulose naturally is not thermoplastic, chemically modified cellulose may have improved processability. A typical cellulose derivative is cellulose acetate, which is thermoplastic [76] and dissolvable in some organic solvents (e.g., acetone, chloroform, *N,N*-dimethylformamide (DMF), dichloromethane (DCM), methanol, formic acid, and pyridine) [77], so that melt processing, solution casting, and electrospinning can be used for the processing of cellulose acetate. Moreover, many cellulose derivatives are soluble in water, for example, nonionic methyl cellulose (MC), hydroxypropyl methyl cellulose (HPMC), hydroxyethyl cellulose (HEC), hydroxypropyl cellulose (HPC), and anionic sodium carboxymethyl cellulose (NaCMC) [78–80]. Some cellulose derivatives such as HPMC have thermal gelling behavior [81–84].

Notably, natural cellulose can be converted into cellulose nanofibrils (CNFs) or cellulose nanocrystals (CNCs), which are referred to as "nanocellulose". CNFs are long particles of cellulose elementary fibrils that have both crystalline and non-crystalline regions and have a high aspect ratio (particle length-to-diameter ratio). The disintegration of the cellulose fibers to obtain CNFs can be achieved by mechanical methods (homogenization or microfluidization), nanogrinding or ultrasonication [85]. On the other hand, CNCs are mainly crystalline and can be obtained by acid hydrolysis or enzyme hydrolysis of cellulose pulp [85]. Nanocellulose has excellent strength, a high Young's modulus, biocompatibility, tunable self-assembly, and thixotropic and photonic properties [70]. Additionally, various surface chemical modifications can be carried out to introduce either charged or hydrophobic moieties to nanocellulose, and these include amidation, esterification, etherification, silylation, polymerization, urethanization, sulfonation, and phosphorylation [70]. Nanocellulose has found a broad array of applications, such as functional paper, optoelectronics, and antibacterial coatings, packaging, mechanically reinforced polymer composites, tissue scaffolds, drug delivery, biosensors, energy storage, catalysis, environmental remediation, and electrochemically controlled separation [70]. For reinforcing plastics, a study [86] indicated that CNFs led to a higher strength and modulus than CNCs did, due to CNFs' larger aspect ratio and fiber entanglement, but to a lower strain at failure because of their relatively large fiber agglomerates.

6.2.3 Chitin and Chitosan

Chitin is a polysaccharide composed of *N*-acetylglucosamine connected by a β-(1→4)-linkage (see Figure 6.1). It is a widespread substance that has even been said to be one of the planet's most common biomolecules, with an estimated annual production of 10^{11}–10^{14} tons [87]. A variety of biomasses, including the exoskeletons of crustaceans and the cell walls of fungus and diatoms, have solid structures made of chitin [87]. Chitin is less hydrophilic than other polysaccharides because of its hydrophobic acetyl groups.

Chitin is insoluble in common solvent systems. Solvents that have been found to dissolve chitin include ILs, DMAc/LiCl, $CaCl_2$/methanol, DESs (e.g., choline chloride/thiourea), highly polar fluorinated solvents (e.g., hexafluoroisopropanol and hexafluoroacetone sesquihydrate), strong acids (e.g., trichloroacetic acid (TCA) and dichloroacetic acid (DCA)), inorganic salt aqueous solutions (e.g., LiCNS, $Ca(CNS)_2$, CaI_2, $CaBr_2$, and $CaCl_2$), and alkali/urea aqueous solutions [88, 89].

Like cellulose, as a semicrystalline polymer composed of nanofibrous units, native chitin is an excellent starting material to fabricate nanomaterials in the form of nanofibers and nanocrystals with a large aspect ratio and specific area [88]. Chitin nanocrystals and nanofibers can be obtained by acid hydrolysis, 2,2,6,6-tetramethylpiperidine 1-oxyl (TEMPO)-mediated oxidation, mechanical treatment (e.g., grinding, homogenization, high-speed blending, aqueous counter collision, pulsed ultrasonication, and steam explosion), and partial deacetylation [88, 90]. This is called the "top-down" approach to preparing nanochitin. On the other hand, using the solvents mentioned above, chitin can be dissolved and then nanochitin can be obtained by methods such as electrospinning and self-assembly, which is the "bottom-up" approach [88, 90]. Nanochitin can then be made into different material forms, such as film, hydrogel, aerogel, fiber, and different composites [88], which can be applied in biomedical (e.g., tissue engineering, wound dressing, drug delivery, and biological adhesives), food, water-treatment, green-electronics, enzyme-immobilization, cosmetics, and other application areas [90].

Additionally, using the solvents mentioned above, chitin can be dissolved and then regenerated to constitute different forms of materials (hydrogel, aerogel, film, membrane, fiber, and microsphere) [6, 91]. For example, Zhang and co-workers have demonstrated the use of aqueous NaOH/urea to prepare high-strength, transparent chitin films with gas-barrier ability and strong mechanical properties [5, 6]. Thin, flexible chitin films with minimal dimensional shrinkage and maximum flexibility were prepared using DMAc/LiCl as a solvent and by a cold-press process [92]. Rogers and co-workers demonstrated that ILs can be used to extract chitin from crustacean shells and to subsequently produce chitin films and fibers [93, 94].

The deacetylation of chitin generates chitosan, which is a polysaccharide made up of β(1,4)-linked *N*-acetyl-D-glucosamine units. Chitosan can be obtained by treating chitin with a strong alkaline compound such as sodium hydroxide. The degree of deacetylation is used to define chitosan products with various percentages of deacetylated groups (amine groups) because, typically, 100% deacetylation is not accomplished. Since chitosan is protonated, positively charged, and hence soluble in mild acids, processing it is significantly easier than processing chitin. The water-solubility of chitosan is also related to its molecular mass. While low-molecular-mass chitosan is soluble at neutral pH and generates a less viscous solution, high-molecular-mass chitosan is insoluble above a pH of 6.3 [95]. The solubility of chitosan allows it to be prepared in films and other different forms. The pH-dependent solubility of chitosan also makes it advantageous for some applications (e.g., controlled release). The positive charge of chitosan under acidic conditions is an appealing property that allows it to interact with negatively charged substances such as some nanomaterials (e.g., graphene oxide) and polymers (e.g., alginate, pectin, protein, and carboxymethyl cellulose), leading to composite materials with outstanding properties [96–102].

Chitosan has a number of enticing qualities, including its ability to prevent the growth of a wide range of fungi, yeasts, and bacteria – but only when it is dissolved and positively charged in an acidic solution [14]. One theory put forth explains how chitosan's antibacterial action is caused by positively charged molecules interacting with negatively charged residues on bacterial surfaces [14]. Additionally, chitosan is superoleophobic, which is a useful property for oil–water separation [103].

Some chitosan derivatives show improved properties such as solubility in water irrespective of pH and greater antimicrobial activity. For example, the acetylation of chitosan with the introduction of small functional groups such as methyl, acetyl, carboxymethyl, and phosphoryl groups improves its solubility in a neutral-pH or alkaline medium without influencing its polycationic character [104]. Carboxymethyl chitosan compounds are water-soluble [105]. The synthesis of trimethyl carboxymethyl chitosan derivatives was performed in order to achieve an enhanced antibacterial property [105]. Chitosan-*O*-polyethylene glycol graft copolymers were prepared and shown to be water-soluble over a range of pHs, with a highly reduced viscosity, similar to that of MPEG-2000 [106]. Moreover, quaternized ammonium salts of chitosan are soluble over a broad range of pHs (either acidic or basic), with antimicrobial activity maintained and even enhanced [107, 108].

It has been demonstrated that chitosan, blended with low contents of an acid solution and plasticizers, can be processed by high-viscosity thermomechanical processing, namely, batch mixing and compression molding [96–98, 109–114]. For chitosan processing, glycerol has been suggested to be a more effective plasticizer than other polyols or ionic liquids [113, 115, 116].

6.2.4 Alginate

Alginate is a polysaccharide derived from brown seaweed known as Phaeophyceae, by treatment with aqueous alkali solutions, typically with NaOH [117]. Alginate is a whole family of linear copolymers containing blocks of (1,4)-linked β-D-mannuronate (M) and α-L-guluronate (G) residues (see Figure 6.1). The blocks are composed of consecutive G residues, consecutive M residues, or alternating M and G residues. The source from which alginate is extracted determines the contents of M and G and the size of the blocks [117]. Their negative charge is a result of the carboxylic groups of uronic acids [118].

An important functional property of alginate is that its aqueous solution undergoes a sol-gel transition with the presence of Ca^{2+} and many other multivalent metal cations, useful for food, biomedical, and pharmaceutical applications [119]. The interaction with these metal cations is an important cross-linking mechanism to realize alginate-based hydrogels [120, 121]. It is also useful for the fabrication of alginate-based films with improved mechanical properties and water-resistance but reduced ductility [118, 122–125]. Ionic cross-linking can also help the development of 3D printing processes [126, 127].

Alginate materials have been widely studied for food packaging/coating [128], biomedical [117, 129], and environmental [130] applications.

In the reported studies, alginate films were normally prepared by solution casting. Alginates possess good film-forming properties, but tend to be quite brittle after being dried; however, plasticizers such as glycerol or sorbitol may improve the flexibility of alginate films [131]. The permeability of alginate films to oxygen and water vapor rises with increasing glycerol content, whereas sorbitol had no effect on the films' barrier qualities against either [131]. Alginate films are impervious to oils and fats but are poor moisture barriers [132]. As mentioned above, cross-linking alginate with Ca^{2+} ions could lead to increased tensile strength and water-resistance and reduced ductility [122–124]. The incorporation of oregano essential oil [124] or garlic oil [133] could increase the antimicrobial and water vapor barrier properties of alginate films. Studies [122, 134, 135] also suggested that a micro- or nanocellulose addition might significantly improve the mechanical characteristics of alginate films and reduce their water vapor permeability (WVP) and water-solubility.

Besides using solution methods, the thermomechanical processing and plasticization of alginate using batch mixing and compression molding has been reported, leading to transparent, soft, and easy-to-handle alginate films [136, 137]. Compared with glycerol-plasticized chitosan or starch processed under the same conditions, glycerol-plasticized alginate had a higher equilibrated moisture content, probably due to the more hydrophilic nature of alginate, and was stiffer [136].

Glycerol is a more efficient plasticizer for thermomechanically processed alginate than sorbitol, resulting in lower mechanical strength, Young's modulus, and glass transition temperature as well as higher water-sensitivity and transparency. The best material properties were obtained when glycerol and sorbitol mixtures were used [137].

6.2.5 Pectin

Pectin is a family of complex polysaccharides that contain 1,4-linked α-D-galactosyluronic acid (GalpA) residues (see Figure 6.1), extracted from higher plant cell walls, consisting mostly of variably methylated galacturonan segments separated by rhamnose residues, some of which may be linked to short neutral sugar side chains [138]. Different pectins vary significantly in terms of the degree of esterification (DE), which is measured as the proportion of carboxyl groups that are esterified and present in the pectin structure. Due to the hydrophobic nature of esters with long hydrocarbon chains, a larger DE decreases pectin's water-solubility while increasing the gelation rate [139]. The gelling ability of pectin makes it widely used in food, pharmaceutical, health promotion, and cosmetics applications [140]. Ionic cross-linking of low-methoxyl pectin occurs with an appropriate concentration of $CaCl_2$ [141, 142].

Pectin is soluble in pure water but insoluble in organic solvents. Thus, casting has been used commonly to prepare pectin-based films [139, 143]. Extrusion has also been used but mainly for pectin blended with other polymers [143]. The most used plasticizer for pectin is glycerol. When plasticizers are added to pectin, the barrier qualities of the resultant pectin films against moisture, gas, and scent chemicals are considerably diminished [143]. It has been found that the addition of citric acid improves the tensile strength, Young's modulus, and water vapor permeability (WVP) of pectin films and also decreases film solubility, likely due to a cross-linking effect [144]. It was also discovered that cast pectin films become insoluble in water and have exceptional tensile strength when submerged in aqueous solutions containing multivalent cations [145]. Pectin composites such as those incorporated with chitosan nanoparticles [146] or chitosan/pectin fibers [147] were reported to have increased mechanical properties and reduced WVP.

6.3 Strategies to Formulate Polysaccharide-Based Materials

6.3.1 Dissolution and Plasticization

Unlike synthetic thermoplastic polymers, polysaccharides, although they do not have a covalently cross-linked structure, will not melt naturally at a high temperature due to the significantly hydrogen-bonded molecular network. As a result, in order to process polysaccharides, solvents that function as hydrogen-bond breakers or plasticizers need to be added. Different solvents, plasticizers, or additives that are instrumental to the dissolution or processing of different polysaccharides were introduced in Section 6.2. For high-viscosity thermomechanical processing of polysaccharides, the processing aids used should be those that can dissolve the polysaccharide when employed in a large amount. For example, for the thermomechanical processing of starch or alginate, a small amount of water should suffice, while thermomechanically processed chitosan can be obtained with acids. Additionally, glycerol is the most used plasticizer for polysaccharides, as it is cheap, nonvolatile, and stable during processing and post-processing. Other plasticizers that are suitable for polysaccharides include glycol, sorbitol, xylitol, maltitol, ethylene glycol, propylene glycol, butanediol, sucrose, fructose, mannose, and fatty acids (such as myristate or palmitate) [148]. It is noteworthy that small-molecule plasticizers like glycerol in polymers may migrate out of the material during storage or usage, leading to material property changes.

In recent years, ILs and DESs have been regarded as novel solvents or plasticizers for polysaccharide extraction, modification, and processing [149, 150]. DESs have many of the same properties as ILs, including low vapor pressure, chemical and thermal stability, nonflammability, strong dissolving ability, low melting point, and tailorability, yet they are less costly, water-neutral, low- or nontoxic, and usually biodegradable. Additionally, DESs may be made from chemicals that are readily available and/or naturally occurring in an easy, practical way by heating and mixing [149].

6.3.2 Polysaccharide-Based Blends

Polysaccharides can lack versatility in terms of processability, and materials based on only polysaccharides may have some shortcomings in properties or functionality for applications. Therefore, the development of blends and composites of different polysaccharides or polysaccharides with other polymers, especially biodegradable, has been widely pursued. Combining various polysaccharides and polysaccharides/

proteins is an effective way to create materials with balanced characteristics and integrated functionalities that are particularly helpful, for example, for biomedical applications [151, 152]. However, it is noteworthy that proteins and polysaccharides, and even different polysaccharides with each other, may not always be compatible due to differences in factors such as hydrophilicity/hydrophobicity, viscosity, thermal properties, gelation behavior, and electrical charge. In the studies by Jebalia et al. [153, 154], extrusion-processed pea starch/pea protein isolate films showed incompatibility and possibly interfacial debonding, which weakened the mechanical properties. Chen et al. [155] found that, in the potato-starch–whey-protein-fibril mixed gel, the phase separation of the protein fibrils and the starch dominated at neutral pH, whereas the two biopolymer components more closely correlated under acidic conditions, increasing the gel storage modulus. Solution-cast films based on HPMC, which gelates at an increased temperature, and hydroxypropyl starch (HPS), which gelates at a reduced temperature, showed phase separation between the two chemically modified polysaccharides [81–84]. Thermomechanically processed chitosan/alginate films also displayed a phase-separated structure observable using transmission electron microscopy (TEM) [111].

Research has been widely carried out to develop blends of polysaccharides, especially starch, with biodegradable polymers, especially polyesters such as PLA, PHAs, polycaprolactone (PCL), polybutylene adipate terephthalate (PBAT), and polybutylene succinate (PBS) [30, 31]. Starch-based blends have been a major type of commercialized bioplastics. In this case, there should be suitable strategies in place to increase the compatibility between hydrophilic polysaccharides and hydrophobic polyesters. For starch/PLA blends, for example, two processes are often used to create the compatibilizers, namely, grafting a polyester monomer onto the starch backbone by polymerization and modifying polyester using maleic anhydride (MA) [156]. Huneault and Li [157] studied the interfacial modification of starch/MA-grafted-PLA blends prepared by twin-screw extrusion. Plasticized starch was produced using the first half of a twin-screw extrusion line, and molten PLA was added halfway down the line using a single-screw extruder as a side feeder. Before PLA was added, the water in the plasticized starch was devolatilized to lessen the breakdown of PLA in the presence of water. It was discovered that the dispersed phase size was significantly lower, in the 1–3 m range, suggesting that this technique could improve the compatibility of starch and PLA at their interface [157]. Additionally, many researchers have used isocyanate-derived compatibilizers such as methylenediphenyl diisocyanate (MDI) and phenyl diisocyanate (PDI) to cross-link PLA and starch molecules. Isocyanate functional groups can generate urethane linkages and improve interfacial adhesion by interacting with the starch hydroxyl group or the terminal carboxyl group of PLA [158]. Yu and co-workers [159–161] found that controlling the distribution of the MDI compatibilizer across multiple extrusion runs is a practical means of enhancing the compatibility of starch and polyester as well as the final product characteristics of the starch/polyester blends.

Polyvinyl alcohol (PVA) is an interesting polymer to be hybridized with polysaccharides as it is hydrophilic, water-soluble, nontoxic, and biodegradable. PVA/polysaccharide-based hydrogels, electrospun nanofibers, films, and coatings have been widely studied for biomedical, food packaging, and other applications [162–165]. This is because the combination of PVA and polysaccharides could allow suitable processability and material properties to be achieved. Moreover, considering that bioinks based on single or simple components cannot meet all the requirements for bioprinting large, complex biomimetic tissues and organs, Zou et al. [166] developed a bioink made of three polysaccharides including nanocellulose, agarose, and sodium alginate with seeded cells and PVA as a sacrificial material to construct structures that did not collapse.

6.3.3 Polysaccharide-Based Nanocomposites

Polysaccharide-based nanocomposites have been widely studied [167, 168], and the inclusion of nanofillers into polysaccharide formations may result in improved properties and added functions. For example, chitosan/silver-nanoparticles nanocomposites showed strong bactericidal activity without significant cytotoxicity [169, 170]. It was proposed that silver nanoparticles immobilized in the polysaccharide gel matrix can exert their antibacterial action by simple contact with the bacterial membrane, while they cannot be taken up and internalized by cells [170]. Svagan et al. [171] prepared nanocomposites made of amylopectin, glycerol, and a high content of microfibrillated cellulose (MFC) by solution casting. Due to the biomimetic structure like plant cell walls, the nanocomposite with 70 wt % MFC showed high tensile strength, high modulus, and very high work of fracture (toughness).

For nanocomposites development, the compatibility between the nanofiller and the polysaccharide matrix and the dispersion of the nanofiller in the polysaccharide matrix are highly important. Most nanofillers that are widely used for developing polymer nanocomposites, such as carbonaceous nanofillers and metal nanoparticles, are not as hydrophilic as polysaccharides, and thus can only form weak interfacial interactions with the polysaccharide matrix. In contrast, hydrophilic nanofillers such as clays and polysaccharide nanofillers (e.g., cellulose nanofibers and chitin nanowhiskers) can be more easily dispersed in polysaccharide matrices. Graphene is hydrophobic and not easy to be dispersed in a polysaccharide matrix, whereas graphene oxide, which has oxygen-containing groups (e.g., –OH and –COOH) and is negatively charged, can interact with polysaccharides (especially positively charged chitosan) more strongly. In situ synthesis of metal nanoparticles (e.g., silver nanoparticles) in polysaccharides by chemical reduction is an effective strategy to create polysaccharide nanocomposites with excellently dispersed metal nanoparticles. In most studies of polysaccharide nanocomposites, solution casting and sonication were used for easier processing of polysaccharides and also for better dispersion of nanoparticles [172]. To a much lesser extent, thermomechanical processing (e.g., batch mixing and extrusion) has been employed to prepare polysaccharide-based nanocomposites [96, 97, 110, 112, 114, 173–176].

6.4 Processing of Polysaccharide-Based Materials

6.4.1 Wet Processing

In most cases, wet processing is used to process polysaccharide into materials. Polysaccharides are dissolved or dispersed using a large amount of solvent, which allows for the creation of various polysaccharide materials such as film or membrane, hydrogel, aerogel, particles, sponge, and fibers.

Most frequently, solution casting, during which the excess solvent is removed by drying, is used to prepare polysaccharide films. Many polysaccharides such as chitosan, starch, alginate, pectin, and carboxymethyl cellulose have an excellent film-forming ability and the resulting films have mechanical properties comparable to those of synthetic polymers, effective gas barrier properties, and oil-impermeability, despite a poor WVP [177, 178].

Layer-by-layer (LbL) deposition has been used to create multilayer polysaccharide films based on the electrostatic interactions of oppositely charged polysaccharides (e.g., positively charged chitosan and negatively charged biopolymers) [179–187]. LbL methods include dip-coating LbL assembly [181, 185], spray-assisted LbL assembly [179], LbL drop-casting [186], and LbL electrostatic deposition [187].

Hydrogel, which has a three-dimensional network structure and contains a large amount of water, is an important form of biopolymer material, especially useful for biomedical applications. To form a hydrogel, dissolved polysaccharide chains need to be cross-linked by chemical and/or physical cross-linking. For chitosan, chemical cross-linkers such as glutaraldehyde and genipin are useful [188, 189], while a phosphate or adipate bridge will cross-link starch [59]. Typical kinds of physical cross-linking include ionic cross-linking and hydrogen-bonding cross-linking. Ionic cross-linking is using multivalent metal cations such as Ca^{2+} to ionically cross-link alginate as discussed in Section 6.2.4. This property of alginate is especially useful for the fabrication of hydrogels, beads, and microspheres and for 3D printing. Additionally, it has been found that aqueous $CaCl_2$ can gelate high-amylose starch to form a hydrogel [57, 58], presumably due to the interaction of Ca^{2+} cations with the hydroxyl groups of starch.

Drying hydrogel microspheres with supercritical carbon dioxide ($scCO_2$) can result in aerogels; these are different from xerogels generated by drying chitosan hydrogels in the air, which cannot maintain the fibrous structure or functional groups [190]. Freeze-drying hydrogels leads to a continuous three-dimensional porous structure generated by ice crystals, which is usually called cryogel or sponge.

Beads or microspheres can also be formed by cationic chitosan interacting spontaneously with anionic polysaccharides in aqueous solutions, and this process is known as coacervation. This allows the generation of particles with a core-shell structure. Further coating this kind of particle based on polyelectrolyte complexation can generate multiple layers on the particle.

Micro- or nanofibers can be prepared via self-assembly through polyelectrolyte complexation (e. g., chitosan and xanthan gum [191]) or electrospinning.

6.4.2 Thermomechanical Processing

Polysaccharides do not have real thermoplasticity as most conventional thermoplastic polymers. However, as discussed in Section 6.2, some polysaccharides such as starch, chitosan, and alginate have been demonstrated to be thermomechanically processable when they are mixed with water, plasticizers, and/or other additives that can function as hydrogen-bond breakers, and techniques such as extrusion, batch mixing, and compression molding have been used to process these polysaccharides. Starch is the polysaccharide that is most suitable for extrusion processing as it behaves like a thermoplastic after it is gelatinized (or "melted"), and that is the reason why it has been widely investigated to develop bioplastics.

While starch is versatile for chemical modification, a lot of research on reactive extrusion (REX) of starch has been carried out to prepare modified starch products – by, for example, grafting, esterification, phosphorylation, and cross-linking – and compatibilized starch blends [192–195]. For example, Hablot et al. [196] prepared maleated thermoplastic starch (MTPS) through REX of starch with MA as an esterification agent and glycerol as a plasticizer. MTPS was then blended with PBAT by REX. This allowed the grafting of glycerol on the starch backbone and, further, the grafting of PBAT onto glycerylated starch via transesterification reactions [196]. Xiao et al. [197] prepared a starch-based superabsorbent by reactive mixing. During this process, acrylamide was successfully grafted onto starch, leading to a porous structure beneficial for the controlled release of urea. In contrast to that of starch, the reactive processing of other polysaccharides has been reported scarcely.

Formulations based on only polysaccharides (e.g., starch mixed with limited amounts of water/plasticizers) still show quite high viscosity, which may cause issues for thermomechanical processing. Starch with a higher amylose content particularly has a higher viscosity [32] and requires a higher mechanical energy input to be processed [34], although it is favorable for generating materials with better mechanical properties [34]. Premixing polysaccharides with water/plasticizers, followed by sufficient time for equilibrium, is helpful for improving their processability [34].

When polysaccharides are mixed with other polymers (as discussed in Section 6.3.2), the processing will be easier, as the overall rheological properties can be mostly determined by the other polymers. However, it is noteworthy that in this case, polysaccharides may not achieve a highly thermoplastic state as when they are processed alone. Therefore, it may be necessary to first process polysaccharides and then blend them with other polymers.

6.4.3 3D Printing

3D printing has recently emerged as a versatile technology to manufacture materials with complex three-dimensional structures. The 3D printing of polysaccharide-based materials has also been studied to some extent, especially in food and biomedical applications [198]. For biopolymers, extrusion-based 3D printing has been used to the greatest extent. Inks based on polysaccharides are usually in the form of paste, which requires the use of non-melting extrusion-based printing techniques, typically like direct ink writing (DIW). After the ink is extruded out of the nozzle and deposited on the stage, it needs to be solidified by cooling, drying, or physical (e.g., ionic) or chemical cross-linking. DIW is also an important technique for the 3D printing of foods, especially starch-based ones [198–204].

Especially important for the DIW process is that ink formulations must exhibit suitable rheological properties. Shear-thinning behavior is important for easy extrusion out of submillimeter-diameter nozzles; once deposited on the stage, the ink must possess a high storage modulus (G′), yield stress (τ_y), and low-shear viscosity, for shape retention (to prevent the distortion of printed objects) [205, 206]. Liu et al. [201] showed that the τ_y (from 44.41 Pa to 883.19 Pa) of a potato-starch paste increased with a higher starch concentration (from 10 wt % to 30 wt %), and a higher τ_y was favorable for higher printing accuracy and higher strength of the printed structure. In addition, it is essential for the material to swiftly regain its initial viscosity during extrusion (shearing) in order to achieve optimal form fidelity, which is determined by the recovery time. Kim et al. [207] observed that for alginate/κ-carrageenan/$CaSO_4$ hydrogels, the viscosity increased with increasing κ-carrageenan concentration while the shear-thinning behavior was preserved and the recovery time was reduced. Similarly, cellulose nanofiber dispersions included with κ-carrageenan could have improved viscoelastic properties, allowing the 3D printing of self-supporting structures [208].

As mentioned before, alginate can be ionically cross-linked (gelated) with the presence of multivalent cations such as Ca^{2+}, which is an important feature that can be utilized for 3D printing. Patrício et al. [126] reported that, as a viscoelastic fluid gel with shear-thinning behavior, an alginate-based xanthan-gum solution containing Ca^{2+} can function as a continuous pseudoplastic matrix and a supporting material to facilitate the fast manufacturing of complicated three-dimensional structures like bifurcate tubes. When alginate was printed into the supporting bath, Ca^{2+} was employed to gelate it [126].

The most typical kind of chemical cross-linking is photo-cross-linking. For free-radical polymerization, biopolymers can be functionalized with methacrylate groups. For extrusion 3D printing, the polysaccharides alginate [209, 210], chitosan [211, 212], and hyaluronic acid [213] have been functionalized with methacrylate moieties to be UV-curable. For example, according to Cao et al. [209], a 3D printed methacrylated alginate (SA-MA) hydrogel was cured during printing under UV light, and a subsequent ion exchange with Ca^{2+} ions and ionic interaction with chitosan could stepwise raise the cross-linking density, increase the mechanical properties, and cause a volume contraction of the hydrogel. The 3D printed hydrogel was able to demonstrate different deformations by two-step ionic immersion because different portions of the material were irradiated to varying degrees during printing, creating a gradient pattern of cross-linking. A bilayer of SA-MA hydrogel was also created, with different chemical cross-linking densities, and following Ca^{2+} soaking, it took on an anisotropic structure.

There are also other chemical cross-linking strategies used for the solidification of biopolymer-based inks. For example, Heidenreich et al. [214] showed that the precursor stage addition of *N*-hydroxysuccinimide (NHS)/1-ethyl-3-(3-dimethylaminopropyl)carbodiimide (EDC) was a promising method for mechanically reinforcing the finished construct. Since NHS/EDC enables the formation of covalent connections between carboxyl (-COOH) or phosphate groups ($-OPO(OH)_2$) and primary amines ($-NH_2$), covalent amide linkages might be created between collagen chains as well as between collagen and chitosan chains.

To achieve continuous printing of internally micropatterned alginate/gelatin-methacryloyl (GelMA) hydrogel fibers for regulating cellular orientation in tissue engineering, Samandari et al. [210] created a DIW 3D printer. Specifically, the printer included a static mixer that striated various hydrogels and a coaxial microfluidic device where $CaCl_2$ ionically cross-linked alginate striations to create fiber. UV light was then used to chemically cross-link the GelMA striations in the fiber. While the interior filaments were on a microscale, the fiber created in this manner is measured in millimeters. It was shown that the internal patterned structure, which could be modified by changing the flow parameters while printing, could result in great cell spreading and alignment and rapid cell proliferation and differentiation, leading to the production of mature tissue engineering constructs.

Besides extrusion-based 3D printing, inkjet 3D printing has also been studied for the fabrication of polysaccharide-based biomaterials, but to a much lesser extent. For example, Xu et al. [215] reported employing a bioink made of sodium alginate (2% w/v) combined with a cell suspension at a 1:1 (v/v) ratio to 3D inkjet print complicated 3D structures like zigzag tubes. The printed structure was created on a platform that could be moved vertically while the printhead, which could move horizontally, jetted the bioink into a bath containing a $CaCl_2$ solution for ionically cross-linking the alginate. Fibroblast (3T3 cell)-laden tubular and overhang-structured constructs with high post-printing 3T3 cell viability (above 82% after an incubation period of 72 h) were obtained [215]. In another study, a customized thermal inkjet printer was used by Xu et al. [216] to create complex-shaped and heterogeneous tissue constructs by simultaneously ejecting several kinds of cells. Separate ink cartridges for the printer were each filled with a 0.1 M $CaCl_2$ solution containing canine smooth muscle cells (dSMCs), bovine aortic endothelial cells (bECs), and human amniotic fluid-derived stem cells (hAFSCs), respectively. Collagen and sodium alginate were combined to create a solution, and $CaCl_2$ was used to cross-link the sodium alginate to create hydrogel, which served as the printing support. Evaluations conducted in vitro and in vivo revealed that the cells in the printed construct had a good level of vitality, were able to multiply well, and could form tissues with adequate vascularization [216].

6.4.4 Electrospinning

In order to electrospin cellulose, it must be dissolved first, but as cellulose solvents are often not volatile, this solvent must be removed from the electrospun fibers by a coagulation step [217]. Cellulose derivatives such as cellulose acetate, hydroxypropyl cellulose, hydroxypropyl methyl cellulose, and ethyl-cyanoethyl cellulose can be more readily dissolved in volatile solvents, making their electrospinning easier [217]. Among these, cellulose acetate has been studied more extensively due to its excellent solubility and application-related properties [218]. Additionally, cellulose nanofibers can be mixed with other polymers for electrospinning. For example, Li et al. [219] developed a reusable, biodegradable, and antibacterial mask fabricated by the electrospinning of PVA, polyethylene glycol (PEG), and CNF, followed by esterification and the deposition of a nitrogen-doped TiO_2 (N-TiO_2) and TiO_2 mixture. They found that the formulation of PVA, PEG, and CNF enhanced the electrospinnability and resulted in an excellent mechanical performance of the resultant mask [219].

Fibrous materials based on only chitosan have been successfully prepared by electrospinning. For example, Ohkawa et al. [220] demonstrated that uniform (not interconnected) chitosan fibers with a mean diameter of 330 nm can be created using a trifluoroacetic acid (TFA)/dichloromethane (DCM) (70:30 (v/v)) mixture as a chitosan solvent (chitosan concentration: 8 wt %). Electrospun chitosan nanofibrous membranes made with a TFA/DCM-combination solvent, after being neutralized in a Na_2CO_3 aqueous solution, could preserve their fibrous structure even after continuous immersion in phosphate-buffered saline (pH = 7.4) or distilled water for 12 weeks [221]. Geng et al. [222] demonstrated that a high concentration of acetic acid (over 30%) is necessary for dissolving chitosan for electrospinning, and the optimum conditions are 7% chitosan in an aqueous 90% acetic-acid solution and an electrical field of 4 kV/cm. They also suggested that a suitable molecular mass of chitosan (106,000 g/mol) and electrical field (4 kV/cm) are necessary to produce bead-free chitosan nanofibers [222]. Homayoni et al. [223] first used an alkali treatment to hydrolyze and partially degrade chitosan molecules, which was for reducing the viscosity of chitosan and increasing its spinnability. Excellent electrospun nanofibers could be produced from solutions of the hydrolyzed chitosan in aqueous 70–90% acetic acid, with the mean diameter of the nanofiber increasing with reduced acetic acid concentration in the solvent. The optimum nanofiber with a mean diameter of 140 nm was achieved with the hydrolyzation of chitosan for 48 hours (molecular mass: 2.94×10^5 g/mol).

Chitin nanofibers prepared by electrospinning have also been studied, but the first step is still to dissolve chitin in a solvent. It was shown that, to increase its solubility, chitin can be depolymerized using gamma radiation and then electrospun using 1,1,1,3,3,3-hexafluoro-2-propanol (HFIP) as the spinning solvent [224, 225]. Additionally, ILs have shown to be an effective solvent for chitin electrospinning. Rogers and co-workers [226] developed a one-pot process of using [C_2mim][OAc] for the extraction of high-molecular-mass chitin from dried shrimp shell and subsequent electrospinning. This process also allowed the electrospinning of chitin together with cellulose to produce composite nanomats [227]. Chemically modified chitin may be dissolved more easily for electrospinning; for example, chitin propionate dissolved in water–ethanol mixtures [228].

Blending chitosan/chitin with synthetic polymers such as PVA [229–232], PEG [228], and biopolymers such as gelatin [233] is a more viable means to create electrospun fibers. The quality and diameter of the PVA/chitosan nanofibers were found to be associated with the PVA/chitosan ratio and the PVA/chitosan concentration [229]. When chitin nanowhiskers were mixed in other polymers such as PVA [232] or PCL [234] for electrospinning, the former nanoparticles just acted as a reinforcement filler for the polymer matrix. Chitosan/gelatin polyelectrolyte-complex nonwoven nanofibrous mats were produced by electrospinning, and the electrospinnability and fiber diameter were found to be strongly influenced by the chitosan/gelatin ratio, the solution concentration, and the electric field [233].

Generally, the electrospinning of starch is quite difficult due to its branched molecular structure, and the strategies employed in this area include the use of a suitable solvent (e. g., DMSO, formic acid, acetic acid, chloroform, and ILs), starch modification (e. g., acetylated starch, oxidized starch, cationic starch, and hydroxypropyl starch), and mixing starch with other polymers (e. g., PVA, PEG, and PCL) [235, 236]. The fabrication of pure starch fibers by electrospinning has been reported [237–241]. A high-amylose starch (Gelose 80, containing about 80% amylose) dissolved in a 95% DMSO solution was electrospun into fibers, which were deposited into a coagulating bath containing ethanol, followed by cross-linking using glutaraldehyde. It was proposed that for this high-amylose starch, 1.2–2.7 times the entanglement concentration is required for effective electrospinning, while starches with lower amylose contents require a higher critical entanglement concentration or are even not electrospinnable [238]. For electrospinning of a high-amylose starch (HYLON® VII, containing about 70% amylose) dissolved in H_2O/DMSO, it was found that a high water content resulted in gelatinization and poor electrospinnability as water negatively impacted hydrogen bonding [242]. Also for HYLON® VII-based electrospun fibers, glutaraldehyde vapor-phase cross-linking could reduce water-sensitivity and result in a tenfold increase in tensile strength [243].

Alginate and pectin have insufficient electrospinnability; thus, it is common to blend alginate with other polymers for electrospinning, such as PEG [244–247] and PVA [246, 248]. It has to be noted that PEG generally is resistant to biodegradation, although there has been evidence showing the long-term oxidative degradation of PEG under biologically relevant conditions [249]. Moreover, constructing polyelectrolyte complexes between reversely charged biopolymers (e. g., chitosan–pectin [250] and chitosan–alginate [245]) can improve the properties of electrospun fibers. For example, Lin et al. [250] introduced a chitosan/pectin/PVA formulation in a concentrated acetic acid solution to prepare electrospun fibers, which benefited from the polyelectrolyte complexation between chitosan and pectin, exhibiting much better mechanical properties, decreased swelling ratio in a saline solution, and a quicker release of preloaded tetracycline HCl, compared with the chitosan/PVA sample.

Chang et al. [251] reported an electrospun nanofibrous mat with a structure of alginate core and chitosan sheath, which was prepared by electrospinning of an alginate/glycerol solution into a coagulating bath containing a chitosan/ethanol solution. They claimed that, compared with the alginate-based electrospun nanofiber coagulated with Ca^{2+}, this core–sheath structured nanofiber could avoid structure damage due to the extraction of calcium in the physiological environment [251].

For enhancing the properties (e.g., wet-stability and mechanical properties) of polysaccharide-based electrospun fibers, a subsequent cross-linking may be necessary [244, 252, 253]. For chitosan-based nanofibers, the commonly used cross-linking strategies include chemical (using, e.g., glutaraldehyde, epichlorohydrin, and genipin) and physical ones (using, e.g., tripolyphosphate, tannic acid, and glycerol phosphate) [252]. Qin et al. [254] demonstrated that for chitosan/pullulan electrospinning nanofibers, the water-resistance and mechanical and thermal properties could be enhanced by thermally induced cross-linking by Maillard and Schiff base reactions with the presence of cinnamaldehyde (as a substitute of the toxic glutaraldehyde). Cui et al. [253] used a dual cross-linking strategy to treat electrospun pectin nanofibers; they found that adipic acid dihydrazide (ADH) cross-linking enhanced the mechanical strength and inhibited degradation, and nanofibers based on high-methoxylated pectin facilitated cell adhesion and proliferation.

Electrospun polysaccharide nanofibers have been found to be highly useful especially for biomedical applications such as wound healing, tissue engineering, drug delivery, and pharmaceutical applications as reviewed elsewhere [218, 235, 236, 246, 252, 255, 256].

6.5 Concluding Remarks

Polysaccharides as a special group of polymers have their unique properties related to processing. They require the use of water or other solvents for disrupting the native hydrogen-bonded structure (resulting in a solution, paste, dispersion, or "melt"), which is the prerequisite for various ways of processing polysaccharides into different forms of material. While techniques to process polysaccharides have been well established for their traditional usage, continuous development in this area keeps generating new methods of polysaccharide engineering.

However, there are still challenges that need to be addressed in future research. Firstly, it is always interesting to develop new, environmentally friendly, low-/non-toxic, and even easily recyclable solvation systems for the dissolution, extraction, modification, and processing of polysaccharides in a cost-effective and sustainable way. Secondly, new, especially "green", chemistries are highly desirable for obtaining new polysaccharide derivatives with improved processability and enhanced material properties or functionality, as well as enabling greater interaction/compatibility of polysaccharides with other polymers and fillers to enlarge the family of polysaccharide-based blends and composites. Additionally, it is also interesting to study the inherent properties (thermal/cooling gelation behavior, polyelectrolyte complexation, self-assembly, gelling ability, water-holding capability, and gas barrier properties) of polysaccharides and develop processes to make use of or magnify such properties to boost applicational outcomes. Moreover, new engineering technologies are worth to be explored to enable new structures and functions of polysaccharide-based materials and composites, to enhance their applicability and allow them to be used for emerging applications (e.g., smart materials, actuation, and bioelectronics). Last but not least, interrogating the mechanisms of the underlying processing–structure–property/functionality relationships will expand our capabilities of developing high-performance polysaccharide-based materials for demanding applications.

6.6 Abbreviations Used in This Chapter

CNC	cellulose nanocrystal
CNF	cellulose nanofibril
DES	deep eutectic solvent
DMAc	*N,N*-dimethylacetamide
DMSO	dimethyl sulfoxide
HPMC	hydroxypropyl methyl cellulose
IL	ionic liquid
MA	maleic anhydride
PCL	polycaprolactone
PEG	polyethylene glycol
PVA	polyvinyl alcohol
WVP	water vapor permeability

References for Chapter 6

[1] R.M. Cywar, N.A. Rorrer, C.B. Hoyt, G.T. Beckham, and E.Y.X. Chen, "Bio-based polymers with performance-advantaged properties", *Nature Reviews Materials* 7(2) (2022), pp. 83–103.

[2] S. Ozilgen and S. Bucak, Chapter 6 – Functional Biopolymers in Food Manufacturing, in: A.M. Grumezescu and A.M. Holban (Eds.), *Biopolymers for Food Design*, Academic Press, 2018, pp. 157–189.

[3] K. Prameela, C. Murali Mohan, and C. Ramakrishna, Chapter 1 – Biopolymers for Food Design: Consumer-Friendly Natural Ingredients, in: A.M. Grumezescu and A.M. Holban (Eds.), *Biopolymers for Food Design*, Academic Press, 2018, pp. 1–32.

[4] I. Chang, A.K. Prasidhi, J. Im, and G.-C. Cho, "Soil strengthening using thermo-gelation biopolymers", *Construction and Building Materials* 77 (2015), pp. 430–438.

[5] J. Huang, Y. Zhong, L. Zhang, and J. Cai, "Extremely Strong and Transparent Chitin Films: A High-Efficiency, Energy-Saving, and "Green" Route Using an Aqueous KOH/Urea Solution", *Advanced Functional Materials* 27(26) (2017), 1701100.

[6] B. Duan, C. Chang, B. Ding, J. Cai, M. Xu, S. Feng, J. Ren, X. Shi, Y. Du, and L. Zhang, "High strength films with gas-barrier fabricated from chitin solution dissolved at low temperature", *Journal of Materials Chemistry A* 1(5) (2013), pp. 1867–1874.

[7] J. Gómez-Estaca, R. Gavara, R. Catalá, and P. Hernández-Muñoz, "The Potential of Proteins for Producing Food Packaging Materials: A Review", *Packaging Technology and Science* 29(4–5) (2016), pp. 203–224.

[8] P. Forssell, R. Lahtinen, M. Lahelin, and P. Myllärinen, "Oxygen permeability of amylose and amylopectin films", *Carbohydrate Polymers* 47(2) (2002), pp. 125–129.

[9] O. Rompothi, P. Pradipasena, K. Tananuwong, A. Somwangthanaroj, and T. Janjarasskul, "Development of non-water soluble, ductile mung bean starch based edible film with oxygen barrier and heat sealability", *Carbohydrate Polymers* 157 (2017), pp. 748–756.

[10] S. Cheng, Y. Zhang, R. Cha, J. Yang, and X. Jiang, "Water-soluble nanocrystalline cellulose films with highly transparent and oxygen barrier properties", *Nanoscale* 8(2) (2016), pp. 973–978.

[11] J. Wang, D. J. Gardner, N. M. Stark, D. W. Bousfield, M. Tajvidi, and Z. Cai, "Moisture and Oxygen Barrier Properties of Cellulose Nanomaterial-Based Films", *ACS Sustainable Chemistry & Engineering* 6(1) (2018), pp. 49–70.

[12] H. J. Park and M. S. Chinnan, "Gas and water vapor barrier properties of edible films from protein and cellulosic materials", *Journal of Food Engineering* 25(4) (1995), pp. 497–507.

[13] S. Sivakanthan, S. Rajendran, A. Gamage, T. Madhujith, and S. Mani, "Antioxidant and antimicrobial applications of biopolymers: A review", *Food Research International* 136 (2020), 109327.

[14] M. Hosseinnejad and S. M. Jafari, "Evaluation of different factors affecting antimicrobial properties of chitosan", *International Journal of Biological Macromolecules* 85 (2016), pp. 467–475.

[15] I. Shahid-ul, M. Shahid, and F. Mohammad, "Green Chemistry Approaches to Develop Antimicrobial Textiles Based on Sustainable Biopolymers – A Review", *Industrial & Engineering Chemistry Research* 52(15) (2013), pp. 5245–5260.

[16] J. Song, B. Winkeljann, and O. Lieleg, "Biopolymer-Based Coatings: Promising Strategies to Improve the Biocompatibility and Functionality of Materials Used in Biomedical Engineering", *Advanced Materials Interfaces* 7(17) (2020), 2000850.

[17] C. Wang, T. Yokota, and T. Someya, "Natural Biopolymer-Based Biocompatible Conductors for Stretchable Bioelectronics", *Chemical Reviews* 121(4) (2021), pp. 2109–2146.

[18] R. Rebelo, M. Fernandes, and R. Fangueiro, "Biopolymers in Medical Implants: A Brief Review", *Procedia Engineering* 200 (2017), pp. 236–243.

[19] T. G. Sahana and P. D. Rekha, "Biopolymers: Applications in wound healing and skin tissue engineering", *Molecular Biology Reports* 45(6) (2018), pp. 2857–2867.

[20] T. Biswal, "Biopolymers for tissue engineering applications: A review", *Materials Today: Proceedings* 41 (2021), pp. 397–402.

[21] P. Kanmani, J. Aravind, M. Kamaraj, P. Sureshbabu, and S. Karthikeyan, "Environmental applications of chitosan and cellulosic biopolymers: A comprehensive outlook", *Bioresource Technology* 242 (2017), pp. 295–303.

[22] M. Zubair and A. Ullah, Chapter 14 – Biopolymers in environmental applications: industrial wastewater treatment, in: S. Thomas, S. Gopi, and A. Amalraj (Eds.), *Biopolymers and their Industrial Applications*, Elsevier, 2021, pp. 331–349.

[23] K. Palanisamy, A. Jeyaseelan, K. Murugesan, and S. B. Palanisamy, Biopolymer Technologies for Environmental Applications, in: K. M. Gothandam, S. Ranjan, N. Dasgupta, and E. Lichtfouse (Eds.), *Nanoscience and Biotechnology for Environmental Applications*, Springer, Cham, 2019, pp. 55–83.

[24] N.R. Nair, V.C. Sekhar, K.M. Nampoothiri, and A. Pandey, 32 – Biodegradation of Biopolymers, in: A. Pandey, S. Negi, and C.R. Soccol (Eds.), *Current Developments in Biotechnology and Bioengineering*, Elsevier, 2017, pp. 739–755.

[25] E.M.N. Polman, G.-J.M. Gruter, J.R. Parsons, and A. Tietema, "Comparison of the aerobic biodegradation of biopolymers and the corresponding bioplastics: A review", *Science of The Total Environment* 753 (2021), 141953.

[26] J.P. Mua and D.S. Jackson, "Fine Structure of Corn Amylose and Amylopectin Fractions with Various Molecular Weights", *Journal of Agricultural and Food Chemistry* 45(10) (1997), pp. 3840–3847.

[27] P. Chen, L. Yu, G.P. Simon, X. Liu, K. Dean, and L. Chen, "Internal structures and phase-transitions of starch granules during gelatinization", *Carbohydrate Polymers* 83(4) (2011), pp. 1975–1983.

[28] P. Chen, L. Yu, G. Simon, E. Petinakis, K. Dean, and L. Chen, "Morphologies and microstructures of cornstarches with different amylose–amylopectin ratios studied by confocal laser scanning microscope", *Journal of Cereal Science* 50(2) (2009), pp. 241–247.

[29] H. Liu, L. Yu, F. Xie, and L. Chen, "Gelatinization of cornstarch with different amylose/amylopectin content", *Carbohydrate Polymers* 65(3) (2006), pp. 357–363.

[30] H. Liu, F. Xie, L. Yu, L. Chen, and L. Li, "Thermal processing of starch-based polymers", *Progress in Polymer Science* 34(12) (2009), pp. 1348–1368.

[31] F. Xie, P.J. Halley, and L. Avérous, "Rheology to understand and optimize processibility, structures and properties of starch polymeric materials", *Progress in Polymer Science* 37(4) (2012), pp. 595–623.

[32] F. Xie, L. Yu, B. Su, P. Liu, J. Wang, H. Liu, and L. Chen, "Rheological properties of starches with different amylose/amylopectin ratios", *Journal of Cereal Science* 49(3) (2009), pp. 371–377.

[33] S. Tajuddin, F. Xie, T.M. Nicholson, P. Liu, and P.J. Halley, "Rheological properties of thermoplastic starch studied by multipass rheometer", *Carbohydrate Polymers* 83(2) (2011), pp. 914–919.

[34] M. Li, P. Liu, W. Zou, L. Yu, F. Xie, H. Pu, H. Liu, and L. Chen, "Extrusion processing and characterization of edible starch films with different amylose contents", *Journal of Food Engineering* 106(1) (2011), pp. 95–101.

[35] F. Xie, E. Pollet, P.J. Halley, and L. Avérous, "Starch-based nano-biocomposites", *Progress in Polymer Science* 38(10) (2013), pp. 1590–1628.

[36] P. Liu, F. Xie, M. Li, X. Liu, L. Yu, P.J. Halley, and L. Chen, "Phase transitions of maize starches with different amylose contents in glycerol–water systems", *Carbohydrate Polymers* 85(1) (2011), pp. 180–187.

[37] M. Paluch, J. Ostrowska, P. Tyński, W. Sadurski, and M. Konkol, "Structural and Thermal Properties of Starch Plasticized with Glycerol/Urea Mixture", *Journal of Polymers and the Environment* 30(2) (2022), pp. 728–740.

[38] A. Šoltýs, V. Hronský, N. Šmídová, D. Olčák, F. Ivanič, and I. Chodák, "Solid-state ^{1}H and ^{13}C NMR of corn starch plasticized with glycerol and urea", *European Polymer Journal* 117 (2019), pp. 19–27.

[39] Z.Y. Ben, H. Samsudin, and M.F. Yhaya, "Glycerol: Its properties, polymer synthesis, and applications in starch based films", *European Polymer Journal* 175 (2022), 111377.

[40] M. Alee, Q. Duan, Y. Chen, H. Liu, A. Ali, J. Zhu, T. Jiang, A. Rahaman, L. Chen, and L. Yu, "Plasticization Efficiency and Characteristics of Monosaccharides, Disaccharides, and Low-Molecular-Weight Polysaccharides for Starch-Based Materials", *ACS Sustainable Chemistry & Engineering* 9(35) (2021), pp. 11960–11969.

[41] M.R. Area, M. Rico, B. Montero, L. Barral, R. Bouza, J. López, and C. Ramírez, "Corn starch plasticized with isosorbide and filled with microcrystalline cellulose: Processing and characterization", *Carbohydrate Polymers* 206 (2019), pp. 726–733.

[42] M.R. Area, B. Montero, M. Rico, L. Barral, R. Bouza, and J. López, "Properties and behavior under environmental factors of isosorbide-plasticized starch reinforced with microcrystalline cellulose biocomposites", *International Journal of Biological Macromolecules* 164 (2020), pp. 2028–2037.

[43] B. Zhang, F. Xie, T. Zhang, L. Chen, X. Li, R.W. Truss, P.J. Halley, J.L. Shamshina, T. McNally, and R.D. Rogers, "Different characteristic effects of ageing on starch-based films plasticised by 1-ethyl-3-methylimidazolium acetate and by glycerol", *Carbohydrate Polymers* 146 (2016), pp. 67–79.

[44] F. Xie, B.M. Flanagan, M. Li, P. Sangwan, R.W. Truss, P.J. Halley, E.V. Strounina, A.K. Whittaker, M.J. Gidley, K.M. Dean, J.L. Shamshina, R.D. Rogers, and T. McNally, "Characteristics of starch-based films plasticised by glycerol and by the ionic liquid 1-ethyl-3-methylimidazolium acetate: A comparative study", *Carbohydrate Polymers* 111 (2014), pp. 841–848.

[45] D. Domene-López, J.J. Delgado-Marín, I. Martin-Gullon, J.C. García-Quesada, and M.G. Montalbán, "Comparative study on properties of starch films obtained from potato, corn and wheat using 1-ethyl-3-methylimidazolium acetate as plasticizer", *International Journal of Biological Macromolecules* 135 (2019), pp. 845–854.

[46] D. Domene-López, J.J. Delgado-Marín, J.C. García-Quesada, I. Martín-Gullón, and M.G. Montalbán, "Electroconductive starch/multi-walled carbon nanotube films plasticized by 1-ethyl-3-methylimidazolium acetate", *Carbohydrate Polymers* 229 (2020), 115545.

[47] A. Sankri, A. Arhaliass, I. Dez, A.C. Gaumont, Y. Grohens, D. Lourdin, I. Pillin, A. Rolland-Sabaté, and E. Leroy, "Thermoplastic starch plasticized by an ionic liquid", *Carbohydrate Polymers* 82(2) (2010), pp. 256–263.

[48] J. Wang, Y. Liang, Z. Zhang, C. Ye, Y. Chen, P. Wei, Y. Wang, and Y. Xia, "Thermoplastic starch plasticized by polymeric ionic liquid", *European Polymer Journal* 148 (2021), 110367.

[49] M. Zdanowicz, R. Jędrzejewski, and R. Pilawka, "Deep eutectic solvents as simultaneous plasticizing and crosslinking agents for starch", *International Journal of Biological Macromolecules* 129 (2019), pp. 1040–1046.

[50] K. Zhang, X. Cheng, F. Cheng, Y. Lin, M. Zhou, and P. Zhu, "Poly(citrate glyceride): a hyperbranched polyester for starch plasticization", *Polymer International* 67(4) (2018), pp. 399–404.

[51] S. Mateyawa, D.F. Xie, R.W. Truss, P.J. Halley, T.M. Nicholson, J.L. Shamshina, R.D. Rogers, M.W. Boehm, and T. McNally, "Effect of the ionic liquid 1-ethyl-3-methylimidazolium acetate on the phase transition of starch: Dissolution or gelatinization?", *Carbohydrate Polymers* 94(1) (2013), pp. 520–530.

[52] B. Zhang, F. Xie, J.L. Shamshina, R.D. Rogers, T. McNally, P.J. Halley, R.W. Truss, L. Chen, and S. Zhao, "Dissolution of Starch with Aqueous Ionic Liquid under Ambient Conditions", *ACS Sustainable Chemistry & Engineering* 5(5) (2017), pp. 3737–3741.

[53] K. Zan, J. Wang, F. Ren, J. Yu, S. Wang, F. Xie, and S. Wang, "Structural disorganization of cereal, tuber and bean starches in aqueous ionic liquid at room temperature: Role of starch granule surface structure", *Carbohydrate Polymers* 258 (2021), 117677.

[54] B. Zhang, F. Xie, J.L. Shamshina, R.D. Rogers, T. McNally, D.K. Wang, P.J. Halley, R.W. Truss, S. Zhao, and L. Chen, "Facile Preparation of Starch-Based Electroconductive Films with Ionic Liquid", *ACS Sustainable Chemistry & Engineering* 5(6) (2017), pp. 5457–5467.

[55] Y. Li, P. Liu, C. Ma, N. Zhang, X. Shang, L. Wang, and F. Xie, "Structural Disorganization and Chain Aggregation of High-Amylose Starch in Different Chloride Salt Solutions", *ACS Sustainable Chemistry & Engineering* 8(12) (2020), pp. 4838–4847.

[56] M. Lin, X. Shang, P. Liu, F. Xie, X. Chen, Y. Sun, and J. Wan, "Zinc chloride aqueous solution as a solvent for starch", *Carbohydrate Polymers* 136 (2016), pp. 266–273.

[57] C. Ma, F. Xie, L. Wei, C. Zheng, X. Liu, L. Wang, and P. Liu, "All-Starch-Based Hydrogel for Flexible Electronics: Strain-Sensitive Batteries and Self-Powered Sensors", *ACS Sustainable Chemistry & Engineering* 10(20) (2022), pp. 6724–6735.

[58] P. Liu, C. Ma, Y. Li, L. Wang, L. Wei, Y. Yan, and F. Xie, "Facile Preparation of Eco-Friendly, Flexible Starch-Based Materials with Ionic Conductivity and Strain-Responsiveness", *ACS Sustainable Chemistry & Engineering* 8(51) (2020), pp. 19117–19128.

[59] Y.-F. Chen, L. Kaur, and J. Singh, Chapter 7 – Chemical Modification of Starch, in: M. Sjöö and L. Nilsson (Eds.), *Starch in Food* (Second Edition), Woodhead Publishing, 2018, pp. 283–321.

[60] D. Zia-ud, H. Xiong, and P. Fei, "Physical and chemical modification of starches: A review", *Critical Reviews in Food Science and Nutrition* 57(12) (2017), pp. 2691–2705.

[61] N. Masina, Y. E. Choonara, P. Kumar, L. C. du Toit, M. Govender, S. Indermun, and V. Pillay, "A review of the chemical modification techniques of starch", *Carbohydrate Polymers* 157 (2017), pp. 1226–1236.

[62] Q. Chen, H. Yu, L. Wang, Z. ul Abdin, Y. Chen, J. Wang, W. Zhou, X. Yang, R. U. Khan, H. Zhang, and X. Chen, "Recent progress in chemical modification of starch and its applications", *RSC Advances* 5(83) (2015), pp. 67459–67474.

[63] C. Fringant, J. Desbrières, and M. Rinaudo, "Physical properties of acetylated starch-based materials: Relation with their molecular characteristics", *Polymer* 37(13) (1996), pp. 2663–2673.

[64] H. Chi, K. Xu, X. Wu, Q. Chen, D. Xue, C. Song, W. Zhang, and P. Wang, "Effect of acetylation on the properties of corn starch", *Food Chemistry* 106(3) (2008), pp. 923–928.

[65] O. V. López, N. E. Zaritzky, M. V. E. Grossmann, and M. A. García, "Acetylated and native corn starch blend films produced by blown extrusion", *Journal of Food Engineering* 116(2) (2013), pp. 286–297.

[66] Z. Fu, L. Zhang, M.-H. Ren, and J. N. BeMiller, "Developments in Hydroxypropylation of Starch: A Review", *Starch – Stärke* 71(1–2) (2019), 1800167.

[67] A. L. Chaudhary, M. Miler, P. J. Torley, P. A. Sopade, and P. J. Halley, "Amylose content and chemical modification effects on the extrusion of thermoplastic starch from maize", *Carbohydrate Polymers* 74(4) (2008), pp. 907–913.

[68] B. Lindman, G. Karlström, and L. Stigsson, "On the mechanism of dissolution of cellulose", *Journal of Molecular Liquids* 156(1) (2010), pp. 76–81.

[69] S. Wang, A. Lu, and L. Zhang, "Recent advances in regenerated cellulose materials", *Progress in Polymer Science* 53 (2016), pp. 169–206.

[70] B. Medronho and B. Lindman, "Brief overview on cellulose dissolution/regeneration interactions and mechanisms", *Advances in Colloid and Interface Science* 222 (2015), pp. 502–508.

[71] N. Mohd, S. F. S. Draman, M. S. N. Salleh, and N. B. Yusof, "Dissolution of cellulose in ionic liquid: A review", *AIP Conference Proceedings* 1809(1) (2017), 020035.

[72] X. Tan, L. Chen, X. Li, and F. Xie, "Effect of anti-solvents on the characteristics of regenerated cellulose from 1-ethyl-3-methylimidazolium acetate ionic liquid", *International Journal of Biological Macromolecules* 124 (2019), pp. 314–320.

[73] Y. Mao, J. Zhou, J. Cai, and L. Zhang, "Effects of coagulants on porous structure of membranes prepared from cellulose in NaOH/urea aqueous solution", *Journal of Membrane Science* 279(1) (2006), pp. 246–255.

[74] L. Zhang, Y. Mao, J. Zhou, and J. Cai, "Effects of Coagulation Conditions on the Properties of Regenerated Cellulose Films Prepared in NaOH/Urea Aqueous Solution", *Industrial & Engineering Chemistry Research* 44(3) (2005), pp. 522–529.

[75] R. Sescousse, K. A. Le, M. E. Ries, and T. Budtova, "Viscosity of Cellulose–Imidazolium-Based Ionic Liquid Solutions", *The Journal of Physical Chemistry B* 114(21) (2010), pp. 7222–7228.

[76] R. Quintana, O. Persenaire, L. Bonnaud, and P. Dubois, "Recent advances in (reactive) melt processing of cellulose acetate and related biodegradable bio-compositions", *Polymer Chemistry* 3(3) (2012), pp. 591–595.

[77] S. Tungprapa, T. Puangparn, M. Weerasombut, I. Jangchud, P. Fakum, S. Semongkhol, C. Meechaisue, and P. Supaphol, "Electrospun cellulose acetate fibers: effect of solvent system on morphology and fiber diameter", *Cellulose* 14(6) (2007), pp. 563–575.

[78] E. Doelker, Swelling Behavior of Water-Soluble Cellulose Derivatives, in: L. Brannon-Peppas and R. S. Harland (Eds.), *Studies in Polymer Science*, Elsevier, 1990, pp. 125–145.

[79] M. Rincón-Iglesias, E. Lizundia, and S. Lanceros-Méndez, "Water-Soluble Cellulose Derivatives as Suitable Matrices for Multifunctional Materials", *Biomacromolecules* 20(7) (2019), pp. 2786–2795.

[80] C. Clasen and W. M. Kulicke, "Determination of viscoelastic and rheo-optical material functions of water-soluble cellulose derivatives", *Progress in Polymer Science* 26(9) (2001), pp. 1839–1919.

[81] Y. Wang, J. Wang, Q. Sun, X. Xu, M. Li, and F. Xie, "Hydroxypropyl methylcellulose hydrocolloid systems: Effect of hydroxypropy group content on the phase structure, rheological properties and film characteristics", *Food Chemistry* 379 (2022), 132075.

[82] Y. Wang, L. Yu, Q. Sun, and F. Xie, "Hydroxypropyl methylcellulose and hydroxypropyl starch: Rheological and gelation effects on the phase structure of their mixed hydrocolloid system", *Food Hydrocolloids* 115 (2021), 106598.

[83] Y. Wang, L. Yu, F. Xie, S. Li, Q. Sun, H. Liu, and L. Chen, "On the investigation of thermal/cooling-gel biphasic systems based on hydroxypropyl methylcellulose and hydroxypropyl starch", *Industrial Crops and Products* 124 (2018), pp. 418–428.

[84] Y. Wang, L. Yu, F. Xie, L. Zhang, L. Liao, H. Liu, and L. Chen, "Morphology and properties of thermal/cooling-gel bi-phasic systems based on hydroxypropyl methylcellulose and hydroxypropyl starch", *Composites Part B: Engineering* 101 (2016), pp. 46–52.

[85] B. Pereira and V. Arantes, Chapter 9 – Nanocelluloses From Sugarcane Biomass, in: A. K. Chandel and M. H. Luciano Silveira (Eds.), *Advances in Sugarcane Biorefinery*, Elsevier, 2018, pp. 179–196.

[86] X. Xu, F. Liu, L. Jiang, J. Y. Zhu, D. Haagenson, and D. P. Wiesenborn, "Cellulose Nanocrystals vs. Cellulose Nanofibrils: A Comparative Study on Their Microstructures and Effects as Polymer Reinforcing Agents", *ACS Applied Materials & Interfaces* 5(8) (2013), pp. 2999–3009.

[87] L. Bastiaens, L. Soetemans, E. D'Hondt, and K. Elst, Sources of Chitin and Chitosan and their Isolation, in: L. A. M. van den Broek and C. G. Boeriu (Eds.), *Chitin and Chitosan*, Wiley, 2019, pp. 1–34.

[88] B. Duan, Y. Huang, A. Lu, and L. Zhang, "Recent advances in chitin based materials constructed via physical methods", *Progress in Polymer Science* 82 (2018), pp. 1–33.

[89] C. K. S. Pillai, W. Paul, and C. P. Sharma, "Chitin and chitosan polymers: Chemistry, solubility and fiber formation", *Progress in Polymer Science* 34(7) (2009), pp. 641–678.

[90] X. Yang, J. Liu, Y. Pei, X. Zheng, and K. Tang, "Recent Progress in Preparation and Application of Nano-Chitin Materials", *Energy & Environmental Materials* 3(4) (2020), pp. 492–515.

[91] J. L. Shamshina, P. Berton, and R. D. Rogers, "Advances in Functional Chitin Materials: A Review", *ACS Sustainable Chemistry & Engineering* 7(7) (2019), pp. 6444–6457.

[92] N. L. B. M. Yusof, L. Y. Lim, and E. Khor, "Flexible chitin films: structural studies", *Carbohydrate Research* 339(16) (2004), pp. 2701–2711.

[93] Y. Qin, X. Lu, N. Sun, and R. D. Rogers, "Dissolution or extraction of crustacean shells using ionic liquids to obtain high molecular weight purified chitin and direct production of chitin films and fibers", *Green Chemistry* 12(6) (2010), pp. 968–971.

[94] C. King, J. L. Shamshina, G. Gurau, P. Berton, N. F. A. F. Khan, and R. D. Rogers, "A platform for more sustainable chitin films from an ionic liquid process", *Green Chemistry* 19(1) (2017), pp. 117–126.

[95] A. Ali, Annu, and S. Ahmed, 17 – Physical and chemical modification of chitosan-based green materials, in: S. Ahmed (Ed.), *Advanced Green Materials*, Woodhead Publishing, 2021, pp. 379–397.

[96] P. Chen, F. Xie, F. Tang, and T. McNally, "Structure and properties of thermomechanically processed chitosan/carboxymethyl cellulose/graphene oxide polyelectrolyte complexed bionanocomposites", *International Journal of Biological Macromolecules* 158 (2020), pp. 420–429.

[97] P. Chen, F. Xie, F. Tang, and T. McNally, "Thermomechanical-induced polyelectrolyte complexation between chitosan and carboxymethyl cellulose enabling unexpected hydrolytic stability", *Composites Science and Technology* 189 (2020), 108031.

[98] Y. Chen, Q. Duan, L. Yu, and F. Xie, "Thermomechanically processed chitosan:gelatin films being transparent, mechanically robust and less hygroscopic", *Carbohydrate Polymers* 272 (2021), 118522.

[99] L. Meng, F. Xie, B. Zhang, D.K. Wang, and L. Yu, "Natural Biopolymer Alloys with Superior Mechanical Properties", *ACS Sustainable Chemistry & Engineering* 7(2) (2019), pp. 2792–2802.

[100] S. Basu, A. Plucinski, and J.M. Catchmark, "Sustainable barrier materials based on polysaccharide polyelectrolyte complexes", *Green Chemistry* 19(17) (2017), pp. 4080–4092.

[101] H.-C. Chiang, B. Eberle, D. Carlton, T.J. Kolibaba, and J.C. Grunlan, "Edible Polyelectrolyte Complex Nanocoating for Protection of Perishable Produce", *ACS Food Science & Technology* 1(4) (2021), pp. 495–499.

[102] K. Chi and J.M. Catchmark, "Improved eco-friendly barrier materials based on crystalline nanocellulose/chitosan/carboxymethyl cellulose polyelectrolyte complexes", *Food Hydrocolloids* 80 (2018), pp. 195–205.

[103] S. Zhang, F. Lu, L. Tao, N. Liu, C. Gao, L. Feng, and Y. Wei, "Bio-Inspired Anti-Oil-Fouling Chitosan-Coated Mesh for Oil/Water Separation Suitable for Broad pH Range and Hyper-Saline Environments", *ACS Applied Materials & Interfaces* 5(22) (2013), pp. 11971–11976.

[104] H. Sashiwa, N. Kawasaki, A. Nakayama, E. Muraki, N. Yamamoto, and S.-i. Aiba, "Chemical Modification of Chitosan. 14: Synthesis of Water-Soluble Chitosan Derivatives by Simple Acetylation", *Biomacromolecules* 3(5) (2002), pp. 1126–1128.

[105] Z. Shariatinia, "Carboxymethyl chitosan: Properties and biomedical applications", *International Journal of Biological Macromolecules* 120 (2018), pp. 1406–1419.

[106] N. Gorochovceva and R. Makuška, "Synthesis and study of water-soluble chitosan-O-poly(ethylene glycol) graft copolymers", *European Polymer Journal* 40(4) (2004), pp. 685–691.

[107] C. Qin, Q. Xiao, H. Li, M. Fang, Y. Liu, X. Chen, and Q. Li, "Calorimetric studies of the action of chitosan-*N*-2-hydroxypropyl trimethyl ammonium chloride on the growth of microorganisms", *International Journal of Biological Macromolecules* 34(1) (2004), pp. 121–126.

[108] W. Sajomsang, S. Tantayanon, V. Tangpasuthadol, and W.H. Daly, "Quaternization of *N*-aryl chitosan derivatives: synthesis, characterization, and antibacterial activity", *Carbohydrate Research* 344(18) (2009), pp. 2502–2511.

[109] V. Epure, M. Griffon, E. Pollet, and L. Avérous, "Structure and properties of glycerol-plasticized chitosan obtained by mechanical kneading", *Carbohydrate Polymers* 83(2) (2011), pp. 947–952.

[110] D.F. Xie, V.P. Martino, P. Sangwan, C. Way, G.A. Cash, E. Pollet, K.M. Dean, P.J. Halley, and L. Avérous, "Elaboration and properties of plasticised chitosan-based exfoliated nano-biocomposites", *Polymer* 54(14) (2013), pp. 3654–3662.

[111] P. Chen, F. Xie, F. Tang, and T. McNally, "Unexpected Plasticization Effects on the Structure and Properties of Polyelectrolyte Complexed Chitosan/Alginate Materials", *ACS Applied Polymer Materials* 2(7) (2020), pp. 2957–2966.

[112] P. Chen, F. Xie, F. Tang, and T. McNally, "Glycerol plasticisation of chitosan/carboxymethyl cellulose composites: Role of interactions in determining structure and properties", *International Journal of Biological Macromolecules* 163 (2020), pp. 683–693.

[113] M. Matet, M.-C. Heuzey, E. Pollet, A. Ajji, and L. Avérous, "Innovative thermoplastic chitosan obtained by thermo-mechanical mixing with polyol plasticizers", *Carbohydrate Polymers* 95(1) (2013), pp. 241–251.

[114] Q. Duan, Y. Chen, L. Yu, and F. Xie, "Chitosan–Gelatin Films: Plasticizers/Nanofillers Affect Chain Interactions and Material Properties in Different Ways", *Polymers* 14(18) (2022), 3797.

[115] M. Chen, T. Runge, L. Wang, R. Li, J. Feng, X.-L. Shu, and Q.-S. Shi, "Hydrogen bonding impact on chitosan plasticization", *Carbohydrate Polymers* 200 (2018), pp. 115–121.

[116] X. Ma, C. Qiao, X. Wang, J. Yao, and J. Xu, "Structural characterization and properties of polyols plasticized chitosan films", *International Journal of Biological Macromolecules* 135 (2019), pp. 240–245.

[117] K. Y. Lee and D. J. Mooney, "Alginate: Properties and biomedical applications", *Progress in Polymer Science* 37(1) (2012), pp. 106–126.

[118] M. J. Costa, A. M. Marques, L. M. Pastrana, J. A. Teixeira, S. M. Sillankorva, and M. A. Cerqueira, "Physicochemical properties of alginate-based films: Effect of ionic crosslinking and mannuronic and guluronic acid ratio", *Food Hydrocolloids* 81 (2018), pp. 442–448.

[119] C. Hu, W. Lu, A. Mata, K. Nishinari, and Y. Fang, "Ions-induced gelation of alginate: Mechanisms and applications", *International Journal of Biological Macromolecules* 177 (2021), pp. 578–588.

[120] Q. Zhou, H. Kang, M. Bielec, X. Wu, Q. Cheng, W. Wei, and H. Dai, "Influence of different divalent ions cross-linking sodium alginate-polyacrylamide hydrogels on antibacterial properties and wound healing", *Carbohydrate Polymers* 197 (2018), pp. 292–304.

[121] X. Zhang, K. Wang, J. Hu, Y. Zhang, Y. Dai, and F. Xia, "Role of a high calcium ion content in extending the properties of alginate dual-crosslinked hydrogels", *Journal of Materials Chemistry A* 8(47) (2020), pp. 25390–25401.

[122] J. A. Sirviö, A. Kolehmainen, H. Liimatainen, J. Niinimäki, and O. E. O. Hormi, "Biocomposite cellulose-alginate films: Promising packaging materials", *Food Chemistry* 151 (2014), pp. 343–351.

[123] J.-W. Rhim, "Physical and mechanical properties of water resistant sodium alginate films", *LWT – Food Science and Technology* 37(3) (2004), pp. 323–330.

[124] S. Benavides, R. Villalobos-Carvajal, and J. E. Reyes, "Physical, mechanical and antibacterial properties of alginate film: Effect of the crosslinking degree and oregano essential oil concentration", *Journal of Food Engineering* 110(2) (2012), pp. 232–239.

[125] G. Liling, Z. Di, X. Jiachao, G. Xin, F. Xiaoting, and Z. Qing, "Effects of ionic crosslinking on physical and mechanical properties of alginate mulching films", *Carbohydrate Polymers* 136 (2016), pp. 259–265.

[126] S. G. Patrício, L. R. Sousa, T. R. Correia, V. M. Gaspar, L. S. Pires, J. L. Luís, J. M. Oliveira, and J. F. Mano, "Freeform 3D printing using a continuous viscoelastic supporting matrix", *Biofabrication* 12(3) (2020), 035017.

[127] C. C. Piras and D. K. Smith, "Multicomponent polysaccharide alginate-based bioinks", *Journal of Materials Chemistry B* 8(36) (2020), pp. 8171–8188.

[128] M. S. Nair, M. Tomar, S. Punia, W. Kukula-Koch, and M. Kumar, "Enhancing the functionality of chitosan- and alginate-based active edible coatings/films for the preservation of fruits and vegetables: A review", *International Journal of Biological Macromolecules* 164 (2020), pp. 304–320.

[129] K. Varaprasad, T. Jayaramudu, V. Kanikireddy, C. Toro, and E. R. Sadiku, "Alginate-based composite materials for wound dressing application: A mini review", *Carbohydrate Polymers* 236 (2020), 116025.

[130] B. Wang, Y. Wan, Y. Zheng, X. Lee, T. Liu, Z. Yu, J. Huang, Y. S. Ok, J. Chen, and B. Gao, "Alginate-based composites for environmental applications: a critical review", *Critical Reviews in Environmental Science and Technology* 49(4) (2019), pp. 318–356.

[131] V. Jost, K. Kobsik, M. Schmid, and K. Noller, "Influence of plasticiser on the barrier, mechanical and grease resistance properties of alginate cast films", *Carbohydrate Polymers* 110 (2014), pp. 309–319.

[132] M. Lacroix and C. Le Tien, 20 – Edible films and coatings from nonstarch polysaccharides, in: J. H. Han (Ed.), *Innovations in Food Packaging*, Academic Press, London, 2005, pp. 338–361.

[133] Y. Pranoto, V. M. Salokhe, and S. K. Rakshit, "Physical and antibacterial properties of alginate-based edible film incorporated with garlic oil", *Food Research International* 38(3) (2005), pp. 267–272.

[134] T. Huq, S. Salmieri, A. Khan, R. A. Khan, C. Le Tien, B. Riedl, C. Fraschini, J. Bouchard, J. Uribe-Calderon, M. R. Kamal, and M. Lacroix, "Nanocrystalline cellulose (NCC) reinforced alginate based biodegradable nanocomposite film", *Carbohydrate Polymers* 90(4) (2012), pp. 1757–1763.

[135] M. Abdollahi, M. Alboofetileh, R. Behrooz, M. Rezaei, and R. Miraki, "Reducing water sensitivity of alginate bio-nanocomposite film using cellulose nanoparticles", *International Journal of Biological Macromolecules* 54 (2013), pp. 166–173.

[136] C. Gao, E. Pollet, and L. Avérous, "Properties of glycerol-plasticized alginate films obtained by thermo-mechanical mixing", *Food Hydrocolloids* 63 (2017), pp. 414–420.

[137] C. Gao, E. Pollet, and L. Avérous, "Innovative plasticized alginate obtained by thermo-mechanical mixing: Effect of different biobased polyols systems", *Carbohydrate Polymers* 157 (2017), pp. 669–676.

[138] X.Z. Tang, P. Kumar, S. Alavi, and K.P. Sandeep, "Recent Advances in Biopolymers and Biopolymer-Based Nanocomposites for Food Packaging Materials", *Critical Reviews in Food Science and Nutrition* 52(5) (2012), pp. 426–442.

[139] C. Mellinas, M. Ramos, A. Jiménez, and M.C. Garrigós, "Recent Trends in the Use of Pectin from Agro-Waste Residues as a Natural-Based Biopolymer for Food Packaging Applications", *Materials* 13(3) (2020), 673.

[140] J. Chen, W. Liu, C.-M. Liu, T. Li, R.-H. Liang, and S.-J. Luo, "Pectin Modifications: A Review", *Critical Reviews in Food Science and Nutrition* 55(12) (2015), pp. 1684–1698.

[141] V. Vancauwenberghe, L. Katalagarianakis, Z. Wang, M. Meerts, M. Hertog, P. Verboven, P. Moldenaers, M.E. Hendrickx, J. Lammertyn, and B. Nicolaï, "Pectin based food-ink formulations for 3-D printing of customizable porous food simulants", *Innovative Food Science & Emerging Technologies* 42 (2017), pp. 138–150.

[142] V. Vancauwenberghe, P. Verboven, J. Lammertyn, and B. Nicolaï, "Development of a coaxial extrusion deposition for 3D printing of customizable pectin-based food simulant", *Journal of Food Engineering* 225 (2018), pp. 42–52.

[143] P.J.P. Espitia, W.-X. Du, R.d.J. Avena-Bustillos, N.d.F.F. Soares, and T.H. McHugh, "Edible films from pectin: Physical-mechanical and antimicrobial properties – A review", *Food Hydrocolloids* 35 (2014), pp. 287–296.

[144] H.M.C. Azeredo, R. Morrugares-Carmona, N. Wellner, K. Cross, B. Bajka, and K.W. Waldron, "Development of pectin films with pomegranate juice and citric acid", *Food Chemistry* 198 (2016), pp. 101–106.

[145] A.E. Pavlath, A. Voisin, and G.H. Robertson, "Pectin-based biodegradable water insoluble films", *Macromolecular Symposia* 140(1) (1999), pp. 107–113.

[146] M.V. Lorevice, C.G. Otoni, M.R.d. Moura, and L.H.C. Mattoso, "Chitosan nanoparticles on the improvement of thermal, barrier, and mechanical properties of high- and low-methyl pectin films", *Food Hydrocolloids* 52 (2016), pp. 732–740.

[147] H.G.R. Younis, H.R.S. Abdellatif, F. Ye, and G. Zhao, "Tuning the physicochemical properties of apple pectin films by incorporating chitosan/pectin fiber", *International Journal of Biological Macromolecules* 159 (2020), pp. 213–221.

[148] T. Mekonnen, P. Mussone, H. Khalil, and D. Bressler, "Progress in bio-based plastics and plasticizing modifications", *Journal of Materials Chemistry A* 1(43) (2013), pp. 13379–13398.

[149] M. Zdanowicz, K. Wilpiszewska, and T. Spychaj, "Deep eutectic solvents for polysaccharides processing. A review", *Carbohydrate Polymers* 200 (2018), pp. 361–380.

[150] F. Ren, J. Wang, F. Xie, K. Zan, S. Wang, and S. Wang, "Applications of ionic liquids in starch chemistry: a review", *Green Chemistry* 22(7) (2020), pp. 2162–2183.

[151] E.J. Bealer, S. Onissema-Karimu, A. Rivera-Galletti, M. Francis, J. Wilkowski, D. Salas-de la Cruz, and X. Hu, "Protein–Polysaccharide Composite Materials: Fabrication and Applications", *Polymers* 12(2) (2020), 464.

[152] F. Seidi, M. Khodadadi Yazdi, M. Jouyandeh, M. Dominic, H. Naeim, M.N. Nezhad, B. Bagheri, S. Habibzadeh, P. Zarrintaj, M.R. Saeb, and M. Mozafari, "Chitosan-based blends for biomedical applications", *International Journal of Biological Macromolecules* 183 (2021), pp. 1818–1850.

[153] I. Jebalia, M. Kristiawan, M.N. Charalambides, S. Humphry-Baker, G.D. Valle, and S. Guessasma, "Microstructure and local mechanical properties of pea starch/protein composites", *Composites Part C: Open Access* 8 (2022), 100272.

[154] I. Jebalia, J.-E. Maigret, A.-L. Réguerre, B. Novales, S. Guessasma, D. Lourdin, G. Della Valle, and M. Kristiawan, "Morphology and mechanical behaviour of pea-based starch-protein composites obtained by extrusion", *Carbohydrate Polymers* 223 (2019), 115086.

[155] D. Chen, F. Fang, E. Federici, O. Campanella, and O.G. Jones, "Rheology, microstructure and phase behavior of potato starch-protein fibril mixed gel", *Carbohydrate Polymers* 239 (2020), 116247.

[156] F. Xie, B. Zhang, and D.K. Wang, Chapter 7 – Starch Thermal Processing: Technologies at Laboratory and Semi-Industrial Scales, in: M.A. Villar, S.E. Barbosa, M.A. García, L.A. Castillo, and O.V. López (Eds.), *Starch-Based Materials in Food Packaging,* Academic Press, 2017, pp. 187–227.

[157] M.A. Huneault and H. Li, "Morphology and properties of compatibilized polylactide/thermoplastic starch blends", *Polymer* 48(1) (2007), pp. 270–280.

[158] K. Martinez Villadiego, M.J. Arias Tapia, J. Useche, and D. Escobar Macías, "Thermoplastic Starch (TPS)/Polylactic Acid (PLA) Blending Methodologies: A Review", *Journal of Polymers and the Environment* 30(1) (2022), pp. 75–91.

[159] F. Xie, L. Yu, H. Liu, and K. Dean, "Effect of compatibilizer distribution on thermal and rheological properties of gelatinized starch/biodegradable polyesters blends", *Int. Polym. Process.* 21(4) (2006), pp. 379–385.

[160] L. Yu, K. Dean, Q. Yuan, L. Chen, and X. Zhang, "Effect of compatibilizer distribution on the blends of starch/biodegradable polyesters", *Journal of Applied Polymer Science* 103(2) (2007), pp. 812–818.

[161] L. Yu, E. Petinakis, K. Dean, H. Liu, and Q. Yuan, "Enhancing compatibilizer function by controlled distribution in hydrophobic polylactic acid/hydrophilic starch blends", *Journal of Applied Polymer Science* 119(4) (2011), pp. 2189–2195.

[162] E.A. Kamoun, S.A. Loutfy, Y. Hussein, and E.-R.S. Kenawy, "Recent advances in PVA-polysaccharide based hydrogels and electrospun nanofibers in biomedical applications: A review", *International Journal of Biological Macromolecules* 187 (2021), pp. 755–768.

[163] X. Yao, J. Liu, H. Hu, D. Yun, and J. Liu, "Development and comparison of different polysaccharide/PVA-based active/intelligent packaging films containing red pitaya betacyanins", *Food Hydrocolloids* 124 (2022), 107305.

[164] M.V. Cruz, A.C. Jacobowski, M.L.R. Macedo, K.A. Batista, and K.F. Fernandes, "Immobilization of antimicrobial trypsin inhibitors onto cashew gum polysaccharide/PVA films", *International Journal of Biological Macromolecules* 127 (2019), pp. 433–439.

[165] B.R. Moreira, M.A. Pereira-Júnior, K.F. Fernandes, and K.A. Batista, "An ecofriendly edible coating using cashew gum polysaccharide and polyvinyl alcohol", *Food Bioscience* 37 (2020), 100722.

[166] Q. Zou, X. Tian, S. Luo, D. Yuan, S. Xu, L. Yang, M. Ma, and C. Ye, "Agarose composite hydrogel and PVA sacrificial materials for bioprinting large-scale, personalized face-like with nutrient networks", *Carbohydrate Polymers* 269 (2021), 118222.

[167] Y. Zheng, J. Monty, and R.J. Linhardt, "Polysaccharide-based nanocomposites and their applications", *Carbohydrate Research* 405 (2015), pp. 23–32.

[168] C. Rodríguez-González, A.L. Martínez-Hernández, V.M. Castaño, O.V. Kharissova, R.S. Ruoff, and C. Velasco-Santos, "Polysaccharide Nanocomposites Reinforced with Graphene Oxide and Keratin-Grafted Graphene Oxide", *Industrial & Engineering Chemistry Research* 51(9) (2012), pp. 3619–3629.

[169] A. Travan, E. Marsich, I. Donati, M. Benincasa, M. Giazzon, L. Felisari, and S. Paoletti, "Silver–polysaccharide nanocomposite antimicrobial coatings for methacrylic thermosets", *Acta Biomaterialia* 7(1) (2011), pp. 337–346.

[170] A. Travan, C. Pelillo, I. Donati, E. Marsich, M. Benincasa, T. Scarpa, S. Semeraro, G. Turco, R. Gennaro, and S. Paoletti, "Non-cytotoxic Silver Nanoparticle-Polysaccharide Nanocomposites with Antimicrobial Activity", *Biomacromolecules* 10(6) (2009), pp. 1429–1435.

[171] A.J. Svagan, M.A.S. Azizi Samir, and L.A. Berglund, "Biomimetic Polysaccharide Nanocomposites of High Cellulose Content and High Toughness", *Biomacromolecules* 8(8) (2007), pp. 2556–2563.

[172] R. Xiong, A.M. Grant, R. Ma, S. Zhang, and V.V. Tsukruk, "Naturally-derived biopolymer nanocomposites: Interfacial design, properties and emerging applications", *Materials Science and Engineering: R: Reports* 125 (2018), pp. 1–41.

[173] A. Ghanbari, T. Tabarsa, A. Ashori, A. Shakeri, and M. Mashkour, "Preparation and characterization of thermoplastic starch and cellulose nanofibers as green nanocomposites: Extrusion processing", *International Journal of Biological Macromolecules* 112 (2018), pp. 442–447.

[174] K. Dean, L. Yu, and D.Y. Wu, "Preparation and characterization of melt-extruded thermoplastic starch/clay nanocomposites", *Composites Science and Technology* 67(3) (2007), pp. 413–421.

[175] Y. Fourati, A. Magnin, J.-L. Putaux, and S. Boufi, "One-step processing of plasticized starch/cellulose nanofibrils nanocomposites via twin-screw extrusion of starch and cellulose fibers", *Carbohydrate Polymers* 229 (2020), 115554.

[176] V. Nessi, X. Falourd, J.-E. Maigret, K. Cahier, A. D'Orlando, N. Descamps, V. Gaucher, C. Chevigny, and D. Lourdin, "Cellulose nanocrystals-starch nanocomposites produced by extrusion: Structure and behavior in physiological conditions", *Carbohydrate Polymers* 225 (2019), 115123.

[177] L.Z. Wang, L. Liu, J. Holmes, J.F. Kerry, and J.P. Kerry, "Assessment of film-forming potential and properties of protein and polysaccharide-based biopolymer films", *International Journal of Food Science & Technology* 42(9) (2007), pp. 1128–1138.

[178] P. Cazón, G. Velazquez, J.A. Ramírez, and M. Vázquez, "Polysaccharide-based films and coatings for food packaging: A review", *Food Hydrocolloids* 68 (2017), pp. 136–148.

[179] M. Criado, E. Rebollar, A. Nogales, T.A. Ezquerra, F. Boulmedais, C. Mijangos, and R. Hernández, "Quantitative Nanomechanical Properties of Multilayer Films Made of Polysaccharides through Spray Assisted Layer-by-Layer Assembly", *Biomacromolecules* 18(1) (2017), pp. 169–177.

[180] J.G. Fernandez and D.E. Ingber, "Unexpected strength and toughness in chitosan-fibroin laminates inspired by insect cuticle", *Adv. Mater.* 24(4) (2012), pp. 480–484.

[181] G.M. Nogueira, A.J. Swiston, M.M. Beppu, and M.F. Rubner, "Layer-by-Layer Deposited Chitosan/Silk Fibroin Thin Films with Anisotropic Nanofiber Alignment", *Langmuir* 26(11) (2010), pp. 8953–8958.

[182] T. Mohan, R. Kargl, K.E. Tradt, M.R. Kulterer, M. Braćić, S. Hribernik, K. Stana-Kleinschek, and V. Ribitsch, "Antifouling coating of cellulose acetate thin films with polysaccharide multilayers", *Carbohydr. Polym.* 116 (2015), pp. 149–158.

[183] B. Lin, A.C.Y. Yuen, A. Li, Y. Zhang, T.B.Y. Chen, B. Yu, E.W.M. Lee, S. Peng, W. Yang, H.-D. Lu, Q.N. Chan, G.H. Yeoh, and C.H. Wang, "MXene/chitosan nanocoating for flexible polyurethane foam towards remarkable fire hazards reductions", *J. Hazard. Mater.* 381 (2020), 120952.

[184] S. Hroncekova, T. Bertok, M. Hires, E. Jane, L. Lorencova, A. Vikartovska, A. Tanvir, P. Kasak, and J. Tkac, "Ultrasensitive $Ti_3C_2T_X$ MXene/Chitosan Nanocomposite-Based Amperometric Biosensor for Detection of Potential Prostate Cancer Marker in Urine Samples", *Processes* 8(5) (2020), 580.

[185] X. Li, Y. Lu, Z. Shi, G. Liu, G. Xu, Z. An, H. Xing, Q. Chen, R.P.S. Han, and Q. Liu, "Onion-inspired MXene/chitosan-quercetin multilayers: Enhanced response to H_2O molecules for wearable human physiological monitoring", *Sens. Actuators, B* 329 (2021), 129209.

[186] P.K. Kalambate, Dhanjai, A. Sinha, Y. Li, Y. Shen, and Y. Huang, "An electrochemical sensor for ifosfamide, acetaminophen, domperidone, and sumatriptan based on self-assembled MXene/MWCNT/chitosan nanocomposite thin film", *Microchim. Acta* 187(7) (2020), 402.

[187] X. Zhang, D. Liu, T.Z. Jin, W. Chen, Q. He, Z. Zou, H. Zhao, X. Ye, and M. Guo, "Preparation and characterization of gellan gum-chitosan polyelectrolyte complex films with the incorporation of thyme essential oil nanoemulsion", *Food Hydrocolloids* 114 (2021), 106570.

[188] F. Garavand, M. Rouhi, S.H. Razavi, I. Cacciotti, and R. Mohammadi, "Improving the integrity of natural biopolymer films used in food packaging by crosslinking approach: A review", *International Journal of Biological Macromolecules* 104 (2017), pp. 687–707.

[189] N. Reddy, R. Reddy, and Q. Jiang, "Crosslinking biopolymers for biomedical applications", *Trends in Biotechnology* 33(6) (2015), pp. 362–369.

[190] A. El Kadib and M. Bousmina, "Chitosan bio-based organic–inorganic hybrid aerogel microspheres", *Chem. Eur. J.* 18(27) (2012), pp. 8264–8277.

[191] A.C. Mendes, T. Strohmenger, F. Goycoolea, and I.S. Chronakis, "Electrostatic self-assembly of polysaccharides into nanofibers", *Colloids and Surfaces A: Physicochemical and Engineering Aspects* 531 (2017), pp. 182–188.

[192] G. Moad, "Chemical modification of starch by reactive extrusion", *Progress in Polymer Science* 36(2) (2011), pp. 218–237.

[193] F. Xie, L. Yu, H. Liu, and L. Chen, "Starch Modification Using Reactive Extrusion", *Starch - Stärke* 58(3–4) (2006), pp. 131–139.

[194] V.D. Miladinov and M.A. Hanna, "Starch esterification by reactive extrusion", *Industrial Crops and Products* 11(1) (2000), pp. 51–57.

[195] S. Kalambur and S.S.H. Rizvi, "An Overview of Starch-Based Plastic Blends from Reactive Extrusion", *Journal of Plastic Film & Sheeting* 22(1) (2006), pp. 39–58.

[196] E. Hablot, S. Dewasthale, Y. Zhao, Y. Zhiguan, X. Shi, D. Graiver, and R. Narayan, "Reactive extrusion of glycerylated starch and starch–polyester graft copolymers", *European Polymer Journal* 49(4) (2013), pp. 873–881.

[197] X. Xiao, L. Yu, F. Xie, X. Bao, H. Liu, Z. Ji, and L. Chen, "One-step method to prepare starch-based superabsorbent polymer for slow release of fertilizer", *Chemical Engineering Journal* 309 (2017), pp. 607–616.

[198] N. Li, D. Qiao, S. Zhao, Q. Lin, B. Zhang, and F. Xie, "3D printing to innovate biopolymer materials for demanding applications: A review", *Materials Today Chemistry* 20 (2021), 100459.

[199] B. Zheng, Y. Tang, F. Xie, and L. Chen, "Effect of pre-printing gelatinization degree on the structure and digestibility of hot-extrusion 3D-printed starch", *Food Hydrocolloids* 124 (2022), 107210.

[200] H. Chen, F. Xie, L. Chen, and B. Zheng, "Effect of rheological properties of potato, rice and corn starches on their hot-extrusion 3D printing behaviors", *J. Food Eng.* 244 (2019), pp. 150–158.

[201] Z. Liu, H. Chen, B. Zheng, F. Xie, and L. Chen, "Understanding the structure and rheological properties of potato starch induced by hot-extrusion 3D printing", *Food Hydrocolloids* 105 (2020), 105812.

[202] Z. Liu, M. Zhang, B. Bhandari, and Y. Wang, "3D printing: Printing precision and application in food sector", *Trends Food Sci. Technol.* 69 (2017), pp. 83–94.

[203] I. Dankar, A. Haddarah, F.E.L. Omar, F. Sepulcre, and M. Pujolà, "3D printing technology: The new era for food customization and elaboration", *Trends Food Sci. Technol.* 75 (2018), pp. 231–242.

[204] A. Le-Bail, B. Chieregato Maniglia, and P. Le-Bail, "3D printing of foods: recent developments, future perspectives and challenges", *Curr. Opin. Food Sci.* (2020).

[205] R.C. Pack and B.G. Compton, "Material Extrusion Additive Manufacturing of Metal Powder-Based Inks Enabled by Carrageenan Rheology Modifier", *Advanced Engineering Materials* 23(2) (2021), 2000880.

[206] D. Kokkinis, M. Schaffner, and A.R. Studart, "Multimaterial magnetically assisted 3D printing of composite materials", *Nat. Commun.* 6(1) (2015), 8643.

[207] M. H. Kim, Y. W. Lee, W.-K. Jung, J. Oh, and S. Y. Nam, "Enhanced rheological behaviors of alginate hydrogels with carrageenan for extrusion-based bioprinting", *J. Mech. Behav. Biomed. Mater.* 98 (2019), pp. 187–194.

[208] W. Zhou, Z. Wu, F. Xie, S. Tang, J. Fang, and X. Wang, "3D printed nanocellulose-based label for fruit freshness keeping and visual monitoring", *Carbohydrate Polymers* 273 (2021), 118545.

[209] P. Cao, L. Tao, J. Gong, T. Wang, Q. Wang, J. Ju, and Y. Zhang, "4D Printing of a Sodium Alginate Hydrogel with Step-Wise Shape Deformation Based on Variation of Crosslinking Density", *ACS Applied Polymer Materials* 3(12) (2021), pp. 6167–6175.

[210] M. Samandari, F. Alipanah, K. Majidzadeh-A, M. M. Alvarez, G. Trujillo-de Santiago, and A. Tamayol, "Controlling cellular organization in bioprinting through designed 3D microcompartmentalization", *Applied Physics Reviews* 8(2) (2021), 021404.

[211] M. Alizadehgiashi, C. R. Nemr, M. Chekini, D. Pinto Ramos, N. Mittal, S. U. Ahmed, N. Khuu, S. O. Kelley, and E. Kumacheva, "Multifunctional 3D-Printed Wound Dressings", *ACS Nano* 15(7) (2021), pp. 12375–12387.

[212] A. R. Osi, H. Zhang, J. Chen, Y. Zhou, R. Wang, J. Fu, P. Müller-Buschbaum, and Q. Zhong, "Three-Dimensional-Printable Thermo/Photo-Cross-Linked Methacrylated Chitosan–Gelatin Hydrogel Composites for Tissue Engineering", *ACS Appl. Mater. Interfaces* 13(19) (2021), pp. 22902–22913.

[213] L. Ouyang, C. B. Highley, C. B. Rodell, W. Sun, and J. A. Burdick, "3D Printing of Shear-Thinning Hyaluronic Acid Hydrogels with Secondary Cross-Linking", *ACS Biomater. Sci. Eng.* 2(10) (2016), pp. 1743–1751.

[214] A. C. Heidenreich, M. Pérez-Recalde, A. González Wusener, and É. B. Hermida, "Collagen and chitosan blends for 3D bioprinting: A rheological and printability approach", *Polym. Test.* 82 (2020), 106297.

[215] C. Xu, W. Chai, Y. Huang, and R. R. Markwald, "Scaffold-free inkjet printing of three-dimensional zigzag cellular tubes", *Biotechnol. Bioeng.* 109(12) (2012), pp. 3152–3160.

[216] T. Xu, W. Zhao, J.-M. Zhu, M. Z. Albanna, J. J. Yoo, and A. Atala, "Complex heterogeneous tissue constructs containing multiple cell types prepared by inkjet printing technology", *Biomaterials* 34(1) (2013), pp. 130–139.

[217] M. W. Frey, "Electrospinning Cellulose and Cellulose Derivatives", *Polymer Reviews* 48(2) (2008), pp. 378–391.

[218] M. A. Wsoo, S. Shahir, S. P. Mohd Bohari, N. H. M. Nayan, and S. I. A. Razak, "A review on the properties of electrospun cellulose acetate and its application in drug delivery systems: A new perspective", *Carbohydrate Research* 491 (2020), 107978.

[219] Q. Li, Y. Yin, D. Cao, Y. Wang, P. Luan, X. Sun, W. Liang, and H. Zhu, "Photocatalytic Rejuvenation Enabled Self-Sanitizing, Reusable, and Biodegradable Masks against COVID-19", *ACS Nano* 15(7) (2021), pp. 11992–12005.

[220] K. Ohkawa, D. Cha, H. Kim, A. Nishida, and H. Yamamoto, "Electrospinning of Chitosan", *Macromolecular Rapid Communications* 25(18) (2004), pp. 1600–1605.

[221] P. Sangsanoh and P. Supaphol, "Stability Improvement of Electrospun Chitosan Nanofibrous Membranes in Neutral or Weak Basic Aqueous Solutions", *Biomacromolecules* 7(10) (2006), pp. 2710–2714.

[222] X. Geng, O.-H. Kwon, and J. Jang, "Electrospinning of chitosan dissolved in concentrated acetic acid solution", *Biomaterials* 26(27) (2005), pp. 5427–5432.

[223] H. Homayoni, S. A. H. Ravandi, and M. Valizadeh, "Electrospinning of chitosan nanofibers: Processing optimization", *Carbohydrate Polymers* 77(3) (2009), pp. 656–661.

[224] B.-M. Min, S. W. Lee, J. N. Lim, Y. You, T. S. Lee, P. H. Kang, and W. H. Park, "Chitin and chitosan nanofibers: electrospinning of chitin and deacetylation of chitin nanofibers", *Polymer* 45(21) (2004), pp. 7137–7142.

[225] H.K. Noh, S.W. Lee, J.-M. Kim, J.-E. Oh, K.-H. Kim, C.-P. Chung, S.-C. Choi, W.H. Park, and B.-M. Min, "Electrospinning of chitin nanofibers: Degradation behavior and cellular response to normal human keratinocytes and fibroblasts", *Biomaterials* 27(21) (2006), pp. 3934–3944.

[226] P.S. Barber, C.S. Griggs, J.R. Bonner, and R.D. Rogers, "Electrospinning of chitin nanofibers directly from an ionic liquid extract of shrimp shells", *Green Chemistry* 15(3) (2013), pp. 601–607.

[227] J.L. Shamshina, O. Zavgorodnya, H. Choudhary, B. Frye, N. Newbury, and R.D. Rogers, "In Search of Stronger/Cheaper Chitin Nanofibers through Electrospinning of Chitin-Cellulose Composites Using an Ionic Liquid Platform", *ACS Sustainable Chemistry & Engineering* 6(11) (2018), pp. 14713–14722.

[228] T. Zhong, W. Liu, and H. Liu, "Green electrospinning of chitin propionate to manufacture nanofiber mats", *Carbohydrate Polymers* 273 (2021), 118593.

[229] Y.-T. Jia, J. Gong, X.-H. Gu, H.-Y. Kim, J. Dong, and X.-Y. Shen, "Fabrication and characterization of poly (vinyl alcohol)/chitosan blend nanofibers produced by electrospinning method", *Carbohydrate Polymers* 67(3) (2007), pp. 403–409.

[230] Y. Zhang, X. Huang, B. Duan, L. Wu, S. Li, and X. Yuan, "Preparation of electrospun chitosan/poly(vinyl alcohol) membranes", *Colloid and Polymer Science* 285(8) (2007), pp. 855–863.

[231] K.T. Shalumon, N.S. Binulal, N. Selvamurugan, S.V. Nair, D. Menon, T. Furuike, H. Tamura, and R. Jayakumar, "Electrospinning of carboxymethyl chitin/poly(vinyl alcohol) nanofibrous scaffolds for tissue engineering applications", *Carbohydrate Polymers* 77(4) (2009), pp. 863–869.

[232] J. Junkasem, R. Rujiravanit, and P. Supaphol, "Fabrication of α-chitin whisker-reinforced poly(vinyl alcohol) nanocomposite nanofibres by electrospinning", *Nanotechnology* 17(17) (2006), pp. 4519–4528.

[233] S. Haider, W.A. Al-Masry, N. Bukhari, and M. Javid, "Preparation of the chitosan containing nanofibers by electrospinning chitosan–gelatin complexes", *Polymer Engineering & Science* 50(9) (2010), pp. 1887–1893.

[234] Y. Ji, K. Liang, X. Shen, and G.L. Bowlin, "Electrospinning and characterization of chitin nanofibril/polycaprolactone nanocomposite fiber mats", *Carbohydrate Polymers* 101 (2014), pp. 68–74.

[235] G. Liu, Z. Gu, Y. Hong, L. Cheng, and C. Li, "Electrospun starch nanofibers: Recent advances, challenges, and strategies for potential pharmaceutical applications", *Journal of Controlled Release* 252 (2017), pp. 95–107.

[236] T. Hemamalini and V.R. Giri Dev, "Comprehensive review on electrospinning of starch polymer for biomedical applications", *International Journal of Biological Macromolecules* 106 (2018), pp. 712–718.

[237] L. Kong and G.R. Ziegler, "Fabrication of pure starch fibers by electrospinning", *Food Hydrocolloids* 36 (2014), pp. 20–25.

[238] L. Kong and G.R. Ziegler, "Role of Molecular Entanglements in Starch Fiber Formation by Electrospinning", *Biomacromolecules* 13(8) (2012), pp. 2247–2253.

[239] L. Kong and G.R. Ziegler, "Quantitative relationship between electrospinning parameters and starch fiber diameter", *Carbohydrate Polymers* 92(2) (2013), pp. 1416–1422.

[240] L. Kong and G.R. Ziegler, "Formation of starch-guest inclusion complexes in electrospun starch fibers", *Food Hydrocolloids* 38 (2014), pp. 211–219.

[241] H. Wang, L. Kong, and G.R. Ziegler, "Aligned wet-electrospun starch fiber mats", *Food Hydrocolloids* 90 (2019), pp. 113–117.

[242] W. Wang, H. Wang, X. Jin, H. Wang, T. Lin, and Z. Zhu, "Effects of hydrogen bonding on starch granule dissolution, spinnability of starch solution, and properties of electrospun starch fibers", *Polymer* 153 (2018), pp. 643–652.

[243] W. Wang, X. Jin, Y. Zhu, C. Zhu, J. Yang, H. Wang, and T. Lin, "Effect of vapor-phase glutaraldehyde crosslinking on electrospun starch fibers", *Carbohydrate Polymers* 140 (2016), pp. 356–361.

[244] D. McCune, X. Guo, T. Shi, S. Stealey, R. Antrobus, M. Kaltchev, J. Chen, S. Kumpaty, X. Hua, W. Ren, and W. Zhang, "Electrospinning pectin-based nanofibers: a parametric and cross-linker study", *Applied Nanoscience* 8(1) (2018), pp. 33–40.

[245] T. C. Mokhena and A. S. Luyt, "Electrospun alginate nanofibres impregnated with silver nanoparticles: Preparation, morphology and antibacterial properties", *Carbohydrate Polymers* 165 (2017), pp. 304–312.

[246] M. A. Taemeh, A. Shiravandi, M. A. Korayem, and H. Daemi, "Fabrication challenges and trends in biomedical applications of alginate electrospun nanofibers", *Carbohydrate Polymers* 228 (2020), 115419.

[247] C. A. Bonino, K. Efimenko, S. I. Jeong, M. D. Krebs, E. Alsberg, and S. A. Khan, "Three-Dimensional Electrospun Alginate Nanofiber Mats via Tailored Charge Repulsions", *Small* 8(12) (2012), pp. 1928–1936.

[248] Y. Xiao, F. Xie, H. Luo, R. Tang, and J. Hou, "Electrospinning SA@PVDF-HFP Core-Shell Nanofibers Based on a Visual Light Transmission Response to Alcohol for Intelligent Packaging", *ACS Applied Materials & Interfaces* 14(6) (2022), pp. 8437–8447.

[249] J. Ulbricht, R. Jordan, and R. Luxenhofer, "On the biodegradability of polyethylene glycol, polypeptoids and poly(2-oxazoline)s", *Biomaterials* 35(17) (2014), pp. 4848–4861.

[250] H.-Y. Lin, H.-H. Chen, S.-H. Chang, and T.-S. Ni, "Pectin-chitosan-PVA nanofibrous scaffold made by electrospinning and its potential use as a skin tissue scaffold", *Journal of Biomaterials Science, Polymer Edition* 24(4) (2013), pp. 470–484.

[251] J.-J. Chang, Y.-H. Lee, M.-H. Wu, M.-C. Yang, and C.-T. Chien, "Preparation of electrospun alginate fibers with chitosan sheath", *Carbohydrate Polymers* 87(3) (2012), pp. 2357–2361.

[252] C. Cui, S. Sun, S. Wu, S. Chen, J. Ma, and F. Zhou, "Electrospun chitosan nanofibers for wound healing application", *Engineered Regeneration* 2 (2021), pp. 82–90.

[253] S. Cui, B. Yao, M. Gao, X. Sun, D. Gou, J. Hu, Y. Zhou, and Y. Liu, "Effects of pectin structure and crosslinking method on the properties of crosslinked pectin nanofibers", *Carbohydrate Polymers* 157 (2017), pp. 766–774.

[254] Z. Qin, X. Jia, Q. Liu, B. Kong, and H. Wang, "Enhancing physical properties of chitosan/pullulan electrospinning nanofibers via green crosslinking strategies", *Carbohydrate Polymers* 247 (2020), 116734.

[255] P. G. Shikhi-Abadi and M. Irani, "A review on the applications of electrospun chitosan nanofibers for the cancer treatment", *International Journal of Biological Macromolecules* 183 (2021), pp. 790–810.

[256] R. Augustine, S. R. U. Rehman, R. Ahmed, A. A. Zahid, M. Sharifi, M. Falahati, and A. Hasan, "Electrospun chitosan membranes containing bioactive and therapeutic agents for enhanced wound healing", *International Journal of Biological Macromolecules* 156 (2020), pp. 153–170.

7 Foam Injection Molding of Poly(lactic acid) Based on Physical Blowing Agents

Valentina Volpe, Fabiana Foglia, and Roberto Pantani

This chapter presents a review of the technologies concerning the foam injection molding of PLA and methods to control the foaming of this biodegradable polymer by injection molding.

Foam injection molding of PLA is a key technology to produce rigid biodegradable molded parts using less material while reducing energy consumption. PLA presents some drawbacks due to its slow crystallization kinetics, low melt strength, and low toughness. These drawbacks are discussed in the present chapter. The best results were obtained by adding nucleating agents and controlling the process variables. The process that seems to allow the largest density reduction is the high-pressure technique with mold opening. A further improvement in the control of the process variables seems to be the best way to reach even better performances.

7.1 Introduction

Structural foams comprise a cellular core and a solid skin, and they can be obtained using an injection molding machine (IMM) with a resin containing a blowing agent (BA). The foaming agent is mixed with the molten polymer inside the cylinder of the IMM under high pressure; this is followed by injection into the mold. Inside the cavity, a sudden pressure drop causes the expansion of the BA's generated gas with formation of cells in the polymer. The foam injection molding process presents many advantages, such as reduced product weight with a stiffness-to-weight ratio comparable to that of unfoamed parts. Moreover, no packing step is needed, resulting in a faster production cycle.

This process is of interest for biodegradable polymers in general and for poly(lactic acid) (PLA) in particular, for several reasons, including the following: First, PLA is more expensive than traditional plastics; hence, the possibility to obtain lower-weight products is of economic importance. Second, one of the drawbacks of PLA is related to its narrow processing window because of PLA's thermal sensitivity: the presence of gas in the polymer leads to a reduction of the PLA viscosity, allowing the processing of PLA at lower temperatures and pressures.

This chapter presents a review of the technologies concerning the foam injection molding of PLA and methods to control the foaming of this biodegradable polymer by injection molding.

7.2 A General Description of Injection Molding

Injection molding is one of the most widespread processes to produce polymer parts. It is a cyclic process consisting of feeding the polymer pellets, plasticizing the pellets, filling the mold, packing the polymer under pressure, and cooling, before part ejection. The melt in the mold must be kept under high pressure for packing. The packing step should be maintained to avoid defects like sink marks, poor surface quality, voids, and distortion. The cooling step starts at the end of the packing step, when the cavity gate solidifies or is shut by an external control. During the cooling step, the screw plasticizes the polymer for the next shot. The cooling step lasts until the polymer solidifies enough to allow the part to preserve its shape. At the end of the cooling step, the mold is opened to eject the molded part (Figure 7.1).

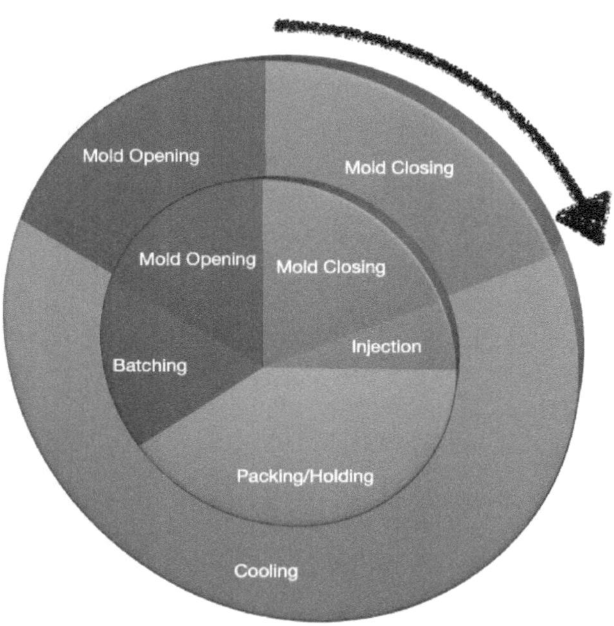

Figure 7.1
Molding cycle. The overall cycle duration depends on the material and the geometry

7.3 Foam Injection Molding

In foam injection molding, the traditional injection molding process is applied to obtain a well-defined shaped part, with less material and limited loss of mechanical properties: the same volume is obtained with a lower amount of plastic by foaming. The resin is mixed with a blowing agent (BA) and fed under high pressure into the mold; the BA forms gas bubbles, which results in an expansion of the material and the complete filling of the cavity. The resulting voids make it possible to reduce weight and blunt crack tips, thus enhancing toughness. The gas acts as a lubricant for the resin, reducing the viscosity and the glass transition temperature [1]. The viscosity decrease allows the use of lower temperatures and pressures, leading to a substantial reduction of the clamp tonnage requirement for mold closure. Furthermore, as the expansion of the foam allows a complete filling of the cavity, the packing phase is eliminated, which results in a significant shortening of the cycle time.

In foam injection molding, four steps can be identified:

- Gas dissolution – a polymer/gas solution is obtained in the cylinder
- Nucleation – many nucleation sites are formed by the pressure drop occurring as the material enters the cavity
- Cell growth – the foam's cells grow and coalesce during filling and post-filling of the cavity
- Shaping – the cavity geometry is reproduced after the solidification of the foamed resin

7.3.1 Structural Foam

The parts obtained by foam injection molding are called "structural foams" or "integral foams" [2, 3], and are characterized by a foamed core and a solid skin (Figure 7.2). The overall density reduction with respect to a solid unfoamed part that has the same volume is usually in the range of 15–40% [4].

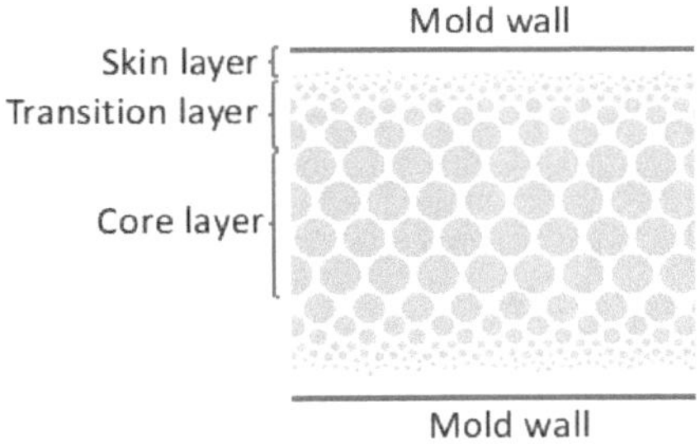

Figure 7.2
Schematic image of a structural foam

Three main regions can be identified in structural foam [5]. The first region is a compact skin layer at the part surface. The second region is the transition layer, in which the cell diameters gradually increase. The third region, at the midplane of the part, is the core layer and is characterized by a homogeneous large cells' diameter.

The thickness of the skin layer can be limited when the structural foams are used for scaffolds in which a high porosity is essential. Surface treatments, such as laser surface texturing [6], can be applied in these cases.

Although structural foams are generally characterized by closed cells, some techniques reported in the literature make it possible to obtain a significant number of open cells in the final part [7, 8].

7.3.2 Surface Appearance

The main drawback of foam injection molding is the poor surface finishing [9]. Swirl marks appear on the part surface and make the surface quality often unacceptable (Figure 7.3) for most applications, except for interior parts.

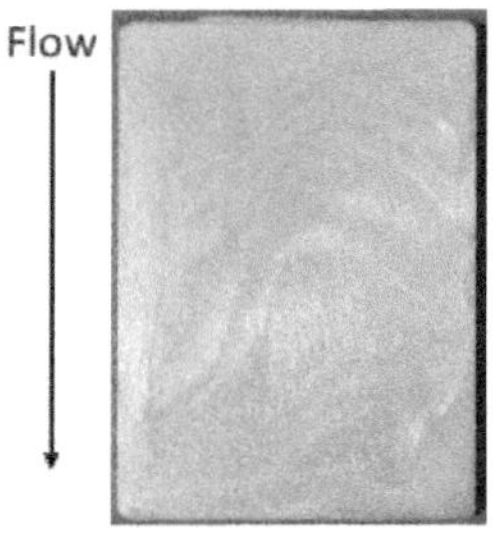

Figure 7.3
Surface appearance of a foamed sample made of PLA. The sample shown is 100 mm long, 70 mm wide, and 5 mm thick [10]

The cause of flow marks on the surface of structural foams is not completely clear. According to some publications [11], they are the result of gas trapped on the mold surface as the plastic/gas mixture begins to solidify. Silver streaks result from macroscopic gas bubbles, which are sheared on the cavity wall during the filling phase [12]. Their fragmentation leads to many microscopic gas bubbles causing craters on the surfaces [13].

Most of the studies on foam injection molding look for an improvement of the surface appearance. Special processes like in-mold decoration or co-injection molding [14] and secondary operations like surface polishing resolve the problem. The control of cell growth during early filling uses breathing molds and gas counterpressure. Furthermore, keeping the cavity surface temperature above the glass transition temperature or the melting temperature during the filling step can have a positive effect [11]. Several methods have been proposed to keep a high mold temperature during filling without unacceptably increasing the cycle time [15]. These methods include induction heating [16] and the addition of an insulating layer on the cavity surface, which induces a heat transfer delay, resulting in a higher surface temperature [17].

■ 7.4 Processing of Structural Foams

7.4.1 Low-Pressure Process

Compared to other foam injection molding methods, low-pressure ones are the most common. The polymer/BA-gas solution under pressure is injected into the cavity to partially fill the mold (realizing a so-called "short shot"). Thus, at the inlet of the cavity, the polymer/gas solution experiences a sudden reduction of the pressure, which allows the expansion of the pressurized gas, to fill the cavity with the foam as can be seen in Figure 7.4.

The main process variables are:

- The injected volume compared to the cavity volume
- The melt temperature
- The mold temperature
- The type and concentration of the BA
- The imposed flow rate

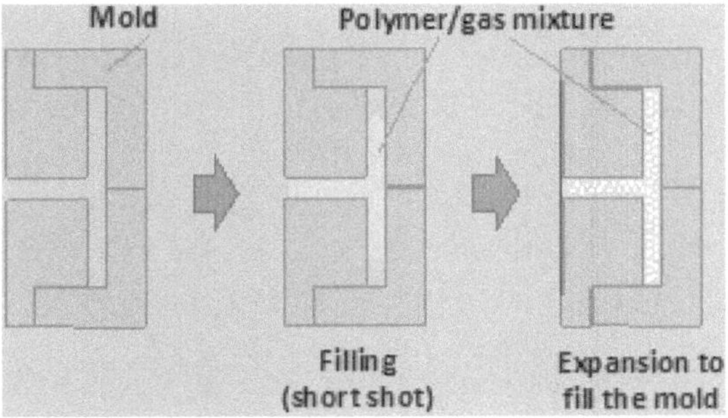

Figure 7.4 Scheme of the low-pressure technique

7.4.2 Low-Pressure with Co-Injection

In this case, two compatible plastics are co-injected using two injection cylinders [14]. A partial filling of one material, without a BA, is injected from one cylinder into the mold. Shortly following the partial filling, a measured shot of another material containing a BA is injected from the second cylinder. After the pressure drop takes place at the end of the injection, the foamed core material will force the unfoamed material to fill the cavity and form a smooth solid skin (Figure 7.5).

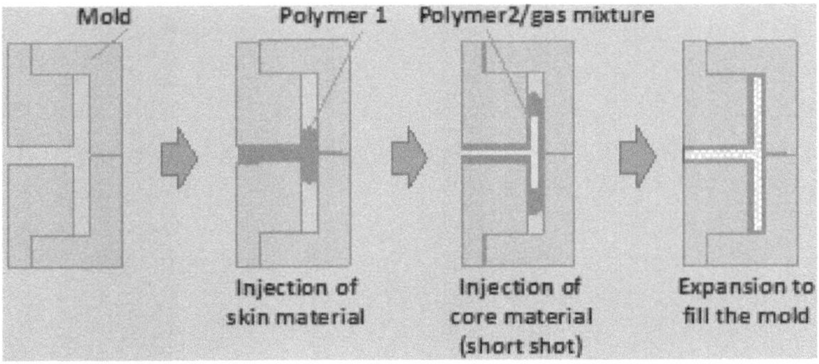

Figure 7.5 Scheme of the low-pressure with co-injection process

7.4.3 Gas Counterpressure

The gas-counterpressure (GCP) method was developed [12, 18] to control the foaming during the early filling stages. During a typical foam injection molding process, the low pressure at the melt front leads to immediate foaming, and the bubbles are forced to the part surface. Distinctively, in the GCP method, the mold is sealed and pressurized with inert gas at a pressure high enough to inhibit foaming when the polymer/gas mixture enters the mold cavity. At the end of filling, the mold pressure is released in a controlled manner, allowing the formation of a foamed core between the already formed skins (Figure 7.6). The GCP process can improve the part surface appearance significantly [19]. The pressure levels needed are between 40 bar and 100 bar, depending on the injection temperature and the polymer/gas solution pressure.

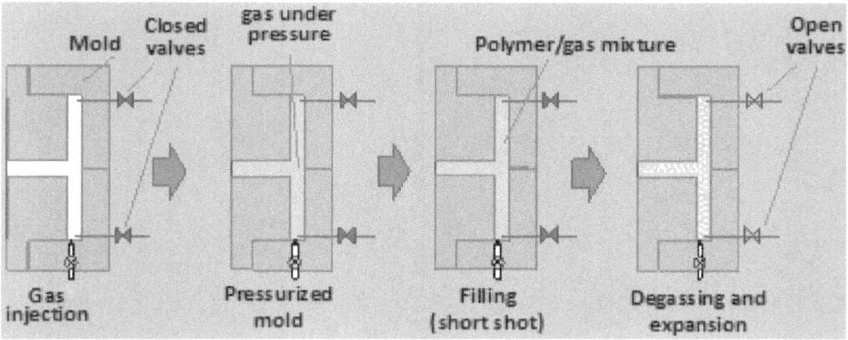

Figure 7.6 Scheme of the gas-counterpressure process

7.4.4 High-Pressure Process

In the high-pressure version of the process, the solution of polymer and gas is injected into the mold to completely fill the cavity and reach a pressure high enough to avoid foaming, while solidification starts at the mold surface. As soon as the thickness of the skin reaches the desired value, the cavity mold pressure is decreased to allow the foaming of the melt in the core region. The pressure drop is obtained either by retracting core inserts (Figure 7.7) or by partially opening the mold halves (in this case, the process is sometimes referred to as "mold breathing" [20]).

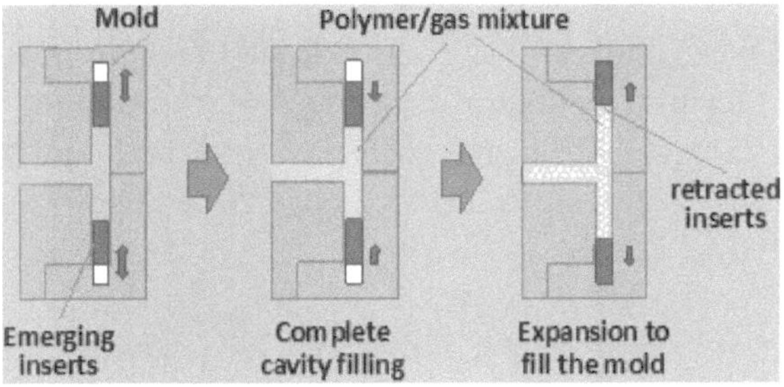

Figure 7.7 Scheme of the high-pressure technique

7.4.5 Mold Opening

In the high-pressure process, the mold-opening (MO) method provides the space for foaming following a programmed delay of the injection time. Thus, the mold is partially opened to a set length. Before demolding the molded part, the subsequent foaming process and cooling take place. The sequence of the mold-opening operation is described in Figure 7.8.

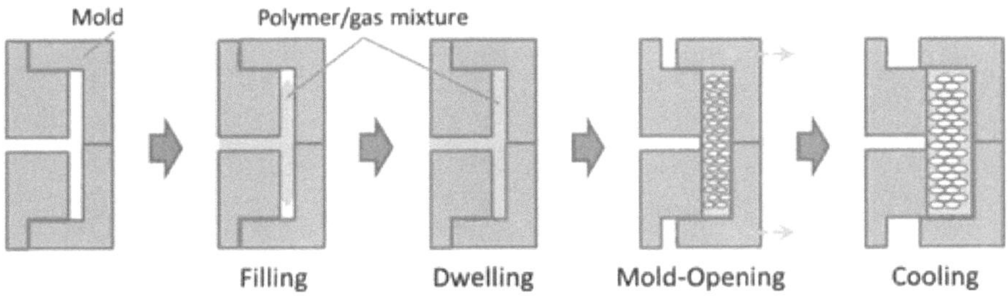

Figure 7.8 Scheme of the mold-opening method [21]

Some processes combine mold opening with the gas-counterpressure method. The process sequence in that case is shown in Figure 7.9. The polymer/gas solution is injected into the cavity previously pressurized by the GCP. After filling the whole cavity, the pressure is reduced by mold opening to a set length.

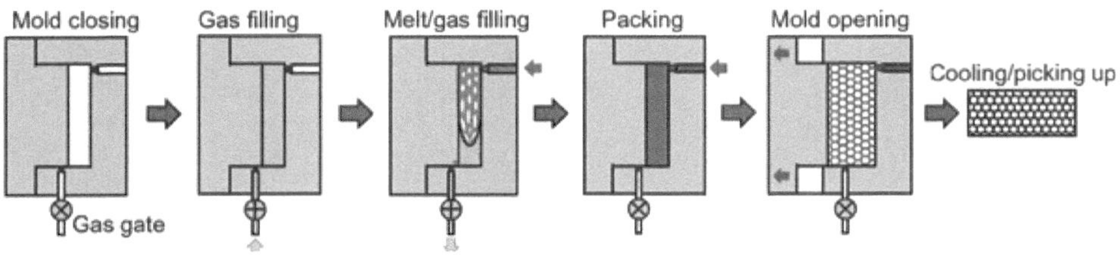

Figure 7.9 Scheme of the mold-opening technique coupled with gas counterpressure [22]

The coupling of MO and GCP reduces the gas diffusion loss in the flow front during the melt-filling process and improves the surface quality of the molded part. Several researchers applied this method to the foam injection molding of polypropylene or polystyrene [23–27], while only a few studies have been done on foam injection molding with MO using PLA.

■ 7.5 Foaming Agents

7.5.1 Chemical Blowing Agents

Chemical blowing agents (CBAs) decompose by the effect of temperature and thus produce gases like carbon dioxide and/or nitrogen [28]. Solid organic or inorganic substances such as azodicarbonamide and sodium bicarbonate are typically used as CBAs. In general, CBAs are classified according to the enthalpy of decomposition, and divided into exothermic and endothermic ones [29]. A combination of exothermic and endothermic CBAs is used for foaming. The main benefits of using CBAs include the possibility of adopting unmodified injection molding machines. Temperature is a critical parameter since it determines the decomposition of the CBA.

7.5.2 Physical Blowing Agents

Physical blowing agents (PBAs) can produce gases because of a physical process, such as evaporation or desorption, due to increased temperatures or reduced pressures. Physical blowing agents adopted for foam injection molding [30] include water, argon, nitrogen, and carbon dioxide. No residues of the blowing agent remain in the part, and thus physically blown parts can be implemented for many applications, from food contact packaging to medical products. Unlike CBAs, PBAs generally need special equipment such as a gas dosage unit and a gas injector. The mainly used physical blowing agents are nitrogen (N_2) and carbon dioxide (CO_2), which allow high degrees of foaming and thus lower density of the foam injection-molded parts. Further advantages of N_2 and CO_2 compared to other physical blowing agents (e. g., pentane) are derived from their nonflammability, nontoxicity, and the low solubility of N_2, which can result in small cells. It should be emphasized that CO_2 at its critical state can enhance the solubility, and thus a larger percentage of CO_2 can be added into the polymer.

7.5.3 Other Foaming Methods

Foaming can be achieved by feeding a conventional injection molding machine with saturated pellets containing a high-pressure physical blowing agent. This kind of pellets can be obtained, for instance, by using an autoclave under high pressure and room temperature [31]. This method involves long sorption times, and then the pellets must be processed in a short time after the preparation. Another possibility is to use an extruder to prepare "gas-laden pellets" [32]. In this case, during extrusion, the PBA is injected and mixed with the polymer. The extruded material is cooled down very quickly to prevent the gas from foaming within the polymer, and then pelletized. The obtained pellets can be processed in a conventional injection molding machine.

A different foaming method to obtain structural foams by CBAs is the adoption of the so-called "expandable thermoplastic microspheres" [33], which are made of microscaled spheres consisting of an outer shell of a polymer encapsulating a hydrocarbon core. When heated, the expanded hydrocarbon swells the microspheres.

This system allows the production of foamed parts with a better surface quality, in comparison with conventional foaming [34], since the outer shell prevents the gas from forming flow marks on the part surface.

7.6 Machine Requirements

As mentioned above, conventional injection molding machines can be adopted for foam injection molding. To use physical blowing agents, a gas injection control unit (Figure 7.10) needs to be added, with one or more gas nozzles that allow the gas injection into the plasticating cylinder.

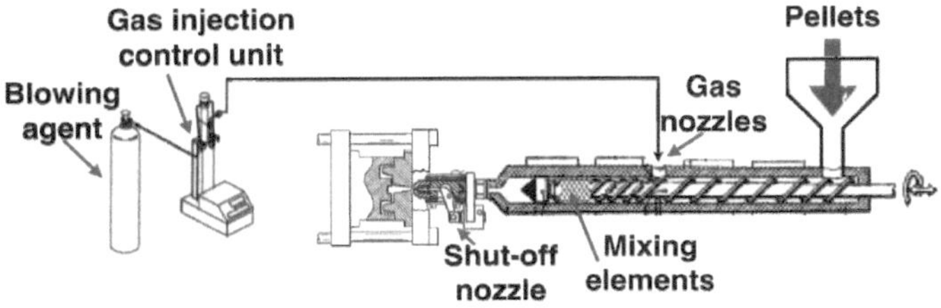

Figure 7.10 Schematics of an injection molding machine for foaming with physical blowing agents

However, some other elements are essential for carrying out the foam injection molding process. The main ones are discussed below.

7.6.1 Control of Back Pressure

After the ejection stage, conventional injection molding machines enter an idle time in which no pressure acts on the screw. During this idle time, the gas exerts pressure on the tip of the screw and thus pushes its back. This expansion can induce a phase separation between the gas and the melt.

A simple way to keep the screw position and thus constant back pressure during the screw's idle time is to place a mechanical brake at the back of the screw. Another possibility, to reach the same goal, is the introduction of a hydraulic control that keeps the final screw position by means of a servo valve. The advantage of this system is the direct control of the pressure of the solution in the injection chamber, which is kept at a value higher than the saturation point at the specific injection temperature.

7.6.2 Screw Modification

The screws of conventional injection molding machines do not have a mixing section long enough to ensure good mixing of the foaming agent and the polymer. Thus, the introduction of a mixing section can be necessary [35]. Rauwendaal [36, 37] used particular mixing elements at the end of a single-screw extruder to increase the stretching and eventual breakup of the bubbles and droplets of liquid additives, obtaining a successful gas dispersion. Park [38] used a static mixer in the nozzle, which creates both a shearing and an elongational effect, which is a much more effective way to promote bubble breakup and, hence, better dispersion of gas bubbles.

A special middle check valve can also be adopted to keep high pressure inside the screw until the solution is injected into the mold [39].

7.6.3 Shut-Off Nozzle

A shut-off nozzle, or a gate valve, is a crucial element for producing foam injection-molded objects. If a conventional open nozzle is implemented, the pressure inside the injection chamber cannot reach a value high enough to keep the gas in the solution inside the melt, causing premature foaming, which produces low-quality products. A shut-off nozzle makes it possible to maintain a high pressure in the injection chamber before injection, thus avoiding premature foaming.

7.6.4 Special Modification of the Injection System

The injection systems of the machine are often modified to allow a better mixing of the gas and the polymer. An additional accumulator after the screw, in which the solution is kept before injection, is often adopted [40], allowing better dispersion of the gas inside the polymer. Xu et al. [41] proposed the incorporation of a gear pump and an accumulator to separate the gas mixing step from the injection and molding stages. This modification allowed the continuous formation of the polymer/gas solution, which is fed to the injection molding machine.

7.7 Process Variables

The significant challenges in foam injection molding include controlling the process variables so that it results in a homogeneous morphology - having fine and uniform microcells - in the foamed part. In addition to the selection of the polymer and the foaming agent, several process variables are of importance, and are divided into two groups:

- Process variables in the preparation of the solution to be injected: screw rotation speed, back pressure, gas-to-polymer ratio, melt temperature, residence time
- Process variables at the injection stage: mold temperature, injection flow rate, gate geometry, ratio of injected to final volume

With reference to the variables in the first group, by increasing the gas-to-polymer ratio, the molded parts possess a smaller cell size and a higher cell density, and the surface quality remains satisfactory until a certain degree of supersaturation is reached [32]. Beyond this threshold, the surface quality starts to worsen as the high cell nucleation rate causes swirl marks on the surface. The melt temperature influences the solubility of the gas, which decreases as the temperature is increased. Thus, supersaturation increases as the melt temperature is decreased. A high rotation speed allows for a better mixing of the gas/polymer solution [38]. A high back pressure increases the mixing time and the concentration of gas in the polymer. A longer residence time of the gas/polymer solution in the injection chamber, while keeping a high pressure to avoid phase separation, can favor the homogeneity of the solution [42].

With regard to the variables in the second group, a thin gate that introduces a significant pressure drop at the cavity entrance can avoid foaming before the melt enters the cavity. A high injection flow rate produces smaller cells with a more homogeneous size distribution in the foamed part [38]. Beyond a certain value of injection flow rate, no further reduction of the cell size is observed and, in some cases, even an increase has been reported [42]. Indeed, when the injection speed is increased, there are larger pressures inside the cavity, significantly closer to the gate, and this can inhibit foaming during filling. When the pressure decreases to the right value, the local temperature can be too low for foaming [10].

7.8 Foam Injection Molding of PLA

In recent years, increasing attention has been directed toward biodegradable polymers due to their potential reduction in the environment. One of the most widely used biodegradable polymers is poly(lactic acid) (PLA). PLA is a biobased, biodegradable, and biocompatible aliphatic polyester that has found many applications in the biomedical and packaging industries, due to its ease of processing and properties, relative to other biodegradable polymers. However, the applications of PLA are limited because of its high costs (compared to polymers derived from nonrenewable sources), its high viscosity, which leads to the use of high pressures and temperatures close to its degradation temperature, and its slow crystallization kinetics.

Foam injection molding appears to remedy some of the shortcomings of PLA to produce high-quality cost-effective parts. The PBA process using a high-pressure supercritical state reduces the viscosity of the polymer melt, thus allowing its processing by injection molding with lower temperatures and pressures. However, PLA's low melt strength and slow crystallization kinetics limit the formation of uniformly distributed fine-celled PLA foams with high void fractions. Thus, many studies relate to the effect of the processing parameters on the foaming behavior of PLA during the injection molding process.

7.8.1 Solubility of Physical Blowing Agents in PLA

Since the products made of PLA are mainly used for food contact or medical applications, it is quite straightforward that mainly PBAs are used for foam injection molding of PLA. As mentioned above, the prerequisite for obtaining a fine cell morphology with well-distributed sizes by foam injection molding of PLA with PBAs is the formation of a single-phase polymer/gas solution. In the single-phase gas/polymer solutions, the dissolved gas increases the free volume of polymeric molecules, thus allowing easier movement of the macromolecules and a reduction of the system's viscosity. The polymer melt viscosity increases once the dissolved gas diffuses out of the melt. The study of the solubility of the gas in the polymer is crucial for selecting the proper process parameters to obtain foams with good quality and properties.

In 2006, Li et al. [43] implemented the Simha-Somcynsky equation of state (SS-EOS) and the Sanchez-Lacombe equation of state (SL-EOS) separately, to determine the solubility of carbon dioxide and nitrogen in a polylactide melt at 28 MPa and temperatures from 180 °C to 200 °C. They used a magnetic suspension balance (MSB) to predict the amount of volume swelling in the PLA/gas mixture due to gas dissolution. Figure 7.11 displays the solubility of CO_2 and N_2 in PLA at 180 °C (a) and 200 °C (b).

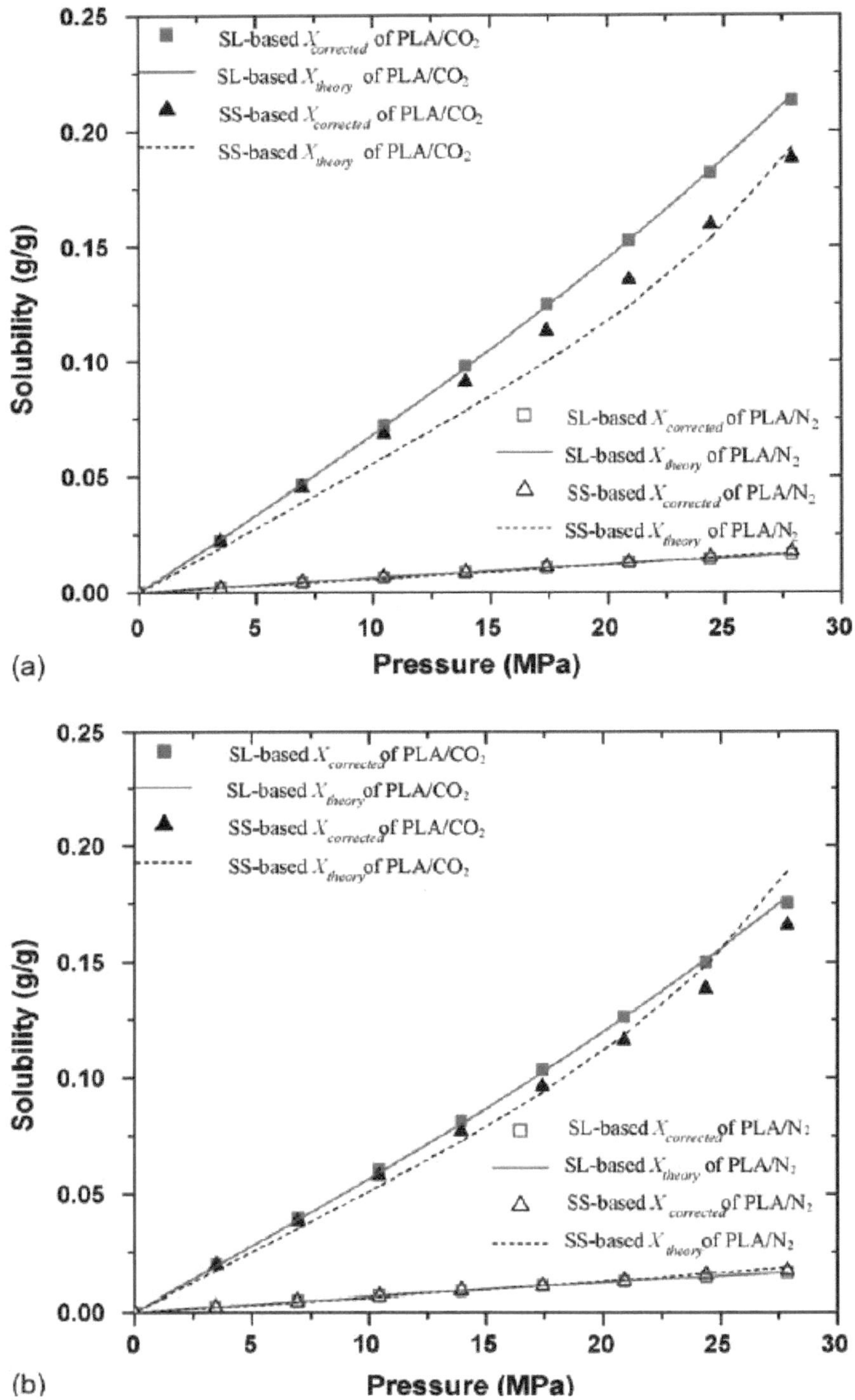

Figure 7.11 Solubility of CO_2 and N_2 in PLA at 180 °C (a) and 200 °C (b) [43]

Significant differences in the solubilities of carbon dioxide and nitrogen in PLA melts can be noted. In particular, the authors confirmed that the solubility of carbon dioxide in PLA is significantly higher than that of nitrogen. At the same temperature, however, nitrogen shows higher diffusivity.

The effect of the processing variables in foam injection molding of PLA is discussed in the following sections.

7.8.2 Mold Temperature

The mold temperature is a crucial parameter that controls the morphology of the foam injection-molded parts. However, only limited studies have been dedicated to the mold temperature of PLA, because this temperature is limited by PLA's relatively low glass transition temperature (55–60 °C) [10]. A high temperature of the mold wall allows better cell density and cell size characteristics. During the filling stage, the polymer/gas solution is injected into the mold cavity and the pressure suddenly drops, inducing the expansion of the gas into the polymer melt. The contact of the polymer/gas solution with the cold wall of the cavity leads to a fast solidification of the material before nucleation and the formation of a solid skin layer. In particular, longer cooling times lead to longer foaming times and the formation of a thinner solid surface layer [44].

Lee and Cha [45] calculated the temperature profile of the polymer near the mold wall (Figure 7.12). They proposed that the variation in skin layer thickness is related to the temperature profile of the polymer near the mold surface, assuming that T_m is the foaming temperature, that is, the temperature at which skin layer is formed. Therefore, the thickness of the solid skin layer depends strongly on the mold temperature.

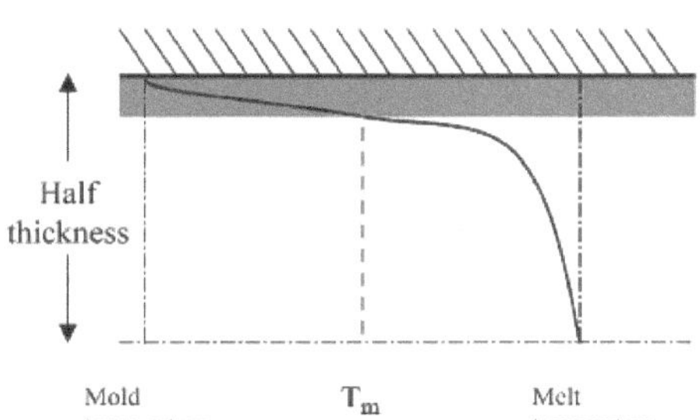

Figure 7.12
Temperature profile of a polymer near the mold wall [45]

In the case of PLA, it can be assumed that the material solidifies when the temperature is below 70 °C, namely, about 5–10 °C above the glass transition temperature. By increasing the mold temperature, the difference between the mold and the melt temperature reduces, the temperature profile flattens, and the thickness of the solid skin layer decreases. The mold temperature is strongly related to the density reduction and morphology of the foamed part. It was demonstrated [46, 47] that by increasing the mold temperature from 25 °C to 40 °C, a more homogeneous morphology can be observed, with a reduction of the core section – characterized by a larger cell size – and a consequent increase of the transition layer (marked with dashed lines in Figure 7.13), characterized by a large number of smaller cells.

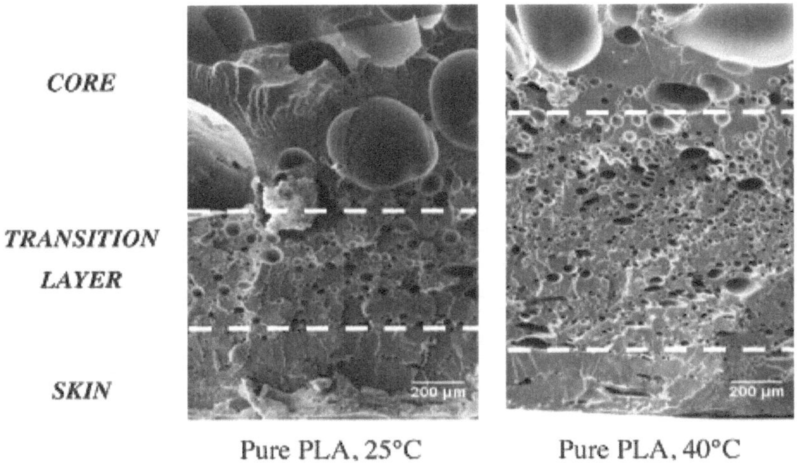

Figure 7.13 Half-thickness SEM (scanning electron microscopy) micrographs of a PLA foamed part molded with a mold temperature of 25 °C (left) and 40 °C (right) [47]

The difference in morphology can be related to the pressure variation due to the increased mold temperature. An increase in mold temperature from 25 °C to 40 °C includes a sharp reduction of the pressure in the cavity, as can be observed on the pressure evolution in Figure 7.14. This reduction involves a more significant pressure drop, determining the formation of a larger number of small cells.

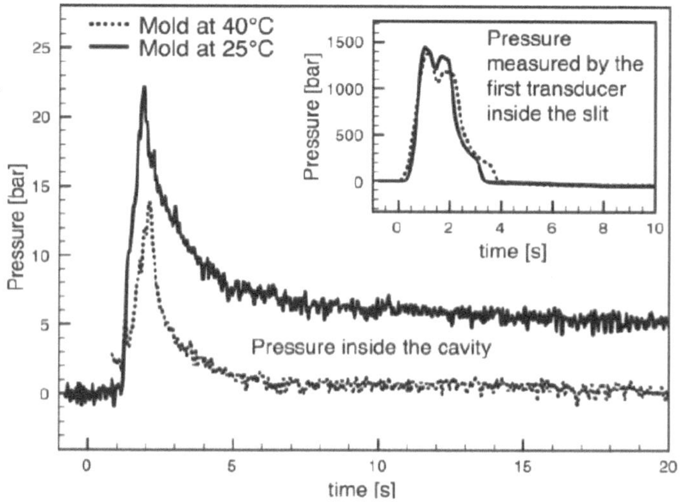

Figure 7.14 Pressure profiles in the cavity with mold temperatures of 25 °C and 40 °C [46]

7.8.3 Injection Flow Rate

Another important parameter for controlling the morphology and the surface appearance of foam injection-molded parts is the injection flow rate. With an increase of the injection flow rate, a shorter time is allocated for the injected volume to cool down before foaming, and, thus, a more homogeneous temperature profile exists when foaming starts. Furthermore, a high injection flow rate results in a lower pressure in the cavity and, consequently, greater pressure drops during injection. The pressure drop causes supersaturation of the polymer/gas solution. The more homogeneous temperature distribution and the lower pressure in the cavity result in a greater number of nucleation sites that lead to a uniform distribution of the cells, with smaller cell sizes and greater cell density [48–50]. Moreover, a high injection flow rate and a short injection time result in homogeneity of pressure and temperature along the sample length. Hence, from the injection point to the end of the product, a small and uniform skin layer and a homogeneous morphology are expected. Ameli et al. [50] described this phenomenon in their study using a mold that contained a rectangular cavity with a fan gate. Figure 7.15 shows the evolution of the skin layer thickness and the cell density with different injection flow rates. From this figure, it can be concluded that both the skin layer thickness and the cell density become similar along the sample length when a high injection flow rate is applied.

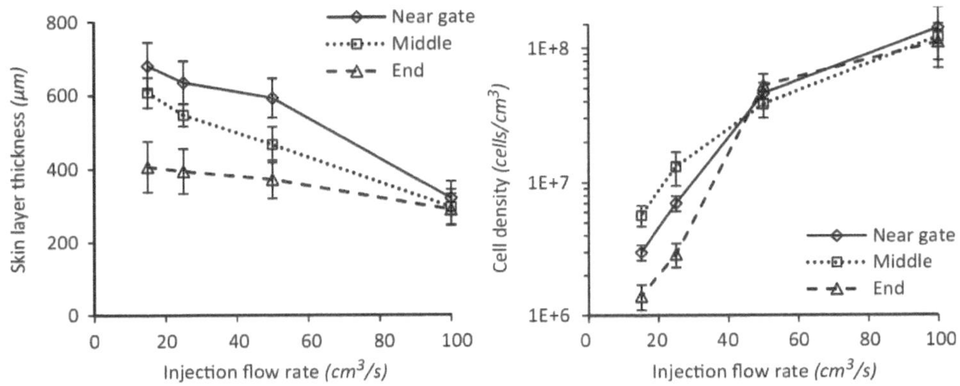

Figure 7.15 Dependence of skin layer thickness and cell density on injection flow rate [50]

Pantani and Volpe [46, 49] also studied the morphology of injection-molded PLA samples obtained with two mold temperatures at different injection flow rates. They analyzed the length of rectangular specimens molded at different injection flow rates, finding a strong dependence on the compressibility, β of the polymer/gas mixture, which is defined as:

$$\beta = \frac{1}{\rho}\frac{\partial \rho}{\partial P} \tag{7.1}$$

where ρ is the material's density and P is the average pressure in the volume considered.

The compressibility is related to the mass flow rate entering the cavity, $\dot{m}$, assuming isothermal conditions during filling, according to the following equation:

$$\dot{m} = \dot{m}_i - V(t)\rho\beta\frac{dP}{dt} \tag{7.2}$$

$$\dot{m}_i = \rho \cdot Q \tag{7.3}$$

where $\dot{m}_i$ is the nominal flow rate,

Q is the flow rate, and $V(t)$ is the volume between the tip of the screw and the cavity entrance.

With a pressure increase, the difference between the flow rate entering the cavity and the nominal one is larger when the shot volume and cushion increase or for a large compressibility. When the pressure decreases – for instance, when the screw stops – this difference decreases.

7.8.4 Back Pressure

An essential parameter in the formation of the polymer/gas solution is the back pressure, i. e., the pressure imposed at the back of the screw when it is returning to prepare a new shot of material to be injected. During this phase of the injection molding cycle, named batching phase, the gas is injected inside the cylinder simultaneously with the loading of the polymer in the form of pellets. The duration of this phase determines the mixing time of the polymer and the gas in the cylinder.

The amount of polymer conveyed from the cylinder toward the nozzle strongly depends on the back pressure: a decrease of the back pressure allows the physical blowing agent, injected in the cylinder, to push back the screw, reducing the amount of polymer conveyed in the injection chamber. Together with the back pressure, the gas injected, the friction along the screw, and the pressure at the tip of the screw (which in turn depends on the melt accumulating in the injection chamber) contribute to the amount of polymer fed to the injection chamber. This phenomenon was well analyzed by Volpe and Pantani [51]. They measured the amount of gas and the volume of the PLA/nitrogen solution injected at different back pressures of the hydraulic system (Figure 7.16).

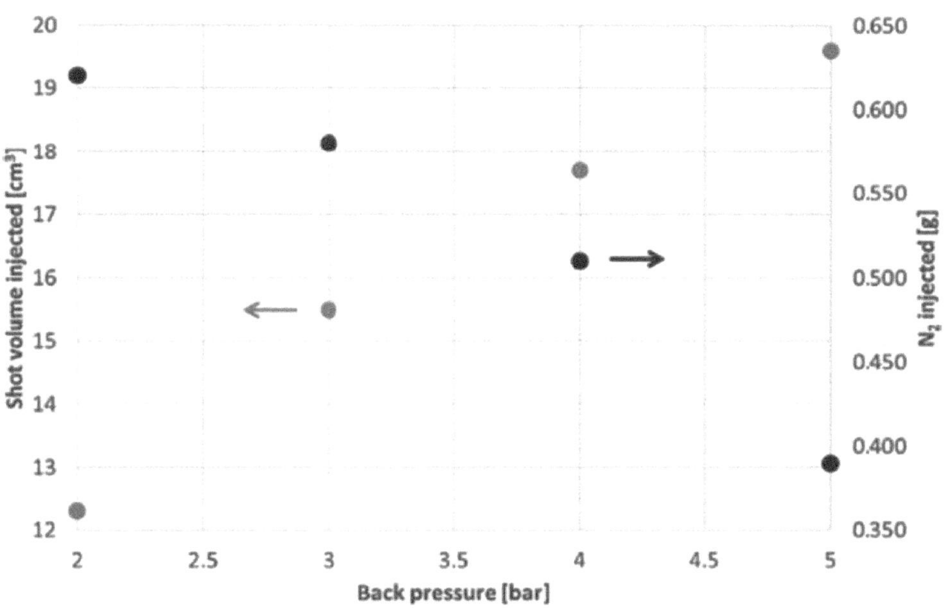

Figure 7.16 Amount of gas and volume of PLA/nitrogen solution injected at different back pressures of the hydraulic system [51]

When the back pressure is decreased, the difference between the gas pressure and the pressure in the cylinder during the gas injection increases, leading to a higher amount of gas injected into the cylinder. A low back pressure also results in a lower viscosity, which means a low pressure required for injection. However, a high back pressure leads to a more efficient mixing, due to the increase of the batching time. For the same injected volume, when the back pressure is decreased, a larger amount of gas and a lower amount of polymer are injected into the cavity, due to a greater compressibility. The foamability and, consequently, the morphology are strongly related to the back pressure: when the back pressure is increased, the pressure in the cavity increases (see Figure 7.17), inhibiting the foaming due to a lower pressure drop between the gate and the cavity.

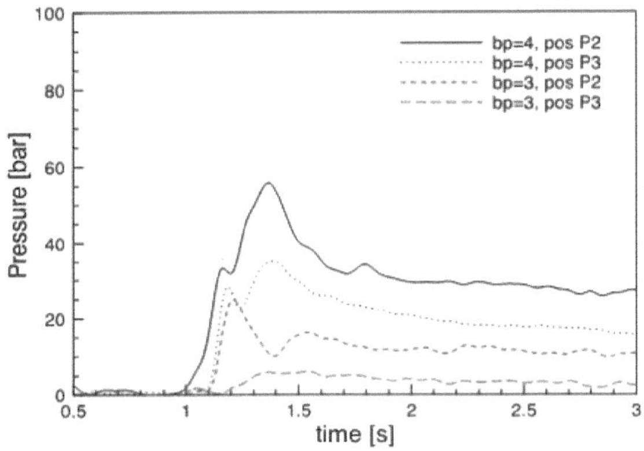

Figure 7.17 Position P2: at 15 mm from the gate; position P3: at 60 mm from the gate [51]

The effect of high back pressure is clearly visible in Figure 7.18, which shows the difference in morphology between samples molded by applying back pressures of 3 bar and 5 bar. With high back pressure, the foaming mainly occurs at the tip of the sample, which presents a small, slightly foamed transition layer and a core region completely unfoamed or with some big voids. A different morphology can be observed for low back pressure: the foaming is more homogeneous in the whole sample, with a skin layer thickness that increases toward the gate and a well-foamed transition layer and core.

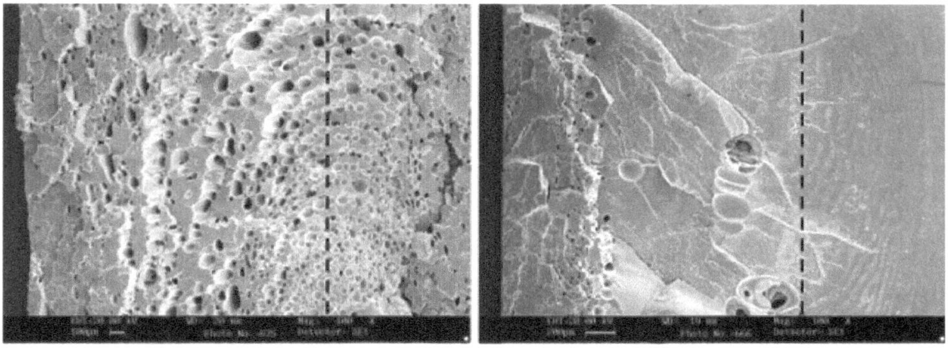

Figure 7.18 SEM micrographs of the cross section of PLA foamed parts molded with a back pressure of 3 bar (a) and 5 bar (b) [51]

The resulting reduction in density with respect to the corresponding unfoamed part is shown in Figure 7.19. With the increase of the back pressure, the density reduction decreases. However, while with high back pressure the reduction in density is almost the same near the gate and at the tip of the specimens, with low back pressure, the reduction in density is enhanced when moving away from the tip of the sample. The optimal back pressure is at an intermediate level that allows the injection of a good polymer-to-gas ratio, resulting in a homogeneous foaming from the gate to the tip of the sample.

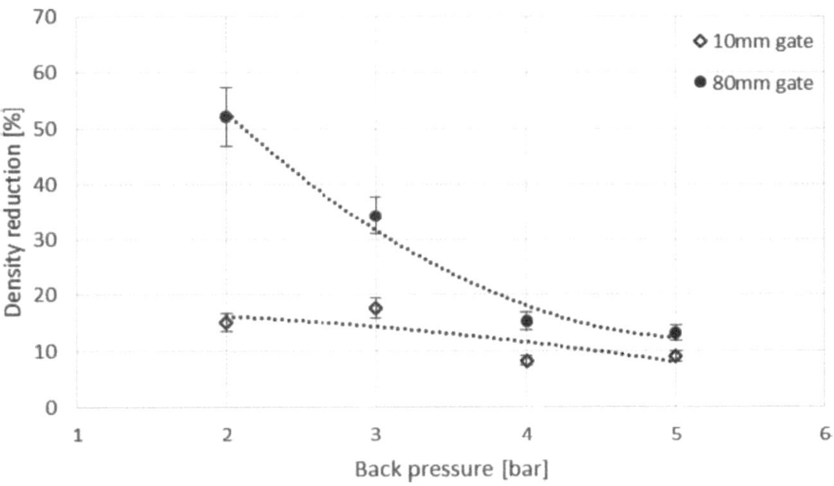

Figure 7.19 Reduction of density versus back pressure near the gate and at the tip of the sample [51]

Volpe and Pantani [51] also studied the mechanical properties of foamed parts. They demonstrated that these properties depend on the type of loading. In the case of flexural loading, the most important part of the sample is the unfoamed skin layer, while in the case of tensile loading, the presence of large voids reduces the effective load-bearing area. Hence, a normalized modulus E_N can be defined:

$$E_N = \frac{E_F}{E_0} \cdot \frac{\rho_0}{\rho_F} \tag{7.4}$$

where E is the modulus, ρ is the density, and the subscripts "F" and "0" indicate the foamed sample and the unfoamed sample, respectively. The normalized moduli derived from flexural and tensile tests are reported in Figure 7.20(a) and (b), respectively.

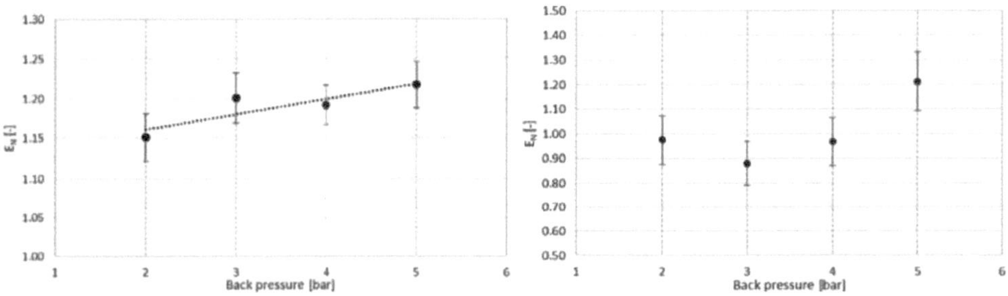

Figure 7.20 Dependence of normalized modulus on back pressure, derived from flexural tests (a) and tensile tests (b) [51]

The normalized modulus derived from flexural tests is larger than 1, indicating that the decrease in flexural modulus is less relevant than the decrease in density. In contrast, the normalized modulus derived from tensile measurements is smaller than 1, showing that the modulus of the microcellular samples is lower than that of their solid counterparts. An exception is obtained for the sample molded with a back pressure of 5 bar, reaching levels close to those of unfoamed specimens. Recently, Kastner et al. [52] proposed a novel approach using reduced back pressure in foam injection molding, based on the compressibility of polymer/gas mixtures. Their method allows the production of components with improved mechanical flexural properties and lower densities, resulting in enhanced sustainability.

7.8.5 Foam Injection Molding with Mold Opening: Effect of Opening Rate

The properties of the foam injection molding process with MO depend on the opening length and the opening rate. Volpe et al. [53] demonstrated that, in the case of PLA, the density reduction strongly depends on the mold-opening length. By increasing the mold-opening length, the same amount of material is injected in an increased volume, causing a significant reduction of the density - from 30% in the case of standard foam injection molding to more than 50% with the MO technique. Figure 7.21 compares the density reduction obtained with MO for two grades of PLA, which differ in the D-lactide content. Both PLA grades lead to a similar density reduction by varying the mold-opening length.

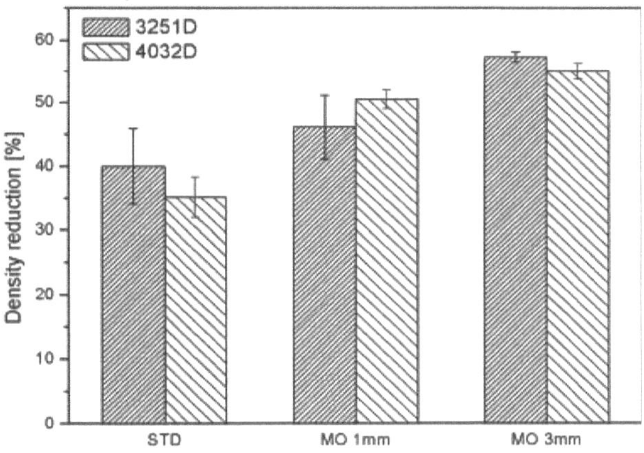

Figure 7.21 Density reduction of two PLA grades obtained by standard foam injection molding and foam injection molding with two different mold-opening lengths [53]

The density reduction also significantly affects the mechanical properties. The same authors assess that the mechanical properties, particularly the normalized modulus, increase with the mold-opening length.

In 2018, Xie et al. [54] studied the effect of the MO rate on the cell structure and, consequently, on the mechanical properties. They demonstrated that the thickness of the skin layer is not affected by the changes in the MO rate. On the contrary, by decreasing the mold-opening rate, the pressure drop rate generated by the MO decreases, resulting in a more uniform cell size distribution both in the transition and the core zones. Figure 7.22 shows the dependence of the average cell size and the average cell density on the MO rate.

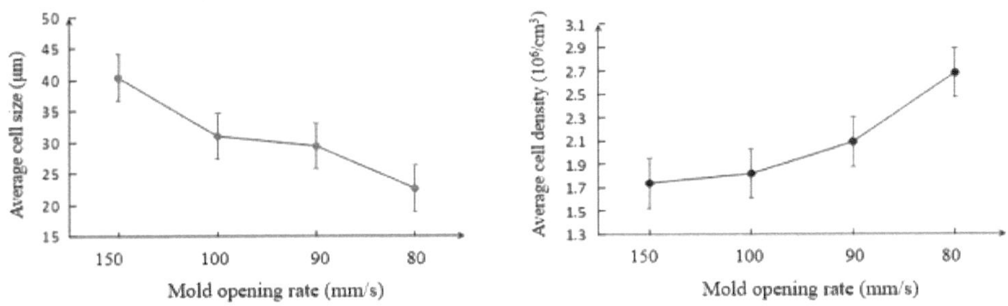

Figure 7.22 Dependence of average cell size and average cell density on the mold-opening rate [54]

As in the case of the MO length, the MO rate also significantly influences the mechanical properties of the molded parts. Both the tensile strength and the modulus increase by decreasing the mold-opening rate, due to the uniform distribution of small cells, as could be observed (for the modulus) in Figure 7.23. However, the mold-opening rate did not significantly affect the impact resistance of the foamed parts.

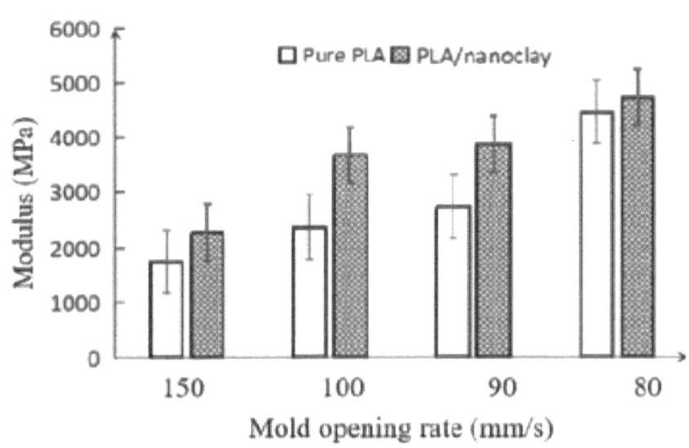

Figure 7.23 Dependence of the modulus on the mold-opening rate [54]

The modulation of the MO rate also allows the regulation of the surface roughness of the PLA foam injection-molded parts. A low mold-opening rate involves a smaller pressure drop and, consequently, the formation of finer cells, affecting the decrease in surface roughness of the molded parts, as can be seen in Figure 7.24.

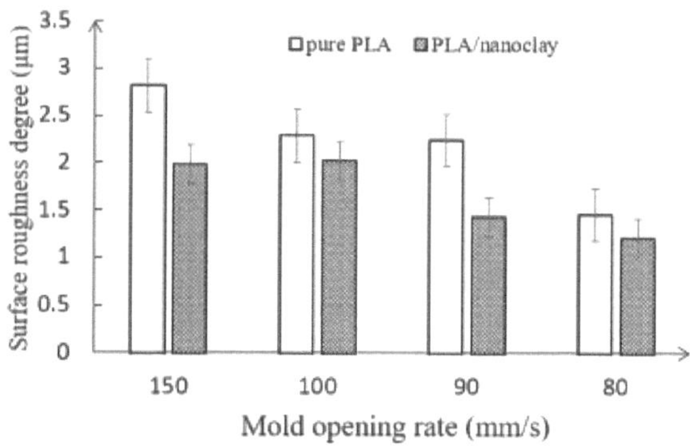

Figure 7.24 Dependence of the surface roughness degree on the mold-opening rate [54]

7.9 Foam Injection Molding of PLA: Additives and Compounds

7.9.1 Chemical Modification of PLA

One of the main shortcomings of foaming PLA is its low melt strength [55]. One of the possible routes to enhance the foamability of PLA is the introduction of chemical modifiers. Multifunctional epoxy chain extenders, such as JONCRYL®, have been shown [56] to significantly improve the cellular structure of foam injection-molded PLA parts. Peroxides were used [57] to induce cross-linking of PLA. In foam injection molding of PLA, peroxides are mainly used to improve phase compatibilization and interfacial adhesion, thus enhancing the mechanical properties of blends with other polymers. This will be discussed in the following sections.

7.9.2 Foam Injection Molding of PLA with Talc

Another shortcoming of PLA is its slow crystallization rate, limiting its mechanical, thermal, and barrier properties. To increase the crystallization rate of PLA, different studies considered the incorporation of additives as potential nucleating agents. Sodium stearate, calcium lactate, and especially talc seem to be efficient nucleating agents for enhancing the crystallinity level and, therefore, the final part's properties. It has been observed that PLA loaded with 1 wt % of talc demonstrates a reduction in crystallization half-time by more than one order of magnitude [58].

Several investigations were conducted on talc-filled PLA composites, reporting significant improvements in mechanical properties [59–61]. The advantages deriving from the addition of talc to PLA are also valid in the case of foam injection molding. Many authors [46, 50] demonstrated the improvement in the morphology and mechanical properties of PLA/talc compounds with respect to the neat PLA. The incorporation of additives with PLA is carried out generally by melt compounding, causing a reduction in the molecular weight of PLA [62, 63]. To minimize the thermal degradation effect, a masterbatch with PLA and talc can be prepared. This masterbatch can then be blended with pure PLA to reach the desired talc concentration during the foam injection molding process. Ameli et al. [50] compared the morphologies of a neat PLA (PLA-A), a processed neat PLA (PLA-P) (compounded without talc), and PLA processed with 5 wt % talc (PLA-T). They reported that the crystallinity increased, first due to the compounding process and further due to the addition of talc (Figure 7.25).

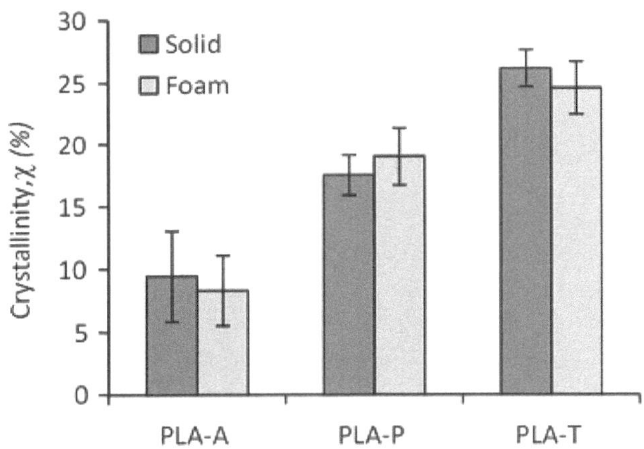

Figure 7.25 Crystallinity of solid and foamed parts of neat PLA (PLA-A), processed neat PLA (PLA-P), and PLA processed with 5 wt % talc (PLA-T) [50]

The same effect was observed by Pantani et al. [47] using differential scanning calorimetry on the core region of composites made of PLA with 3 wt % of talc. The thermogram (Figure 7.26) of the composite shows a smaller and anticipated cold crystallization peak with respect to that of the pure PLA. During the cooling stage, the PLA with 3 wt % of talc reaches almost 45% crystallinity, while the thermogram of the pure PLA does not show any crystallization peak.

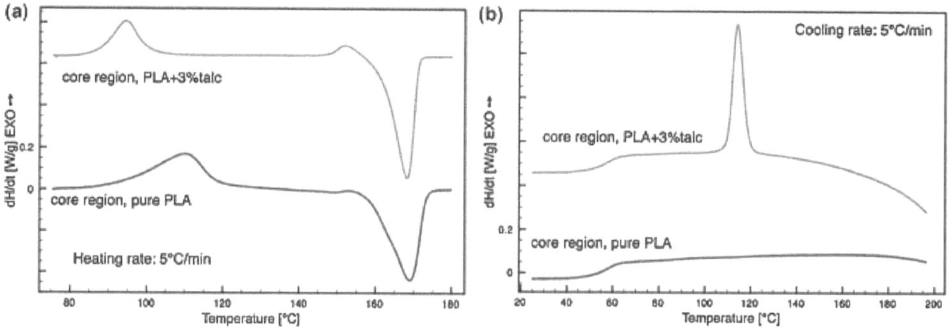

Figure 7.26 Thermograms of pure PLA and PLA with 3 wt % talc during the heating (a) and the cooling (b) stage [47]

The higher crystallinity in the PLA/talc composites and the consequent increase in melt strength allow the achievement of more homogeneous foams. Although the void fraction does not change, because the same processing parameters are used, the addition of talc involves a change in cell density and cell size. The cell density of the PLA/talc compound increases by an order of magnitude with respect to the neat and compounded PLA, as is possible to distinguish in Figure 7.27(a). Furthermore, the addition of talc significantly reduces the cell size, thus improving the uniformity of the foam (Figure 7.27(b)).

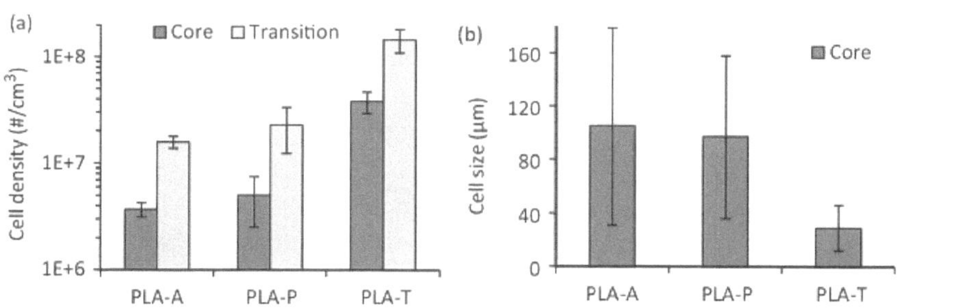

Figure 7.27 Cell density of core and transition zone (a) and cell size of core zone (b) of neat PLA (PLA-A), processed neat PLA (PLA-P), and PLA processed with 5 wt % talc (PLA-T) [50]

The SEM images in Figure 7.28 show the half-thickness of specimens made of pure PLA and PLA with talc, injected with a mold temperature of 40 °C. The PLA/talc composite shows a homogeneous distribution of the cells already close to the gate. Both compact zones and large cells disappear in the core region, making the difference in morphology between the transition and the core layer less visible.

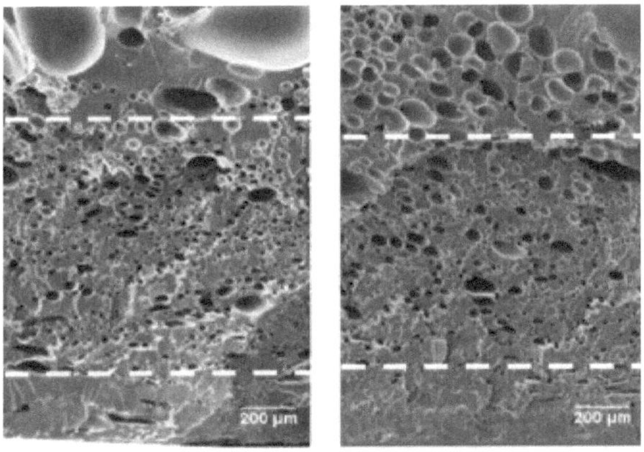

Pure PLA, 40°C PLA+ 3 % talc, 40°C

Figure 7.28 Half-thickness SEM micrographs of neat foamed PLA and PLA with 3% talc [47]

7.9.3 Foam Injection Molding of PLA with Natural Fibers

Many studies have been conducted on the foam injection molding of PLA with additives and natural fibers to enhance the crystallization rate and mechanical properties. Recently, Zhang et al. [64] studied the foam injection molding of PLA/cotton-fiber composites. Despite the difficulty of dispersing the cotton fibers into the PLA matrix, they found that the fibers slightly increase the material's crystallinity but deteriorate the foam uniformity, reducing the tensile strength as a result. Cellulosic fiber is another natural, biodegradable, and biocompostable additive that could improve PLA's crystallization kinetics. Ding et al. [65] demonstrated that the presence of cellulosic fibers significantly increases the degree of crystallinity and decreases the crystallization half-time. During the foam injection molding process, the shear stress and the biaxial stretching allow the alignment of the PLA molecules along the fibers, significantly enhancing the final degree of crystallinity of the foamed part. Furthermore, the PLA/cellulosic-fiber composites also showed improved mechanical and thermal properties.

7.9.4 Foam Injection Molding of PLA-Based Nanocomposites

The research group of Zhang [66] also developed lightweight and environmentally friendly high-performance electromagnetic interference (EMI) shielding materials based on foam injection molding with mold opening of PLA/graphite nanocomposite foams. They demonstrated that nanographite improves both the crystallization and the viscoelastic characteristics of PLA. Upon orientation of the foamed

PLA nanographite flakes during the foaming process, higher electrical conductivities were achieved. Furthermore, the addition of nanographite to PLA improves the mechanical properties of the final parts, particularly the specific strength, rigidity, and toughness.

A promising additive for nanocomposite PLA is halloysite nanotubes (HNTs), a biocompatible and inexpensive filler with a tubular microstructure. HNTs improve the mechanical properties, flame retardancy, and thermal stability of PLA [67, 68]. Different processing methods have been explored to obtain PLA/HNTs nanocomposites. Eryildiz and Altan [6, 69] adopted the foam injection molding process with a chemical blowing agent to produce PLA/HNTs nanocomposite scaffolds. Results of their study demonstrated that the presence of HNTs improves the mechanical properties and cell viability of foamed PLA, which are crucial for the composites used in tissue engineering applications. Furthermore, the high wetting-ability coupled with the high porosity and the large pore diameter of the PLA/HNTs nanocomposite systems facilitate the water entry, allowing an accelerated degradation.

In 2015, other researchers [70] developed a new technique to fabricate high-performance microcellular PLA foams. They used pellets prefoamed with supercritical fluid to disperse nanoparticles and enhance the foaming process. The resulting nanocomposite foams had small cell sizes and high cell densities, with well-separated cells. The use of prefoamed pellets seems to improve the mechanical properties of PLA/clay samples and enhance the PLA crystallization without significantly changing the molecular weight of PLA.

7.9.5 PLA Blends and Compounds

Polymer compounding offers an efficient and cost-effective way to broaden the applications of PLA. In particular, concerning the foamability of PLA, the main goal of compounding is to increase the melt strength and toughness.

Several studies [71–73] demonstrated that thermoplastic polyurethane (TPU) acts as a toughening agent in PLA, preventing the PLA/TPU blends from breaking. Mi et al. [74] characterized PLA/TPU parts produced by foam injection molding for tissue engineering scaffolds. Xu et al. [75] fabricated PLA foams using microcellular foaming technology with TPU and CO_2 as blowing agents. They assessed that TPU acted as a nucleating agent, with decreased cell size and increased cell density and expansion ratio of the PLA/TPU composite foams.

PLA is often compounded with other biodegradable polymers to obtain fully biobased blends with tailored material properties and improved foam morphologies. Haibin Zhao et al. [76, 77] applied foam injection molding to blends of PLA with poly(3-hydroxybutyrate-*co*-3-hydroxyvalerate), PHBV. They found that the increase of the PHBV content decreased the cell size and the tensile strength of the specimens while increasing the strain at break.

Other studies adopted the foam injection molding process to prepare PLA/poly(butylene succinate) (PBS)/activated carbon (AC) composites [78]. They demonstrated that the properties of the composites depend strongly on the PLA/PBS ratio: by increasing the PBS concentration in the PLA, an increase in Young's modulus and crystallinity were observed. Furthermore, at a high PBS concentration, the melt viscosity of the composite foam increases, and as a result the cell growth decreases and the cell density increases. Other researchers investigated the thermal and mechanical properties of biodegradable PLA/poly(butylene succinate-*co*-adipate) (PBSA) composites processed by supercritical fluid (SCF)-assisted foam injection molding [79]. Results showed that adding triphenyl phosphite (TPP) as a compatibilizer led to an increased molecular weight and a shift in the melting temperature. They also showed that the compatibilized SCF-foamed blends had higher crystallinity and storage modulus than their physically foamed counterparts. In 2019, Campuzano and Lopez [80] analyzed for the first time the effect of dicumyl peroxide (DCP) as a compatibilizer and foamability enhancer on foam injection-molded PLA/PBS blends with chemical foaming agents. They demonstrated that PBS enhances the thermal behavior of the blend, while DCP contributes to an increase in the melt viscosity and crystallinity.

Another interesting additive for PLA for foam injection molding is polytetrafluoroethylene (PTFE) [81]. The fibrous nature of PTFE allows the formation of a network structure in the polymer melt, also defined as microfibrillar composites (MFCs). The presence of PTFE in PLA seems to enhance its crystallization and create heterogeneous bubble nucleation sites involving a small cell size, high cell density, and high open-cell content.

■ 7.10 Conclusions and Perspectives

Foam injection molding of PLA is a key technology to produce rigid biodegradable molded parts using less material and reducing energy consumption. One of the main advantages of the process is the possibility to implement conventional injection molding with minimal modifications. However, PLA presents some drawbacks for foam injection molding due to its slow crystallization kinetics, low melt strength, and low toughness, and often does not make it possible to obtain a good surface appearance. Most of the solutions to these drawbacks have been discussed in this chapter: we reviewed the process parameters and PLA's composition modifications that efficiently improve the properties of foam injection-molded PLA parts. The best results were obtained by adding nucleating agents and controlling the process variables. The process that seems to allow the largest density reduction is the high-pressure technique with mold opening. A further improvement in the control of the process variables seems to be the best way to reach even better performances.

References for Chapter 7

[1] Di Maio, E., S. Iannace, G. Mensitieri, and L. Nicolais, "A predictive approach based on the Simha-Somcynsky free-volume theory for the effect of dissolved gas on viscosity and glass transition temperature of polymeric mixtures", *Journal of Polymer Science Part B-Polymer Physics*, 2006. 44(13): pp. 1863–1873.

[2] Shutov, F. A., "Morphology of integral (structural) polymer foams", *Makromolekulare Chemie. Macromolecular Symposia*, 1986. 2(1): pp. 113–123.

[3] Müller, N. and G. W. Ehrenstein, "Evaluation and modeling of injection-molded rigid polypropylene integral foam", *Journal of Cellular Plastics*, 2004. 40(1): pp. 45–59.

[4] *https://newsroom.mazda.com/en/publicity/release/2008/200809/080909c.html* (accessed: 30/06/2023).

[5] Bledzki, A. K., H. Kirschling, M. Rohleder, and A. Chate, "Correlation between injection moulding processing parameters and mechanical properties of microcellular polycarbonate", *Journal of Cellular Plastics*, 2012. 48(4): pp. 301–340.

[6] Eryildiz, M., M. Altan, and S. Odabas, "Improvement of Mechanical and Biological Properties of PLA/HNT Scaffolds Fabricated by Foam Injection Molding: Skin Layer Effect and Laser Texturing", *International Polymer Processing*, 2021. 36(5): pp. 564–576.

[7] Kramschuster, A. and L. S. Turng, "An injection molding process for manufacturing highly porous and interconnected biodegradable polymer matrices for use as tissue engineering scaffolds", *Journal of Biomedical Materials Research Part B-Applied Biomaterials*, 2010. 92B(2): pp. 366–376.

[8] Wu, H. B., H. J. Haugen, and E. Wintermantel, "Supercritical CO_2 in injection molding can produce open porous polyurethane scaffolds – a parameter study", *Journal of Cellular Plastics*, 2012. 48(2): pp. 141–159.

[9] Chen, H. L., R. D. Chien, and S. C. Chen, "Using thermally insulated polymer film for mold temperature control to improve surface quality of microcellular injection molded parts", *International Communications in Heat and Mass Transfer*, 2008. 35(8): pp. 991–994.

[10] Pantani, R., A. Sorrentino, V. Volpe, and G. Titomanlio, "Foam injection molding of poly(lactic acid) with physical blowing agents", Proceedings of PPS-29: The 29th International Conference of the Polymer Processing Society – Conference Papers, *AIP Conference Proceedings*, 2014. 1593: pp. 397–400.

[11] Cha, S. W. and J. D. Yoon, "The Relationship of Mold Temperatures and Swirl Marks on the Surface of Microcellular Plastics", *Polymer-Plastics Technology and Engineering*, 2005. 44(5): pp. 795–803.

[12] Michaeli, W. and A. Cramer, Increasing the surface quality of foamed injection molded parts, Annual Technical Conference ANTEC, Conf. Proc., 2006. Charlotte, North Carolina.

[13] Jiang, J., Z. Li, H. Yang, X. Wang, Q. Li, and L. S. Turng, "Microcellular injection molding of polymers: a review of process know-how, emerging technologies, and future directions", *Current Opinion in Chemical Engineering*, 2021. 33: 100694.

[14] Turng, L. S. and H. Kharbas, "Development of a hybrid solid-microcellular co-injection molding process", *International Polymer Processing*, 2004. 19(1): pp. 77–86.

[15] Wang, Y. and G. H. Hu, Research Progress of Improving Surface Quality of Microcellular Foam Injection Parts. In Applied Mechanics and Materials, 2011. Trans. Tech. Publ.

[16] Chen, S.C., Y.W. Lin, R.D. Chien, and H.M. Li, "Variable mold temperature to improve surface quality of microcellular injection molded parts using induction heating technology", *Advances in Polymer Technology: Journal of the Polymer Processing Institute*, 2008. 27(4): pp. 224–232.

[17] Chen, S.C., H.M. Li, S.S. Hwang, and H.H. Wang, "Passive mold temperature control by a hybrid filming-microcellular injection molding processing", *International Communications in Heat and Mass Transfer*, 2008. 35(7): pp. 822–827.

[18] Bledzki, A.K., J. Kühn, H. Kirschling, and W. Pitscheneder, "Microcellular injection molding of PP and PC/ABS with precision mold opening and gas counterpressure", *Cellular Polymers*, 2008. 27(2): pp. 91–100.

[19] Chen, S.C., P.S. Hsu, and S.S. Hwang, "The effects of gas counter pressure and mold temperature variation on the surface quality and morphology of the microcellular polystyrene foams", *Journal of Applied Polymer Science*, 2013. 127(6): pp. 4769–4776.

[20] Stumpf, M., A. Spörrer, H.-W. Schmidt, and V. Altstädt, "Influence of supramolecular additives on foam morphology of injection-molded i-PP", *Journal of Cellular Plastics*, 2011. 47(6): pp. 519–534.

[21] Chu, R.K., L.H. Mark, D. Jahani, and C.B. Park, "Estimation of the foaming temperature of mold-opening foam injection molding process", *Journal of Cellular Plastics*, 2016. 52(6): pp. 619–641.

[22] Wang, G., G. Zhao, G. Dong, Y. Mu, C.B. Park, and G. Wang, "Lightweight, super-elastic, and thermal-sound insulation bio-based PEBA foams fabricated by high-pressure foam injection molding with mold-opening", *European Polymer Journal*, 2018. 103: pp. 68–79.

[23] Heim, H.-P. and M. Tromm, "General aspects of foam injection molding using local precision mold opening technology", *Polymer*, 2015. 56: pp. 111–118.

[24] Michaeli, W., L. Florez, D. Obeloer, and M. Brinkmann, "Analysis of the impact properties of structural foams", *Journal of Cellular Plastics*, 2009. 45(4): pp. 321–351.

[25] Reglero Ruiz, J.A., M. Vincent, J.-F. Agassant, A. Claverie, and S. Huck, "Morphological analysis of microcellular PP produced in a core-back injection process using chemical blowing agents and gas counter pressure", *Polymer Engineering & Science*, 2015. 55(11): pp. 2465–2473.

[26] Sadik, T., C. Pillon, C. Carrot, J.A. Reglero Ruiz, M. Vincent, and N. Billon, "Polypropylene structural foams: Measurements of the core, skin, and overall mechanical properties with evaluation of predictive models", *Journal of Cellular Plastics*, 2017. 53(1): pp. 25–44.

[27] Shaayegan, V., G. Wang, and C.B. Park, "Study of the bubble nucleation and growth mechanisms in high-pressure foam injection molding through in-situ visualization", *European Polymer Journal*, 2016. 76: pp. 2–13.

[28] Nema, A.K., A.V. Deshmukh, K. Palanivelu, S.K. Sharma, and T. Malik, "Effect of exo- and endothermic blowing and wetting agents on morphology, density and hardness of thermoplastic polyurethanes foams", *Journal of Cellular Plastics*, 2008. 44(4): pp. 277–292.

[29] Thompson, M.R., X. Qin, G. Zhang, and A.N. Hrymak, "Aspects of foaming a glass-reinforced polypropylene with chemical blowing agents", *Journal of Applied Polymer Science*, 2006. 102(5): pp. 4696–4706.

[30] Sun, X.F. and L.S. Turng, "Novel injection molding foaming approaches using gas-laden pellets with N_2, CO_2, and $N_2 + CO_2$ as the blowing agents", *Polymer Engineering and Science*, 2014. 54(4): pp. 899–913.

[31] Pfannschmidt, O. and W. Michaeli, Foam injection molding of thermoplastics loaded with carbon dioxide prior to processing, 57th Annual Technical Conference – Society of Plastics Engineers, 1999, Vol. 2, pp. 2100–2103.

[32] Lee, J., L.S. Turng, E. Dougherty, and P. Gorton, "Novel Foam Injection Molding Technology Using Carbon Dioxide-Laden Pellets", *Polymer Engineering and Science*, 2011. 51(11): pp. 2295–2303.

[33] Peng, J., E. Yu, X. Sun, L.S. Turng, and X.F. Peng, "Study of microcellular injection molding with expandable thermoplastic microsphere", *International Polymer Processing*, 2011. 26(3): pp. 249–255.

[34] Peng, J., J. Wang, K. Li, L.S. Turng, and X.F. Peng, Effect of expandable thermoplastic microspheres on microcellular injection molded polylactic acid (PLA): Microstructure, surface roughness, and tensile properties", Annual Technical Conference – Society of Plastics Engineers, ANTEC Conf. Proc. 2011, 3, pp. 2643–2648.

[35] Bravo, V.L. and A.N. Hrymak, "Nozzle injection of physical blowing agents in the injection molding of microcellular foams", *International Polymer Processing*, 2005. 20(2): pp. 149–156.

[36] Rauwendaal, C., T. Osswald, P. Gramann, and B. Davis, New dispersive mixers for single screw extruders, 56th Annual Technical Conference – Society of Plastics Engineers, 1998, Vol. 1, pp. 277–283.

[37] Rauwendaal, C., "New developments in mixing and screw design", *Plastics, Additives and Compounding*, 2008. 10(6): pp. 32–36.

[38] Lee, J.W.S., J. Wang, J.D. Yoon, and C.B. Park, "Strategies to Achieve a Uniform Cell Structure with a High Void Fraction in Advanced Structural Foam Molding", *Industrial & Engineering Chemistry Research*, 2008. 47(23): pp. 9457–9464.

[39] Xu, J., Microcellular Injection Molding, Vol. 9, 2011. John Wiley & Sons.

[40] Pfannschmidt, O. and W. Michaeli, Foam injection molding of thermoplastics loaded with carbon dioxide prior to processing, 57th Annual Technical Conference – Society of Plastics Engineers, 1999, Vol. 2, pp. 2100–2103.

[41] Xu, X., C.B. Park, J.W.S. Lee, and X. Zhu, "Advanced structural foam molding using a continuous polymer/gas melt flow stream", *Journal of Applied Polymer Science*, 2008. 109(5): pp. 2855–2861.

[42] Rizvi, S.J.A. and N. Bhatnagar, "Optimization of Microcellular Injection Molding Parameters", *International Polymer Processing*, 2009. 24(5): pp. 399–405.

[43] Li, G., H. Li, L.S. Turng, S. Gong, and C. Zhang, "Measurement of gas solubility and diffusivity in polylactide", *Fluid Phase Equilibria*, 2006. 246(1–2): pp. 158–166.

[44] Gómez-Monterde, J., M. Sánchez-Soto, and M.L. Maspoch, "Influence of injection molding parameters on the morphology, mechanical and surface properties of ABS foams", *Advances in Polymer Technology*, 2018. 37(8): pp. 2707–2720.

[45] Lee, J.J. and S.W. Cha, Characteristics of the skin layers of microcellular injection molded parts", *Polymer-Plastics Technology and Engineering*, 2006. 45(7): pp. 871–877.

[46] Pantani, R., A. Sorrentino, V. Volpe, and G. Titomanlio, "Foam injection molding of poly(lactic acid) with physical blowing agents", AIP Conf. Proc., 2014. 1593: pp. 397–400.

[47] Pantani, R., V. Volpe, and G. Titomanlio, "Foam injection molding of poly(lactic acid) with environmentally friendly physical blowing agents", *Journal of Materials Processing Technology*, 2014. 214(12): pp. 3098–3107.

[48] Rizvi, S.J.A. and N. Bhatnagar, "Optimization of microcellular injection molding parameters", *International Polymer Processing*, 2009. 24(5): pp. 399–405.

[49] Volpe, V. and R. Pantani, "Effect of processing condition on properties of polylactic acid parts obtained by foam injection molding", *Journal of Cellular Plastics*, 2017. 53(5): pp. 491–502.

[50] Ameli, A., D. Jahani, M. Nofar, P.U. Jung, and C.B. Park, "Processing and characterization of solid and foamed injection-molded polylactide with talc", *Journal of Cellular Plastics*, 2013. 49(4): pp. 351–374.

[51] Volpe, V. and R. Pantani, "Foam injection molding of poly(lactic) acid: Effect of back pressure on morphology and mechanical properties", *Journal of Applied Polymer Science*, 2015. 132(48).

[52] Kastner, C., T. Mitterlehner, D. Altmann, and G. Steinbichler, "Backpressure Optimization in Foam Injection Molding: Method and Assessment of Sustainability", *Polymers*, 2020. 12(11): 2696.

[53] Volpe, V., M. De Filitto, V. Klofacova, F. De Santis, and R. Pantani, "Effect of mold opening on the properties of PLA samples obtained by foam injection molding", *Polymer Engineering & Science*, 2018. 58(4): pp. 475–484.

[54] Xie, P., G. Wu, Z. Cao, Z. Han, Y. Zhang, Y. An, and W. Yang, "Effect of mold opening process on microporous structure and properties of microcellular polylactide-polylactide nanocomposites", *Polymers*, 2018. 10(5): 554.

[55] Lim, L.-T., R. Auras, and M. Rubino, "Processing technologies for poly(lactic acid)", *Progress in Polymer Science*, 2008. 33(8): pp. 820–852.

[56] Najafi, N., M.-C. Heuzey, P.J. Carreau, D. Therriault, and C.B. Park, "Mechanical and morphological properties of injection molded linear and branched-polylactide (PLA) nanocomposite foams", *European Polymer Journal*, 2015. 73: pp. 455–465.

[57] Standau, T., C. Zhao, S. Murillo Castellón, C. Bonten, and V. Altstädt, Chemical modification and foam processing of polylactide (PLA)", *Polymers*, 2019. 11(2): 306.

[58] Ke, T.Y. and X.Z. Sun, "Melting behavior and crystallization kinetics of starch and poly(lactic acid) composites", *Journal of Applied Polymer Science*, 2003. 89(5): pp. 1203–1210.

[59] Jain, S., M. Misra, A.K. Mohanty, and A.K. Ghosh, "Thermal, mechanical and rheological behavior of poly(lactic acid)/talc composites", *Journal of Polymers and the Environment*, 2012. 20(4): pp. 1027–1037.

[60] Jain, S., M.M. Reddy, A.K. Mohanty, M. Misra, and A.K. Ghosh, "A New Biodegradable Flexible Composite Sheet from Poly(lactic acid)/Poly(?-caprolactone) Blends and Micro-Talc", *Macromolecular Materials and Engineering*, 2010. 295(8): pp. 750–762.

[61] Huda, M.S., L.T. Drzal, A.K. Mohanty, and M. Misra, "The effect of silane treated- and untreated-talc on the mechanical and physico-mechanical properties of poly(lactic acid)/newspaper fibers/ talc hybrid composites", *Composites Part B: Engineering*, 2007. 38(3): pp. 367–379.

[62] Pantani, R., F. De Santis, A. Sorrentino, F. De Maio, and G. Titomanlio, Crystallization kinetics of virgin and processed poly(lactic acid)", *Polymer Degradation and Stability*, 2010. 95(7): pp. 1148–1159.

[63] Gorrasi, G. and R. Pantani, "Effect of PLA grades and morphologies on hydrolytic degradation at composting temperature: Assessment of structural modification and kinetic parameters", *Polymer Degradation and Stability*, 2013. 98(5): pp. 1006–1014.

[64] Zhang, X., W. Ding, E. Chang, X. Chen, J. Chen, C.B. Park, and C. Shen, "Foaming Behaviors and Mechanical Properties of Injection-Molded Polylactide/Cotton-Fiber Composites", *Industrial & Engineering Chemistry Research*, 2020. 59(40): pp. 17885–17893.

[65] Ding, W., D. Jahani, E. Chang, A. Alemdar, C.B. Park, and M. Sain, "Development of PLA/cellulosic fiber composite foams using injection molding: Crystallization and foaming behaviors", *Composites Part A: Applied Science and Manufacturing*, 2016. 83: pp. 130–139.

[66] Wang, G., G. Zhao, S. Wang, L. Zhang, and C.B. Park, "Injection-molded microcellular PLA/graphite nanocomposites with dramatically enhanced mechanical and electrical properties for ultra-efficient EMI shielding applications", *Journal of Materials Chemistry C*, 2018. 6(25): pp. 6847–6859.

[67] Dong, Y., J. Marshall, H.J. Haroosh, S. Mohammadzadehmoghadam, D. Liu, X. Qi, and K.-T. Lau, "Polylactic acid (PLA)/halloysite nanotube (HNT) composite mats: Influence of HNT content and modification", *Composites Part A: Applied Science and Manufacturing*, 2015. 76: pp. 28–36.

[68] Prashantha, K., B. Lecouvet, M. Sclavons, M.F. Lacrampe, and P. Krawczak, "Poly(lactic acid)/ halloysite nanotubes nanocomposites: Structure, thermal, and mechanical properties as a function of halloysite treatment", *Journal of Applied Polymer Science*, 2013. 128(3): pp. 1895–1903.

[69] Eryildiz, M. and M. Altan, "Fabrication of polylactic acid/halloysite nanotube scaffolds by foam injection molding for tissue engineering", *Polymer Composites*, 2020. 41(2): pp. 757–767.

[70] Zhao, H., G. Zhao, L.S. Turng, and X. Peng, "Enhancing nanofiller dispersion through prefoaming and its effect on the microstructure of microcellular injection molded polylactic acid/clay nanocomposites", *Industrial & Engineering Chemistry Research*, 2015. 54(28): pp. 7122–7130.

[71] Lai, S.-M. and Y.-C. Lan, "Shape memory properties of melt-blended polylactic acid (PLA)/thermoplastic polyurethane (TPU) bio-based blends", *Journal of Polymer Research*, 2013. 20(5): pp. 1–8.

[72] Jing, X., H.Y. Mi, X.F. Peng, and L.S. Turng, "The morphology, properties, and shape memory behavior of polylactic acid/thermoplastic polyurethane blends", *Polymer Engineering & Science*, 2015. 55(1): pp. 70–80.

[73] Yu, F. and H.-X. Huang, "Simultaneously toughening and reinforcing poly(lactic acid)/thermoplastic polyurethane blend via enhancing interfacial adhesion by hydrophobic silica nanoparticles", *Polymer Testing*, 2015. 45: pp. 107–113.

[74] Mi, H.-Y., M.R. Salick, X. Jing, B.R. Jacques, W.C. Crone, X.F. Peng, and L.S. Turng, "Characterization of thermoplastic polyurethane/polylactic acid (TPU/PLA) tissue engineering scaffolds fabricated by microcellular injection molding", *Materials Science and Engineering: C*, 2013. 33(8): pp. 4767–4776.

[75] Xu, D., K. Yu, K. Qian, and C.B. Park, "Foaming behavior of microcellular poly(lactic acid)/TPU composites in supercritical CO_2", *Journal of Thermoplastic Composite Materials*, 2018. 31(1): pp. 61–78.

[76] Zhao, H., Z. Cui, X. Sun, L.S. Turng, and X. Peng, "Morphology and properties of injection molded solid and microcellular polylactic acid/polyhydroxybutyrate-valerate (PLA/PHBV) blends", *Industrial & Engineering Chemistry Research*, 2013. 52(7): pp. 2569–2581.

[77] Zhao, H. and L.-S. Turng, Mechanical performance of microcellular injection molded biocomposites from green plastics: PLA and PHBV. In *Biocomposites*, M. Misra, J.K. Pandey, A.K. Mohanty (Eds.), 2015. Woodhead Publishing, pp. 141–160.

[78] Hrimchum, K., D. Aussawasathien, and T. Kajornprai, Injection Moldable Poly(Lactic Acid)-Poly (Butylene Succinate)-Activated Carbon Composite Foams: Effects of PLA/PBS Ratios. In Key Engineering Materials, 2019. Trans. Tech. Publ.

[79] Pradeep, S.A., H. Kharbas, L.S. Turng, A. Avalos, J.G. Lawrence, and S. Pilla, "Investigation of thermal and thermomechanical properties of biodegradable PLA/PBSA composites processed via supercritical fluid-assisted foam injection molding", *Polymers*, 2017. 9(1): 22.

[80] Campuzano, J. and I. Lopez, "Study of the effect of dicumyl peroxide on morphological and physical properties of foam injection molded poly (lactic acid)/poly(butylene succinate) blends", *Express Polymer Letters*, 2020. 14(7).

[81] Ishihara, S., Y. Hikima, and M. Ohshima, "Preparation of open microcellular polylactic acid foams with a microfibrillar additive using coreback foam injection molding processes", *Journal of Cellular Plastics*, 2018. 54(4): pp. 765–784.

8 Microfibrillation as a Facile Processing Technique to Produce Improved Biobased Thermoplastic Blends

Adel Ramezani Kakroodi, Yasamin Kazemi, and Chul B. Park

This chapter summarizes the results of two of our studies [1, 2] on the use of the microfibrillation process as a practical and facile technique to enhance the properties of a biobased thermoplastic, namely, poly(lactic acid) (PLA). In this process, a blend of two immiscible thermoplastics is first formed, through extrusion, and then stretched to transform the dispersed phase's domains into fibrils with submicronic diameters. Having a melting point higher than that of the continuous matrix phase, the dispersed phase maintains its fibrillar shape even after post-processing. This chapter provides insight into the processing-structure-property relationships in microfibrillar composites with biobased polymers. Special attention is paid to the effects of the stretched microfibrils on the performance and characteristics of the matrix phase.

8.1 Introduction

Despite the desire to replace petroleum-based thermoplastics with biobased alternatives, the majority of the efforts dedicated to the vast usage of bioplastics have had limited success due to their many limitations. For instance, PLA, which has many promising attributes, also suffers from limitations such as slow crystallization kinetics and poor melt strength, which restrict its application in film blowing and foaming [3, 4]. Its intrinsic brittleness, due to having short and semirigid chains, and its low thermal stability also prevent the large-scale adoption of PLA in the packaging industry. Many methods for the improvement of PLA's properties suffer from one or more drawbacks, such as high cost, environmental and health hazards, and low practicality. Among these methods, blending PLA with other polymers is considered as the most cost-effective and practical way to enhance its properties. However, the ability to control the morphology of such blends as a means to tailor their properties - for example, their mechanical properties, crystallization kinetics, and processability - has been an ongoing challenge. Furthermore, to maintain the green nature of the PLA mixtures, it is essential to keep the concentration of the second polymeric phase in the blends to a minimum.

As noted above, the ability to control the morphology of a polymeric blend is vital. Among all blend morphologies, in situ microfibrillar composites (MFCs) have shown great potential to improve the crystallization behavior, mechanical properties, and melt strength of thermoplastic hosts such as polypropylene (PP) and polyethylene (PE). In microfibrillation, a homogeneous blend of matrix (A) and dispersed phase (B) polymers is first produced using conventional melt blending equipment, where the melting point (T_m) of B must be at least 40 °C higher than that of A and the processing temperature ($T_{processing}$) is $T_{processing1} > T_{mA}, T_{mB}$. The blend is subsequently stretched, either in molten (hot stretching) or frozen (cold drawing) state, which transforms B from spherical domains into highly oriented microfibrils with high aspect ratios. Then, the resulting fibrillated blend is further processed at $T_{processing2}$ ($T_{mA} < T_{processing2} < T_{mB}$) to produce isotropic products with the desired shapes. During this step, the matrix melts (and its molecules relax) and it is reshaped while the dispersed phase microfibrils maintain their microfibrillar shapes. The success of this process is controlled by characteristics of the A and B polymers such as their viscosity ratio, compatibility, melting points, and the processing conditions [1, 2, 5, 6].

The unique procedure by which MFCs are produced has several advantages over the traditional composite production methods. First, in traditional composites, especially nanocomposites, the strength and stiffness are usually restricted by a poor dispersion of fillers in the matrix [7]. In MFCs, on the other hand, the microfibrils are made by the elongation of already well-dispersed spherical B domains, making the microfibrils perfectly dispersed. Naturally, a high level of dispersion of the microfibrils increases their specific surface area, and this substantially benefits their properties. Second, the B microfibrils' high aspect ratio, in conjunction with their flexible nature, causes them to bend and form an entangled network. Such a network has been reported to significantly improve the matrix's melt strength and, consequently, its processability. Third, MFCs eliminate the environmental concerns and health hazards that are associated with the production and use of traditional fibers. Fourth, MFCs are less expensive, lighter, and easier to produce than traditional composites.

In the following sections, the results of two previously published studies, performed by our group, on the effects of microfibrillation on PLA's various properties are provided [1, 2].

8.2 Microstructure of Microfibrillar Composites

As mentioned above, microfibrillar composites are characterized as having a network of flexible and thin thermoplastic microfibrils perfectly dispersed within a sea of a lower-melting-point plastic. Figure 8.1 shows the scanning electron microscopy (SEM) images of a microfibrillated compound of PLA, with a melting point of around 170 °C, and polyamide 6 (PA6), with a melting point of 220 °C [2]. In addition, quantitative analyses of the distribution of diameters of the PA6 microfibrils in each blend are shown. In this work, PA6 was dried at 100 °C for at least 8 h to avoid hydrolysis. PLA was also dried at 70 °C for 12 h in a vacuum oven. PLA/PA6 blends with 3 wt %, 7 wt %, 15 wt %, and 25 wt % of PA6 were produced using a co-rotating twin-screw extruder (Leistritz) with a screw diameter of 27 mm and an *L/D* ratio of 40. The temperature profile in the extruder's zones was from 160 °C to 230 °C. The processing conditions were chosen to provide a high level of mixing while minimizing PLA's degradation. In order to produce MFCs, the extrudates with a melt temperature of around 235 °C were hot-stretched at the die exit using a take-up roller at a high draw ratio (that is, the ratio of the area of the transverse section of the unstretched extrudate to that of the stretched extrudate) of nearly 2000 and were collected in the form of fibers with diameters of 80 µm. The drawn extrudates froze before reaching the take-up roller, which was located 3 m away from the die exit. The produced fibers were then chopped into 1 mm long pieces. All of the produced samples were compression-molded using a laboratory hydraulic press, Carver SC7620, at 180 °C for 5 min. This was followed by cooling at a rate of 40 °C/min, using cooling air, to form isotropic MFCs. This temperature was chosen to maintain the microfibrillar structure of the PA6 domains in the MFCs.

As shown in Figure 8.1, after hot stretching, the PA6 phase was fully extended into fine microfibrillar structures. The average diameter of the PA6 microfibrils in the PLA/PA6 97/3 (w/w) was as low as 198 nm. Increasing the concentration of the PA6 phase to 25 wt % increased the average diameter of the microfibrils to 233 nm. Further, the thin PA6 microfibrils were fully dispersed in the PLA matrix with no trace of agglomeration of the fibers, even at high concentrations. The microfibrils also seemed to have very high aspect ratios. Additionally, it was reported that the generated microfibrils had much larger sizes, and volumes, than their precursor spheres. This was especially the case in blends with 3 wt % and 7 wt % of PA6, and it was attributed to the simultaneous extension and coalescence of the dispersed spheres during the hot-stretching stage.

Figure 8.2, which was taken after the selective etching of the PLA matrix using a water/methanol/sodium hydroxide solution, also demonstrates the long and flexible nature of the PA6 fibrils.

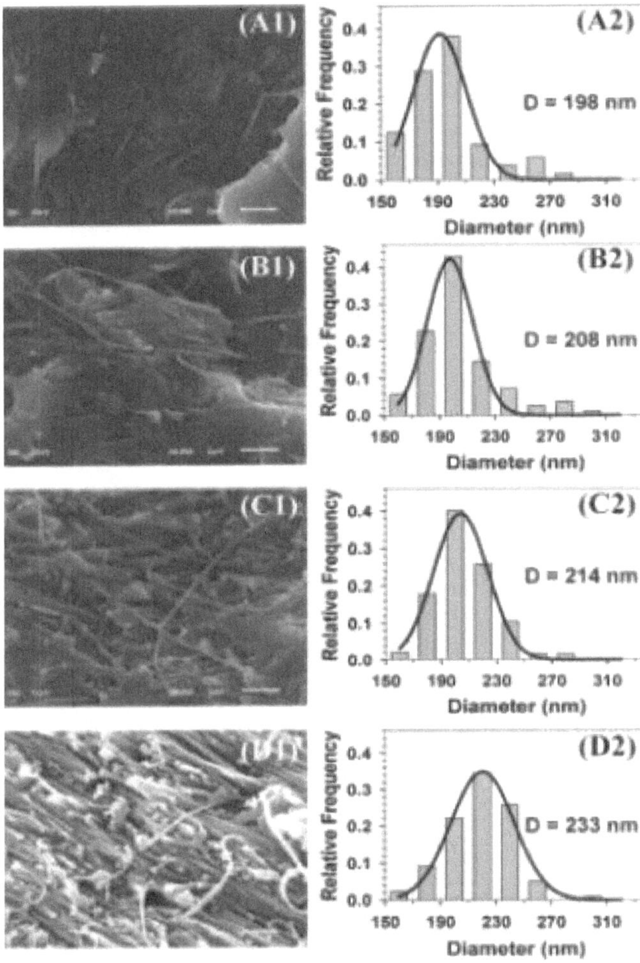

Figure 8.1 SEM micrographs of as-broken surfaces of PLA/PA6 in situ MFCs with 3 wt % (A1), 7 wt % (B1), 15 wt % (C1), and 25 wt % (D1) of PA6 microfibrils (all scale bars: 2 µm). Quantitative analyses of the distributions of diameters of PA6 microfibrils in blends with 3 wt %, 7 wt %, 15 wt %, and 25 wt % PA6 are also shown in (A2), (B2), (C2), and (D2), respectively. *D* stands for the average diameter [2]

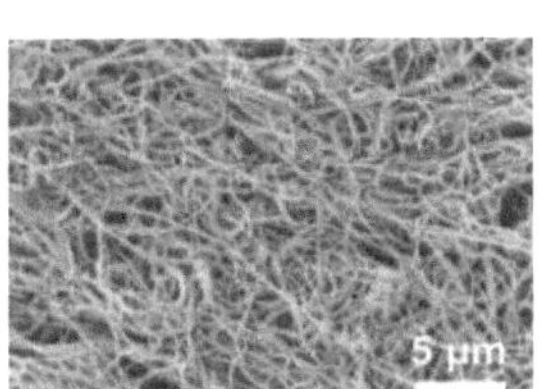

Figure 8.2
SEM micrograph of a fractured surface of PLA/PA6 97/3 (w/w) in situ MFC after the surface etching of the PLA matrix [2]

8.3 Effects of Microfibrillation on Rheological Properties of Bioplastics

The rheological analysis of polymers and their composites is an efficient method to provide insight into their microstructures as well as their processing behaviors [8]. The rheological properties of pure polymers are controlled by the topological interactions (entanglement) of the polymer chains and, in low-molecular-weight polymers, by the friction between the molecules. However, the relaxation dynamics of the polymer molecules in composite melts are significantly affected by several additional factors such as filler size and shape, deformability, level of dispersion, and the level of adhesion between the polymer molecules and the fillers. Figure 8.3 shows the shear responses of a pure PLA and its microfibrillar composite melts with a partially biobased polyamide 610 (PA610) and a fully biobased polyamide 1010 (PA1010) [1].

As shown in Figure 8.3 (A1) and (B1), a classic viscoelastic response was obtained for the pure PLA melt. The storage modulus (G') of this sample was strongly dependent on the frequency in the low frequency range. This suggests a viscous behavior of the melt at low frequencies (long time scales). However, an elastic response was observed at high frequencies (short time scales), where the storage modulus of the PLA melt became increasingly independent of the frequency [9]. The semilogarithmic plot of the phase angle δ versus the complex shear modulus G^* (known as van Gurp plot) of the pure PLA melt is presented in Figure 8.3 (A2) and (B2). A phase angle of nearly 90°, at low values of G^*, was calculated for the PLA melt, which is also a proof of the dominance of the viscous flow (a liquid-like behavior). The liquid-like behavior of the PLA melt can be attributed to PLA's typically short and semirigid chains. Such a characteristic is highly undesirable from a processing point of view, especially in applications such as film forming, foaming, and fiber spinning, where good melt strength is required. Remarkably, Figure 8.3 shows that the in situ generation of even a small number of PA microfibrils resulted in a significant increase in the elasticity of PLA's melt. The addition of only 3 wt % of either PA610 or PA1010 increased the storage modulus of PLA by over three orders of magnitude. Furthermore, the dependence of the samples' G' on frequency, at low frequencies, was also dramatically decreased. The van Gurp plots, Figure 8.3 (A2) and (B2), also demonstrate that the phase angles of the MFCs were significantly smaller than those of pure PLA. This is indicative of a transition from the liquid-like behavior to a solid-like behavior upon the introduction of the microfibrils. The transition in the melt's behavior can be attributed to the topological constraint of motion of the PA phase in the composites due to the formation of a physically entangled network of the long and flexible microfibrils, which is capable of storing the deformation energy over large time scales. The formation of such a stable network is also known as physical gelation or rheological percolation of the microfibrils [10]. It is worth mentioning that after the hot-stretching step in the production of the MFCs, the PA microfibrils generated were highly oriented and incapable of forming a network. During the isotropization step, however, the microfibrils lost their orientation and became interconnected. According to Winter and Chambon [11, 12], at the gel point (that is, the smallest filler concentration at which gelation occurs), G' and G'' (the loss modulus) have similar dependencies on the frequency and, as a result, the loss tangent ($\tan\delta = G''/G'$) becomes independent of the frequency. In Figure 8.3 (A3) and (B3), the loss tangents are plotted as a function of the microfibril content for different frequencies. Using Winter and Chambon's criterion, the gel point of the MFCs would be the concentration at which the lines of the loss tangents for different frequencies intersect. Thus, the gel point of the MFCs has been calculated at 2–3 wt % of PA microfibrils.

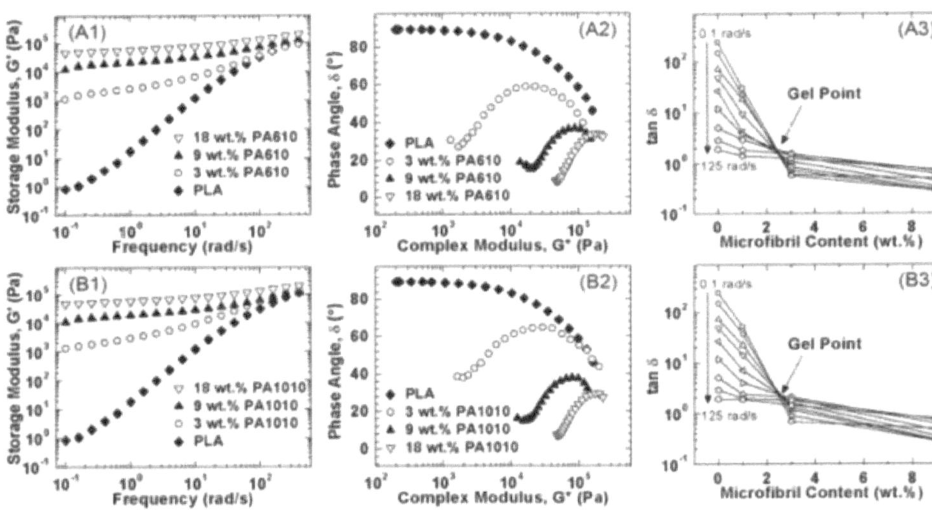

Figure 8.3 Linear viscoelastic behavior of pure PLA compared with those of PLA/PA610 ((A1), (A2), and (A3)) and PLA/PA1010 ((B1), (B2), and (B3)) microfibrillar composites. In (A3) and (B3), different lines represent different frequencies and the gel point appears as the point where the lines intersect [1]

The response of a polymeric melt during an extensional flow is mainly controlled by the physical entanglement of its macromolecules. These entanglements, in turn, depend on molecular structure characteristics such as length, deformability, and degree of branching [13]. For instance, PLA's linear, short, and semirigid chains are known to cause poor performance during the elongational flow [3, 14]. Figure 8.4 [2] shows the responses of pure PLA and PLA/PA6 97/3 (w/w) MFC melts under elongational deformations at the strain rates of 0.01 s^{-1}, 0.1 s^{-1}, and

1 s^{-1}. We have also provided the linear viscoelastic behavior envelope of each sample, given by $\eta_E^+ = 3\eta^+$. Here, η_E^+ is the elongational viscosity and η^+ is the sample's linear viscoelastic shear viscosity, obtained from the start-up of a steady shear flow at a strain rate of 0.003 s−1. The deviation of the experimental values of the elongational viscosities from this envelope is a measure of the strain hardening behavior [12, 15]. As expected, PLA had a very low elongational viscosity and a negligible strain hardening behavior at all strain rates. The PLA's deformability was also small due to the premature breakage of its melt, which was caused by the quick disentanglement of its chains during the flow. It is shown in Figure 8.4 that the inclusion of 3 wt % of PA6 microfibrils led to significant improvements in the PLA melt's elongational characteristics. Strong strain hardening effects (nearly an order of magnitude increase in the elongational viscosity) were observed in the case of the microfibrillar compound, which suggested an increase in the relaxation times of the PLA molecules [13]. The PLA melt's deformability was also significantly improved, especially at a strain rate of 0.01 s−1. This finding is of great importance in the processing of linear polymers. The formation of a three-dimensional network of deformable and well-dispersed PA6 microfibrils, which was also confirmed by the shear viscosity measurements, was responsible for the improved elongational behavior of the PLA melt. The excessively high melt viscosities of the MFCs with a PA6 content of more than 3 wt % made it impossible to perform elongational viscosity measurement experiments.

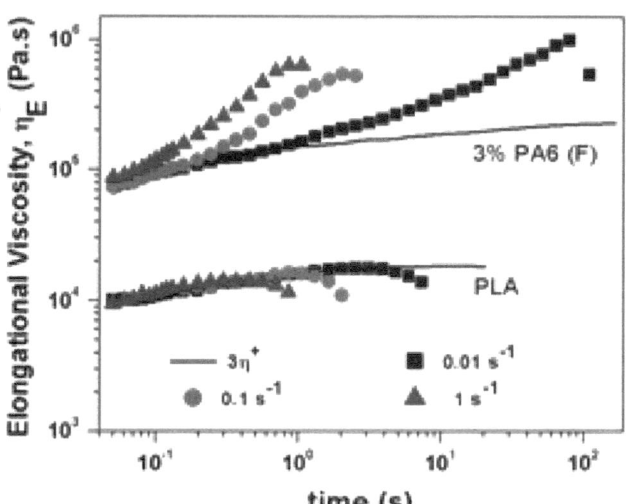

Figure 8.4 Uniaxial elongational responses of PLA and PLA/PA6 97/3 (w/w) MFC melts at strain rates of 0.01 s−1, 0.1 s−1, and 1 s−1 [2]

8.4 Effects of Microfibrillation on Crystallization Behavior of Bioplastics

The non-isothermal differential scanning calorimetry (DSC) heating thermograms of compression-molded films are presented in Figure 8.5 [1]. It is shown that, in the case of pure PLA, a broad exothermic peak can be observed from 90 °C to nearly 155 °C, which results from the cold crystallization of PLA during the heating scan. The number of crystals that were formed during this cold crystallization process nearly equaled the total number of PLA crystals that were later molten from 165 °C to 175 °C during the same heating process. This proves that almost no crystals had been formed within the pure PLA sample during the cooling step (at a cooling rate of 40 °C/min) in the compression molding process. This can be ascribed to the extremely slow crystallization of PLA. Figure 8.5 (A) and (B), on the other hand, show that in the case of the microfibrillar composites with 3 wt % and 9 wt % microfibril content, only a small number of crystals were formed during the heating process in the temperature range of 150 °C to 160 °C. In addition, during the heating of the MFCs with 18 wt % of microfibrils, no cold crystallization occurred. These observations suggest that nearly all of the crystal content of the MFCs was previously formed during the cooling step of the compression molding process. This signals a dramatic improvement in the crystallization kinetics of the PLA upon the in situ generation of the PA microfibrils, which is caused by the heterogeneous crystal nucleating effect of the microfibrils.

The non-isothermal DSC thermograms in Figure 8.5 also provide important information regarding the glass transition temperatures (T_gs) of the samples, displayed by the endothermic peaks at around 60 °C in their DSC scans. It is shown that the T_gs of the MFCs were generally around 4–5 °C higher than that of pure PLA. The increase in the glass transition temperature of PLA can be attributed to the confinement of the segmental movement of the PLA chains (in the amorphous region), caused by two major factors: (i) interactions between the PLA molecules and the (well-dispersed and fine) PA microfibrils – this effect is amplified by the affinity between PLA and PA [16] – and (ii) presence of a large number of small crystals in the MFCs (due to the crystal nucleating effect of the microfibrils) that act as physical cross-links [17].

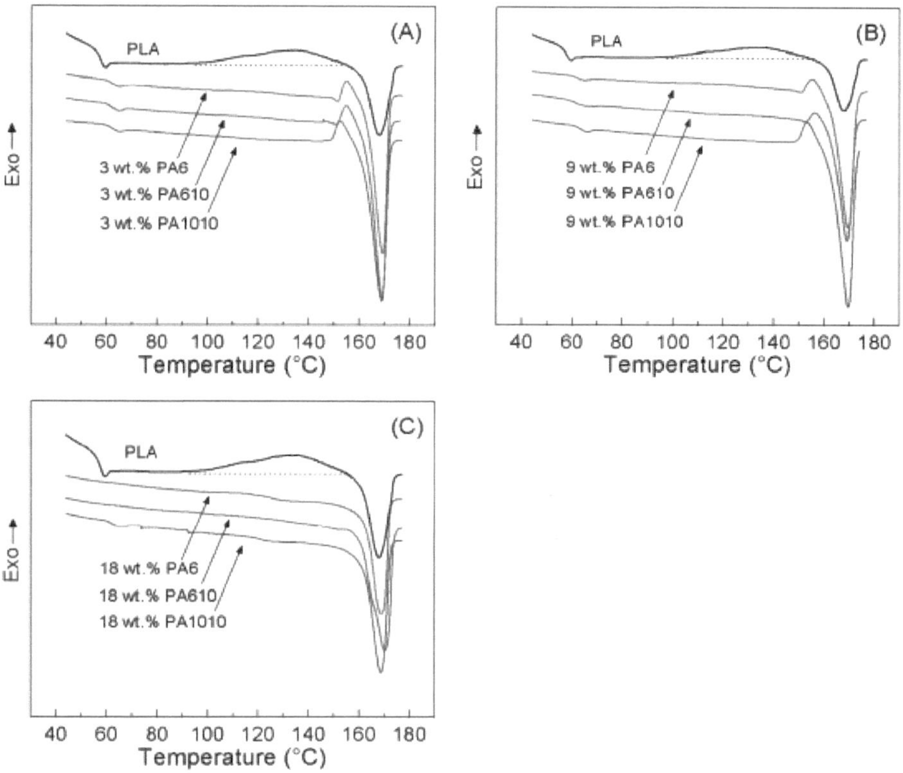

Figure 8.5 Non-isothermal DSC thermograms of PLA and MFCs with microfibril content of 3 wt % (A), 9 wt % (B), and 18 wt % (C) [1]

Polarized optical microscopy (POM) micrographs of the isothermally crystallized PLA and PLA/PA1010 97/3 (w/w) MFC are shown in Figure 8.6. This figure clearly demonstrates that the microfibrils' heterogeneous crystal nucleation effect led to a significant reduction in the average size and an increase in the number of the PLA crystals.

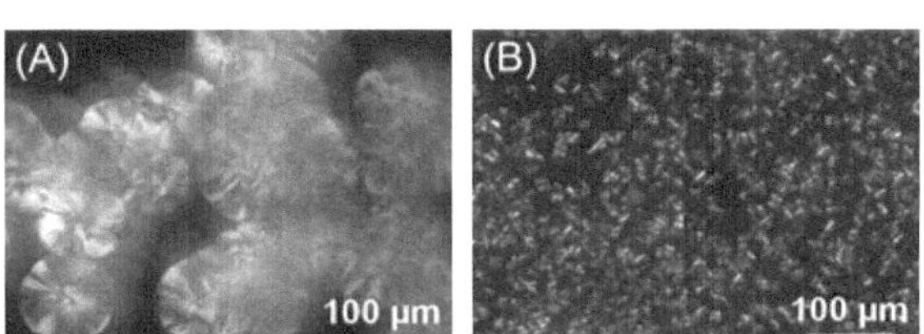

Figure 8.6 POM micrographs of pure PLA (A) and PLA/PA1010 97/3 (w/w) MFC (B) after the completion of their melt crystallizations at 140 °C [1]

Figure 8.7 presents the isothermal crystallization behavior of PLA and the MFCs with PA610 and PA1010 microfibrils under atmospheric pressure and at 97 °C [1]. This figure confirms that the crystallization of the pure PLA was very slow. According to Figure 8.8, the PLA's crystallization half-time ($t_{1/2}$, time required to achieve 50% of the final crystallinity) was 1412 s. The completion of the PLA crystallization took nearly 50 min. Such slow crystallization is extremely unfavorable from a processing point of view. The in situ formation of both PA microfibrils was shown to improve PLA's crystallization kinetics dramatically. The crystallization half-time of PLA was reduced to 181 s and 196 s after the inclusion of only 3 wt % of PA610 and PA1010 microfibrils, respectively. As mentioned before, the improvement in the crystallization kinetics of PLA is a result of the heterogeneous crystal nucleation effect of the microfibrils, which occurs, in part, due to their very fine diameters (that is, large specific surface areas). In other words, the fine microfibrils provide more heterogeneous crystal nucleation sites. Despite the improvement in the PLA's crystallization kinetics after the inclusion of the microfibrils, Figure 8.8 shows that the final crystallinities of the MFCs were lower than that of pure PLA. For instance, the final crystallinities of the MFCs with 18 wt % of PA610 and PA1010 microfibrils were 26.2% and 27.8%, respectively, in contrast with 34.8% for pure PLA. This effect can be ascribed to the severe reduction in the mobility of the PLA chains in the MFCs, caused by the abovementioned reasons, which prevents them from reaching the crystallization sites. It must be noted that although the pure PLA had higher final crystallinity than the MFCs, its crystallization occurred much more slowly. Under the high cooling rates that are normally applied during the processing of polymers, the pure PLA is likely to produce completely amorphous products.

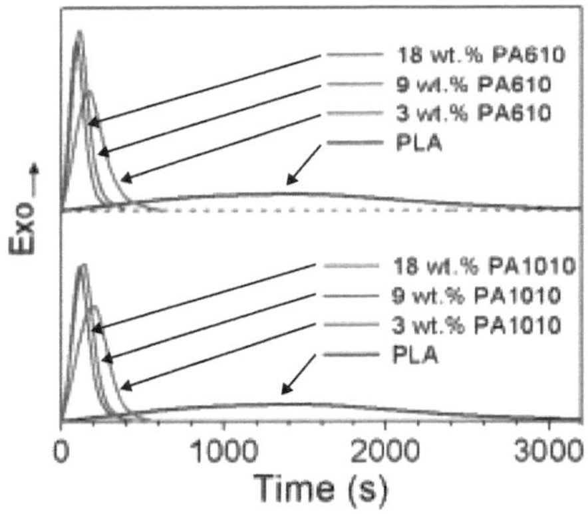

Figure 8.7
Isothermal crystallization behavior of PLA and MFCs with PA610 and PA1010 microfibrils under atmospheric pressure and at 97 °C [1]

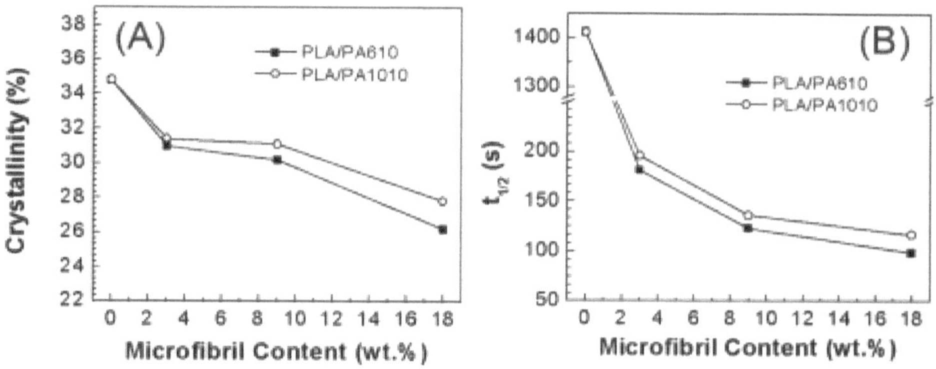

Figure 8.8 Values of the final crystallinities (A) and crystallization half-times ($t_{1/2}$) (B) for isothermal crystallization of pure PLA and MFCs with PA610 and PA1010 microfibrils at 97 °C [1]

8.5 Effects of Microfibrillation on Mechanical Properties of Bioplastics

The improvements in the crystallization kinetics and the crystallinity of PLA after the creation of PA6 microfibrils could dramatically improve PLA's mechanical properties. The tensile properties of the pure PLA and the PLA/PA6 blends with spherical and microfibrillar PA6 domains, both before and after isothermal crystallization at 97 °C for 7 min, are presented in Figure 8.9 [2]. It is shown in Figure 8.9 (A) that the inclusion of PA6 in the form of spherical domains did not improve PLA's tensile modulus significantly even after the isothermal crystallization process. On the other hand, the tensile moduli of the MFCs were shown to be significantly higher than that of PLA. Remarkably, the presence of only 3 wt % of microfibrillar PA6 was sufficient to improve PLA's tensile modulus by 38% after isothermal crystallization for 7 min. The highest tensile modulus was observed for the sample with 15 wt % PA6 microfibrils. However, a further increase in the concentration of PA6 to 25 wt % resulted in a sharp drop in the compound's modulus. Such a behavior was clearly the result of the flawed microstructure of this compound. Figure 8.9 (A) shows that in the isothermally crystallized samples, this reduction in modulus was much less noticeable. This could be ascribed to the presence of crystals or even to possible improvements in the sintering of the PLA/PA6 fibers. Figure 8.9 (B) also shows that the inclusion of small amounts of PA6 in the form of in situ generated microfibrils increased the tensile strain at break of PLA significantly. The strain at break of the compound with 3 wt % microfibrillar PA6 was 82% higher than that of pure PLA. This was a result of the high flexibility of the long PA6 microfibrils and also of their crystal nucleation effect, which led to the formation of a large number of small crystals. The isothermal crystallization of the blends, however, reduced their deformability. The reduction was more noticeable after the crystallization of the MFCs, which was due to their higher crystallinities. Figure 8.9 (C) shows that the tensile strength of the microfibrillar PLA/PA6 97/3 (w/w) compound was 55% (before the isothermal crystallization) and 80% (after the isothermal crystallization) higher than that of the pure PLA. Such a remarkable improvement in PLA's strength was a result of the simultaneous increase in its modulus and deformability after the formation of the PA6 microfibrils.

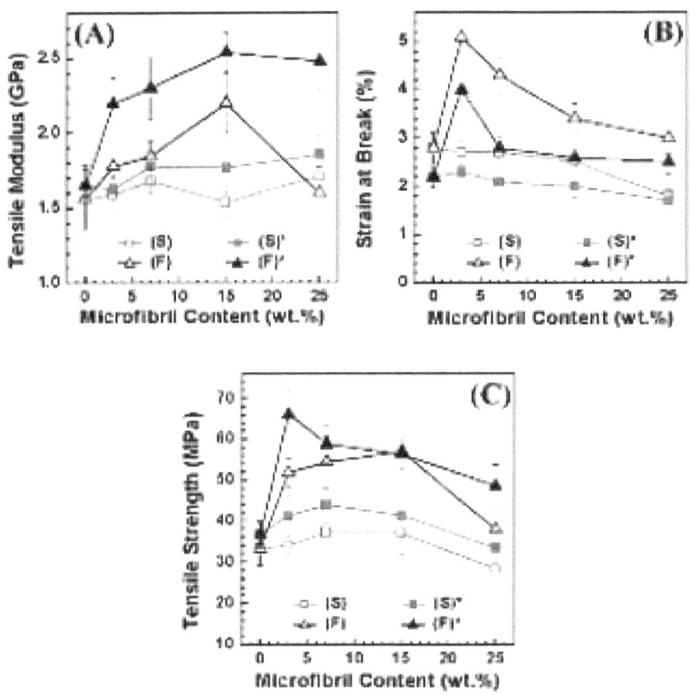

Figure 8.9 Tensile modulus (A), strain at break (B), and tensile strength (C) of pure PLA and blends with spherical and microfibrillar PA6 domains. Isothermally crystallized samples are denoted by asterisks. (S) and (F) stand for blends with spherical and microfibrillar PA6 domains, respectively [2]

8.6 Effects of Microfibrillation on Foaming Ability of Bioplastics

The changes in bioplastics' rheological properties (improved elasticity and strain hardening) after the in situ microfibrillation process promise improvements in their foaming-ability [18, 19]. Batch foaming experiments were performed at 140 °C under a CO_2 atmosphere with a pressure of 13.8 MPa for a 1 h exposure time [2]. Figure 8.10 (A) and (B1) show that the inclusion of the 3 wt % microfibrillar PA6 phase significantly improves the microstructure of the produced foams. The addition of the 3 wt % PA6 microfibrils decreases the PLA foam's average cell size from 70 µm to 23 µm while its cell density increases by a remarkable two orders of magnitude (from 2.4×10^6 cells/cm^3 to 1.8×10^8 cells/cm^3). Such dramatic enhancements in the microstructure of the foams can be primarily ascribed to the effects of the previously noted three-dimensional network of the physically entangled PA6 microfibrils on the elasticity and strength of the PLA melt. The MFC melt's strain hardening behavior prevented the cells from collapsing during the extensional flow, which has been induced in the cell growth stage of the foaming process. Figure 8.10 (B2), on the other hand, shows a very poor microstructure for the foamed samples of PLA/PA6 97/3 (w/w) with spherical PA6 domains. A lack of capability of the spherical PA6 domains to enhance the PLA's rheological properties is responsible for the undesirable foam microstructure. Figure 8.10 (C1) shows that foaming of the compound with 7 wt % PA6 microfibrils is impossible. No cells have been generated in the sample while the escaping gas formed huge pockets just beneath the skin layers. This behavior is the result of an excessive increase in the melt viscosity of the blend. We mentioned earlier that it was impossible to even measure the elongational viscosity of this sample due to its high melt strength. Samples with higher concentrations of PA6 microfibrils showed similar behaviors and were not included in Figure 8.10.

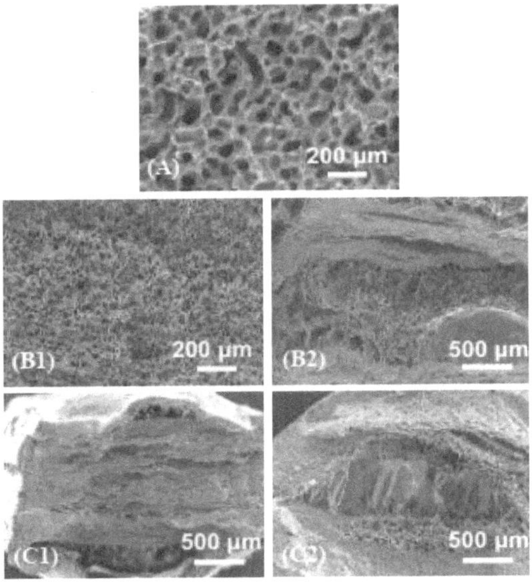

Figure 8.10 SEM micrographs of cryogenically fractured surfaces of foamed samples of pure PLA (A), PLA/PA6 97/3 (w/w) with microfibrillar (B1) and spherical (B2) PA6 domains, and PLA/PA6 93/7 (w/w) with microfibrillar (C1) and spherical (C2) PA6 domains. Note that the magnifications of the (A) and (B1) images are different from the magnifications of (B2), (C1), and (C2) [2]

8.7 Conclusions

The results presented in the studies mentioned in this chapter demonstrate that microfibrillation is a practical, cost-effective, and environmentally friendly method and that it guarantees the attainment of melt blending as a technique to improve PLA's characteristics. After the inclusion of only 3 wt % of a PA6 microfibrillar phase, significant improvements were observed in PLA's characteristics. Thus, we believe that this is an ideal and facile approach to PLA's and other bioplastics' commercialization, turning PLA into a competitive candidate in the commodity thermoplastics market without sacrificing its biodegradability.

References for Chapter 8

[1] A. R. Kakroodi, Y. Kazemi, M. Nofar, and C. B. Park, "Tailoring poly(lactic acid) for packaging applications via the production of fully bio-based in situ microfibrillar composite films," *Chem. Eng. J.*, vol. 308, pp. 772–782, 2017.

[2] A. R. Kakroodi, Y. Kazemi, W. D. Ding, A. Ameli, and C. B. Park, "Poly(lactic acid)-Based in Situ Microfibrillar Composites with Enhanced Crystallization Kinetics, Mechanical Properties, Rheological Behavior, and Foaming Ability," *Biomacromolecules*, vol. 16, no. 12, pp. 3925–3935, 2015.

[3] M. Nofar, A. Tabatabaei, and C. B. Park, "Effects of nano-/micro-sized additives on the crystallization behaviors of PLA and PLA/CO_2 mixtures," *Polymer*, vol. 54, pp. 2382–2391, 2013.

[4] A. Ameli, D. Jahani, M. Nofar, P. U. Jung, and C. B. Park, "Development of high void fraction polylactide composite foams using injection molding: Mechanical and thermal insulation properties," *Compos. Sci. Technol.*, vol. 90, pp. 88–95, 2014.

[5] K. Jayanarayanan, T. Jose, S. Thomas, and K. Joseph, "Effect of draw ratio on the microstructure, thermal, tensile and dynamic rheological properties of insitu microfibrillar composites," *Eur. Polym. J.*, vol. 45, pp. 1738–1747, 2009.

[6] R. Gonzalez-Nunez, B. D. Favis, P. J. Carreau, and C. Lavallée, "Factors influencing the formation of elongated morphologies in immiscible polymer blends during melt processing," *Polym. Eng. Sci.*, vol. 33, pp. 851–859, 1993.

[7] S. Fujisawa, T. Saito, S. Kimura, T. Iwata, and A. Isogai, "Surface engineering of ultrafine cellulose nanofibrils toward polymer nanocomposite materials," *Biomacromolecules*, vol. 14, pp. 1541–1546, 2013.

[8] Y. Li, M. Kröger, and W. K. Liu, "Nanoparticle geometrical effect on structure, dynamics and anisotropic viscosity of polyethylene nanocomposites," *Macromolecules*, vol. 45, pp. 2099–2112, 2012.

[9] D. S. Bangarusampath, H. Ruckdäschel, V. Altstädt, J. K. W. Sandler, D. Garray, and M. S. P. Shaffer, "Rheology and properties of melt-processed poly(ether ether ketone)/multi-wall carbon nanotube composites," *Polymer*, vol. 50, pp. 5803–5811, 2009.

[10] S. R. Raghavan and J. F. Douglas, "The conundrum of gel formation by molecular nanofibers, wormlike micelles, and filamentous proteins: gelation without cross-links?," *Soft Matter*, vol. 8, pp. 8539–8546, 2012.

[11] F. Chambon and H. H. Winter, "Linear Viscoelasticity at the Gel Point of a Crosslinking PDMS with Imbalanced Stoichiometry," *J. Rheol.*, vol. 31, pp. 683–697, 1987.

[12] A. Rizvi and C. B. Park, "Dispersed polypropylene fibrils improve the foaming ability of a polyethylene matrix," *Polymer*, vol. 55, pp. 4199–4205, 2014.

[13] L. Palade, H. J. Lehermeier, and J. R. Dorgan, "Melt Rheology of High L-Content Poly(lactic acid)," *Macromolecules*, vol. 34, pp. 1384–1390, 2001.

[14] Y. Di, S. Iannace, E. Di Maio, and L. Nicolais, "Poly(lactic acid)/organoclay nanocomposites: Thermal, rheological properties and foam processing," *J. Polym. Sci. Part B Polym. Phys.*, vol. 43, pp. 689–698, 2005.

[15] A. Rizvi, C. B. Park, and B. D. Favis, "Tuning viscoelastic and crystallization properties of polypropylene containing *in-situ* generated high aspect ratio polyethylene terephthalate fibrils," *Polymer*, vol. 68, pp. 83–91, 2015.

[16] J. M. Herrera-Alonso, E. Marand, J. C. Little, and S. S. Cox, "Transport properties in polyurethane/clay nanocomposites as barrier materials: Effect of processing conditions," *J. Memb. Sci.*, vol. 337, no. 1–2, pp. 208–214, 2009.

[17] L. S. Loo, R. E. Cohen, and K. K. Gleason, "Chain mobility in the amorphous region of nylon 6 observed under active uniaxial deformation," *Science*, vol. 288, no. 5463, pp. 116–119, 2000.

[18] N. Najafi, M. Heuzey, P. J. Carreau, D. Therriault, and C. B. Park, "Rheological and foaming behavior of linear and branched polylactides," *Rheol. Acta*, vol. 53, pp. 779–790, 2014.

[19] J. Wang, W. Zhu, H. Zhang, and C. B. Park, "Continuous processing of low-density, microcellular poly(lactic acid) foams with controlled cell morphology and crystallinity," *Chem. Eng. Sci.*, vol. 75, pp. 390–399, 2012.

9 Electrospinning of Biodegradable Polymers

Shih-Jung Liu

Electrospinning has become one of the most important methods to manufacture micro-/nanoscale fibers, due to its capability to produce nanostructures with extraordinary features such as high porosity, big surface area, controllable morphology, and high chemical and thermal stability. This chapter aims to provide a general introduction regarding the electrospinning of biodegradable polymers, including natural and synthetic ones. The essential device used for the electrospinning process is introduced and variants of electrospinning, including needleless electrospinning, electrospraying, and others, are also addressed. The interaction between the injected solution and the applied electric voltage is discussed, and insight into the spinning behavior of the fluid is summarized. Moreover, the influence of process parameters on the morphology, dimensions, and properties of electrospun fibers is discussed. The electrospinning of various natural polymers - including deoxyribonucleic acid, silk fibroin, fibrinogens, dextran, chitin, chitosan, alginate, collagen, and gelatin - and synthetic polymers such as polycaprolactone, poly(lactic acid), and poly(lactic-*co*-glycolic acid) is addressed. Finally, the future trend for the electrospinning of biodegradable polymers is discussed.

9.1 Introduction

Nanotechnology comprises the understanding of basic chemistry, physics, biology, and technology at the nanoscale. This technology helps to sustainably promote - if not revolutionize - various other technologies and industry sectors, including, among many others, information technology, transportation, medicine, energy, food safety, and environmental science. Among the distinct nanotechniques reported in the literature, electrospinning has gained considerable attention due to its capability to manufacture nanostructures with extraordinary features such as high porosity, big surface area, controllable morphology, and high chemical and thermal stability [1].

Electrospinning is a simple yet potent process for preparing ultrathin fibers. Since its first use in the early 20th century, it has become one of the most important techniques for producing nanomaterials. Tremendous advances have been achieved in terms of the development of electrospinning technologies and the fabrication of nanofibers. Electrospun nanofibers have found various applications in biomedical technology, such as drug delivery, tissue engineering, wound dressings, and antibacterials, and in industry, including filtration, desalination, protective clothing, and biosensors [1, 2]. The electrospinning process facilitates the preparation of fibers possessing diameters from tens of nanometers to a few micrometers. As shown in Figure 9.1(a), the fundamental components of electrospinning involve a high-voltage power supply (either direct current (DC) or alternating current (AC)), a syringe pump, a spinneret (usually a syringe with a needle), and a grounded collector [1].

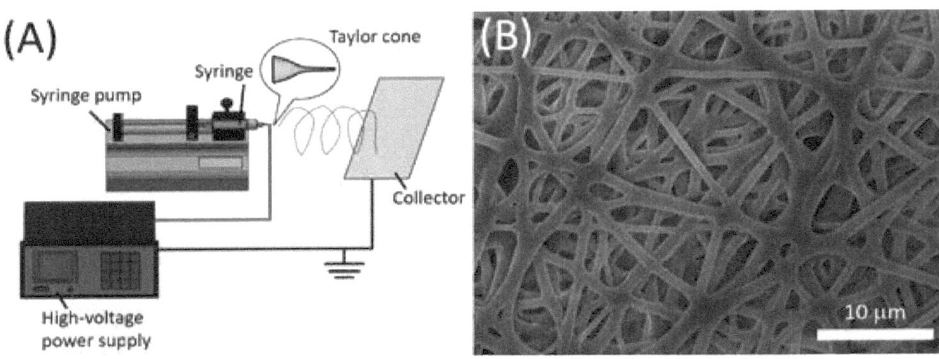

Figure 9.1 (a) Schematic setup of the electrospinning process, (b) microscopic image of electrospun fibers

Electrospinning is essentially an electrohydrodynamic procedure. During the process, a liquid is initially extruded from the spinneret to form a droplet due to the effect of surface tension. After being electrified, the electrostatic repulsion on the surface shapes the droplet into a Taylor cone, from which a charged jet is discharged. The jet first elongates in a straight line and subsequently undertakes robust spiral movements due to bending instabilities. As the jet is extended into thinner diameters, it solidifies rapidly and results in the accumulation of delicate fibers onto the collector (Figure 9.1(b)) [3–5].

■ 9.2 Variants of the Electrospinning Process

9.2.1 Coaxial Electrospinning

Unlike the traditional electrospinning process, coaxial electrospinning [6] involves the use of two or more feeding systems to concurrently deliver multiple polymeric solutions from a coaxial nozzle [7]. As shown in Figure 9.2(a), the spinnerets in the nozzle share the same axis and allow the introduction of one solution into the other at the needle tip to produce continuous core-shell-structured nanofibers. The setup for coaxial electrospinning is almost the same as that for conventional electrospinning, except that two sets of syringe pumps and spinnerets and a special mold (Figure 9.2(b)) accommodating simultaneously the core and shell flows are needed [7]. The coaxial electrospinning scheme possesses several benefits relative to the conventional electrospinning process, particularly in the aspect that it can encapsulate and deliver biomolecules that may lose the bioactivity if dissolved in organic solvents [8, 9].

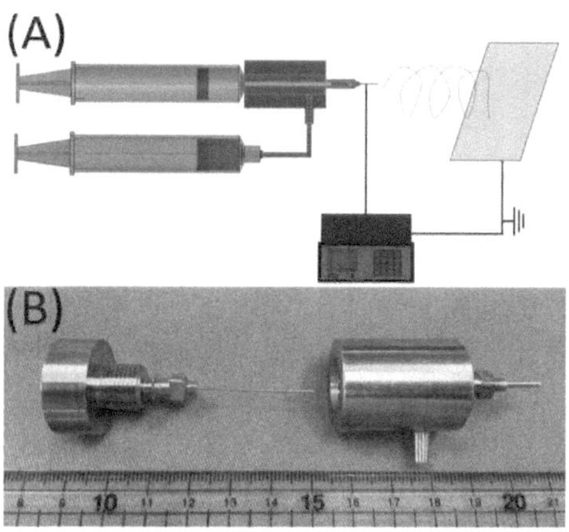

Figure 9.2
(a) (Schematic) Coaxial electrospinning process, (b) photo of the nozzle used for the process

Other advantages of the process include that the two-phase fibers enable the combination of properties in a single fiber [10]. This allows for the integration of electrically spinnable material with non-spinnable material in the construction of polymer-polymer, polymer-inorganic, and inorganic-inorganic coaxial nanofibers. Coaxial electrospinning also provides a novel dimension to the creation of nanoscaled constructions for distinct applications in sensors, energy, catalysis, and textiles as well as biomedical and environmental applications. As an example, Lee et al. [7] developed degradable core-shell nanofibers, with chitosan as the shell material and polylactide (PLA) at the core (Figure 9.3(a)), for the filtration of metallic ions. The average diameter of the electrospun fibers ranged from 234 nm to 562 nm.

With a core-shell structure, the PLA/chitosan nanofibers possess excessive chitosan on their surfaces and exhibit excellent capability in terms of metallic ion removal (Figure 9.3(b)). Additionally, with the progress of electrospinning technology, it can construct more than two layers of coaxially spun nano-/microfibers (Figure 9.4), after removing some of the layers, and manufacture hollow, multi-structured, or functionalized ultrathin fibers.

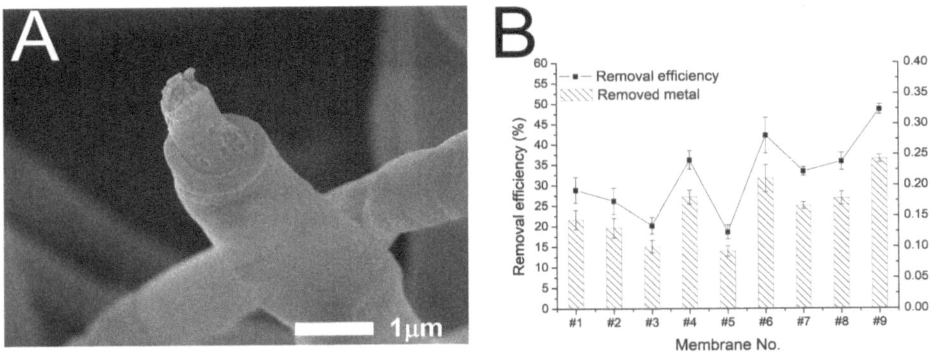

Figure 9.3 (a) Microimage of core-shell-structured nanofiber, (b) uptake capacity of electrospun nanofibrous membranes [7]

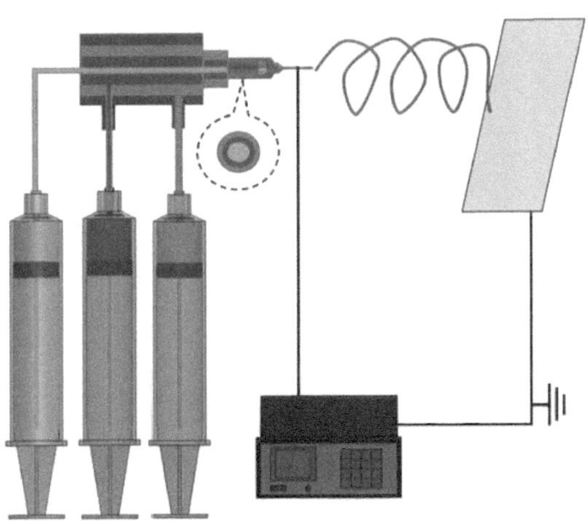

Figure 9.4
Setup for three-layered electrospinning

9.2.2 Other Variants of Electrospinning

Despite the existence of various methods for the fabrication of nanofibers, electrospinning is considered as the most versatile procedure. Materials such as polymers [2, 3] and polymer composites [5, 11, 12] have been made into nanofibers using electrospinning directly or through post-spinning processes. One unique feature of the electrospinning process has been its capability to realize distinct fiber assemblies, which promotes the functionality of products fabricated from nanofibers and enables application-specific modifications. For example, the adoption of a rotating roller, shown in Figure 9.5(a), has enabled the production of highly aligned fibrous mats (Figure 9.5(b)). In general, the degree of alignment of spun nanofibers increases with the rotating speed of the roller.

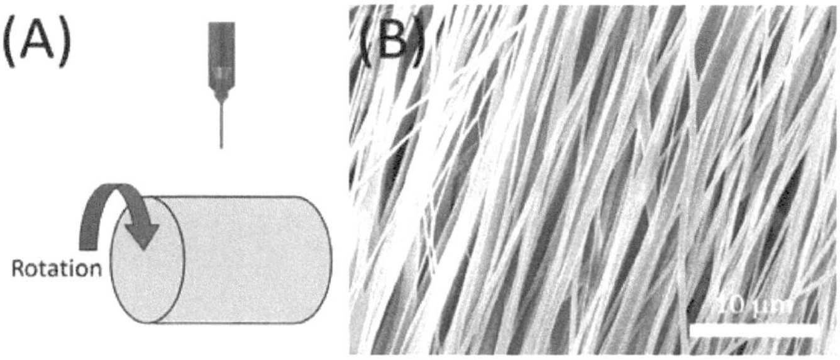

Figure 9.5 (a) Roller for producing aligned nanofibrous mesh, (b) image of highly aligned spun nanofibers

Meanwhile, through the employment of a small-diameter rod or pin as a collector (Figure 9.6), a nanofibrous tube can be obtained by accumulating fibers on the rotating rod/pin. A multilayered tube consisting of different materials can also be spun as a vascular graft. For example, materials such as collagen, which promotes cell growth and attachment but is mechanically weak, can be electrospun as the inner layer, while mechanically stronger synthetic polymers can be employed as the outer layer [13]. Additionally, by adopting degradable polymers of different molecular weights for the inner and outer layers, it is possible to achieve a sequential release of drugs from the nanofibrous tubes [14]. Other variants of electrospinning can be found in the review article by Teo et al. [15].

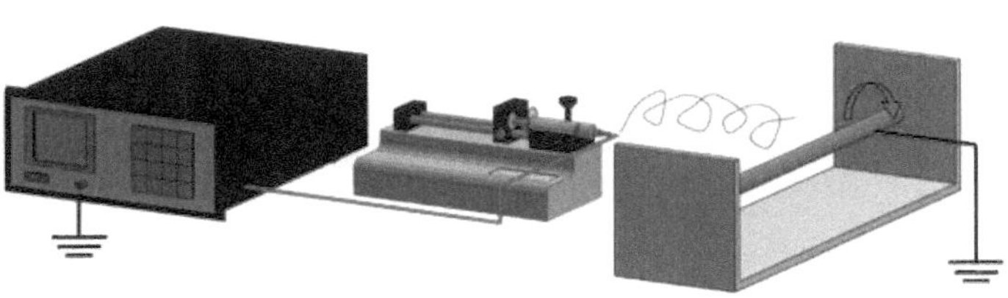

Figure 9.6 The use of a small-diameter rod or pin as a collector to produce a nanofibrous tube

9.2.3 Needleless Electrospinning

One major limitation associated with electrospinning is its low throughput. Mass production of nanofibers could be realized through the creation of multiple jets by employing different needleless electrospinning techniques [16]. Besides the stationary-spinneret electrospinning techniques, needleless electrospinning has also attracted much attention. Needleless electrospinning is the procedure of manufacturing nanofibers directly from an open polymer-solution surface [17]. One kind of needleless electrospinning is the rotary-type spinning, which employs a rotary cylinder as the spinneret (Figure 9.7). The spinneret is partially submerged and rotating in a polymeric solution, allowing the formation of a thin layer of solution on its surface and thus the formation of spikes, which, after the application of an electric field, initiate electrospinning. While in traditional (needle) electrospinning, it is induced by the capillary forces, the initiation of polymeric jets in needleless electrospinning is a self-organized procedure taking place on a free liquid surface. Therefore, the needleless electrospinning process becomes more difficult to control. The spinneret plays a key role affecting the process performance and the final nanofiber quality [18].

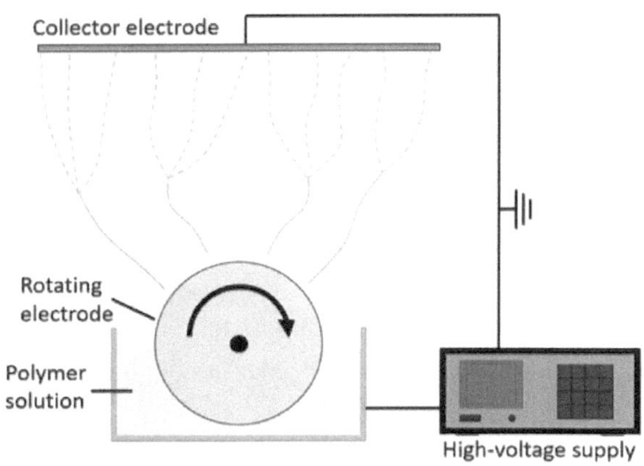

Figure 9.7 (Schematic) Needleless electrospinning process

9.2.4 Electrospraying

Electrospraying is considered as a variant of the electrospinning technique, since both processes employ a high voltage to eject liquid jets. Electrospraying is capable of producing small particulates in micro- and nanodimensions [19]. The setup for electrospraying is almost the same as that for electrospinning (Figure 9.8(a)). In this process, the liquid (or solution) is charged by the nozzle, which is connected to a high-voltage power supply. The electrostatic force incurred by these charges surmounts the surface tension of the solution and splits the solution jet into very small droplets. Tiny droplets with charges resist one another and droplet consolidation can be minimized, which in turn produces droplets with a uniform size distribution (Figure 9.8(b)). The major differences between electrospraying and electrospinning lie in the viscosity and viscoelasticity of the liquid (solution) and the relevant jet behavior. During electrospinning, the polymer concentration should be high enough to trigger the onset of significant chain entanglements in the solutions. The jet can then remain in a continuous shape to generate fibers, while in electrospraying the jet splits into droplets due to the low polymeric concentration for the formation of particles.

Electrospraying has several advantages with respect to traditional mechanical spraying processes [20]. Droplets produced by electrospraying have smaller dimensions and can be as tiny as tens of nanometers (diameter). The size distribution of the droplets is also narrower with a low standard deviation. Electrosprayed biodegradable polymers such as poly-lactide-*co*-glycolide (PLGA) can be found in drug delivery and biomedical engineering [21–25], in addition to other applications. Moreover, core-shell particles, which encapsulate biomolecules at the core, can also be successfully prepared using the co-electrospraying technique [26].

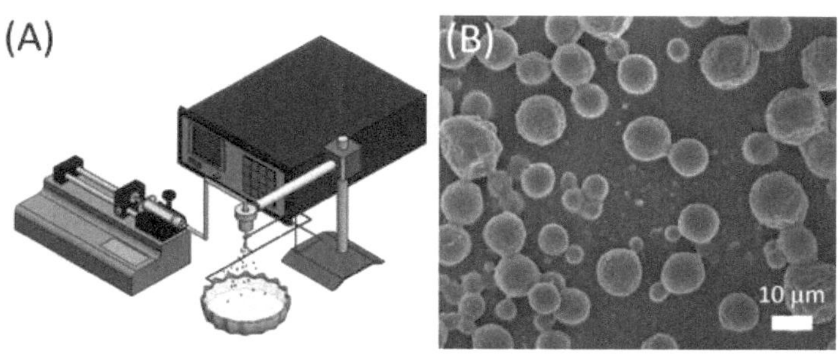

Figure 9.8 (a) Setup of electrospraying process, (b) electrosprayed microparticles

■ 9.3 Effects of Processing Parameters on Electrospinning

The most widely employed materials for electrospinning are organic polymers in the form of melts or, mostly, solutions. Tiny molecules can also be spun into nanofibers if they possess sufficient molecular chain entanglement. The preparation of nanofibers via electrospinning can be affected by several process parameters, including the spinning setup (applied electric voltage, diameter of spinneret, distance between spinneret and collector, solution delivery rate), properties of the solution (polymer concentration, solvent type, solution viscosity and conductivity), and environmental conditions (temperature, relative humidity). Depending on the materials used and the setup, the influence of processing factors on the electrospun nanofibers may be different. For example, Chen et al. [27] employed an orthogonal matrix design to investigate the influence of different process parameters on the electrospinning of poly(D,L)-lactide-*co*-glycolide (PLGA). They evaluated three different solvents, namely, hexafluoro-2-propanol (HFIP), dichloromethane (DCM), and trichloromethane (TCM), and five other parameters, including PLGA concentration, the solution's flow rate, the distance between the spinneret and the collector, and the applied voltage. Their experimental results showed that the type of solvent and the applied voltage are the principal parameters affecting the diameter of spun PLGA fibers. Using response surface methodology, Angel et al. [28] investigated the effect of electrospinning parameters, including cellulose acetate concentration, voltage, and spinning distance, on the electrospinnability of cellulose acetate nanofibers and the fiber diameter. They found that the spun fiber diameter increased with increasing cellulose acetate concentration and was less influenced by the voltage and the spinning distance. Dorati et al. [29] explored the effect of process parameters on the alignment of tubular electrospun nanofibers, and showed that the voltage, combined with the flow rate, has a significant impact on the electrospun fiber orientation.

In the electrospinning process, a high voltage is generally applied to the spinneret to create an electric field. The polarity of the voltage, either positive or negative, affects the allocation of charged molecules in the liquid as well as the type of charge that builds up on the surface of the jet. In addition, the applied voltage also governs the amount of charge conveyed by the jet and the magnitude of the electrostatic repulsion between the charges, as well as the interactive strength between the jet and the exterior electric field. While higher voltages generally favor the formation of thinner fibers, they may also cause more fluid to be ejected, resulting in larger-diameter fibers. Furthermore, although AC has also been employed to electrospin fibers, the jet exhibits a markedly different behavior relative to that employing DC [1].

The molecular weight of the polymer plays a key role in the rheological behavior and electrical properties of the solution. Generally, decreasing the molecular weight is apt to produce beads rather than fibers, due to the limited chain entanglement. Furthermore, the volatility of a solvent controls the speed of evaporation and, therefore, the solidification rate of the jet. An extremely high volatility may not be appropriate for the electrospinning of fibers since the jet solidifies instantly on leaving the spinneret. However, if the volatility is too low, spun fibers may still remain wet when stockpiled on the collector. Meanwhile, the dielectric constant of the solvent determines the amount of electrostatic repulsion between surface charges on the jet. As the dielectric constant increases, the applied voltage required to achieve a stable jet will increase.

With regard to the flow rate of the liquid, any increase will generally lead to the formation of fibers of increased diameter. The working distance between the spinneret tip and the collector determines the unstable phase of jet deposition on the collector. A sufficient distance is required to ensure full extension and curing of the jet to form solid fibers. Generally, as the distance increases, finer fibers are formed. When the distance between the spinneret tip and the collector exceeds a certain range, the fibers will not become thinner with increasing distance due to the coagulation of the jet. In general, it is the complex interplay of all processing parameters that controls fiber morphology and diameter. For example, as the flow rate increases, the critical voltage for fiber generation, as well as the working distance between the spinneret tip and the collector, must be increased to ensure adequate extension and coagulation of the jet. Therefore, it is necessary to optimize all process parameters to control the electrospinning process [1].

9.4 Biodegradable Polymers for Electrospinning

Biodegradable polymers represent a growing field. Owing to their wide-ranging properties, both synthetic and natural polymeric materials perform a vital and ubiquitous role in daily life [30, 31]. Theoretically, all biodegradable polymers can be electrospun into nanofibers as long as they can be dissolved in appropriate solvents. Various biodegradable polymers, including both natural and synthetic polymers, have been investigated for electrospinning to prepare nanofibers.

9.4.1 Natural Polymers

Many natural biopolymers, such as deoxyribonucleic acid (DNA), silk fibroin, fibrinogens, dextran, chitin, chitosan, alginate, collagen, and gelatin, have been reported to be successfully electrospun into nanofibers from their solutions.

9.4.1.1 DNA

DNA is an extensive polymer composed of repeating nucleotides, featuring a combination of adenine (A), thymine (T), cytosine (C), or guanine (G). The structure of DNA is dynamic along its length and is able to coil into tight loops or other geometries. Fang and Reneker [32] electrospan thin fibers of calf thymus Na-DNA from aqueous solutions that have a concentration ranging between 0.3% and 1.5%. The spun DNA fibers possess a diameter of approximately 50 nm to 80 nm, which is 10 times smaller than that of fibers prepared by other methods. They also observed bead-like constructions on many electrospun fibers. In addition, a process named splaying led the jet to split longitudinally into two smaller jets, which split again, repeatedly, until the very-tiny-diameter fibers are created. Bellan et al. [33] successfully electrospan fluorescently labeled DNA molecules into nanofibers with a diameter of approximately 27 nm, by mixing stock λ bacteriophage DNA solution (500 μg/mL) with the intercalating fluorescent dye YOYO-1.

9.4.1.2 Silk Fibroin

Silk fibroin (SF) is a naturally derived biopolymer with many unique properties, including good mechanical properties, versatile processability into different forms, appropriate biocompatibility, very low immunogenicity, and adjustable biodegradability, making it a good material for the preparation of vascular scaffolds [72, 73]. Min et al. [34] employed the electrospinning technique to manufacture SF nonwoven nanofibrous mats for cell culture of normal human keratinocytes and fibroblasts by utilizing formic acid as a solvent. The spun SF nanofibers showed a circular cross section with a smooth surface and possessed an average diameter of 80 nm (ranging from 30 nm to 120 nm).

Liu et al. [35] electrospan nanofibers from native-SF-fibrils solutions with a certain diameter and length by mixing silk with $CaCl_2$/formic acid solvents. When the solution concentration was elevated from 2% to 10%, the diameter of the electrospun SF nanofibers increased from 40 nm to 1.8 μm, demonstrating a good size control of the spun fibers. Biagiotti et al. [36] dissolved Bombyx mori SF films in formic acid for the fabrication of SF scaffolds. They noted that, despite its high volatility, residual formic acid was being trapped into the matrix of the SF nanofibers. Their experimental results also showed that formic acid interacted with SF by forming formyl esters. Yet, the quantity of formic acid that remained on the SF mats after electrospinning was demonstrated to raise no concerns for human health.

9.4.1.3 Fibrinogens

Fibrinogen is a thrombin-coagulable glycoprotein that exists in the blood of vertebrates. The construction of the alpha, beta, and gamma polypeptide chains of human fibrinogen is mainly composed of amino acid and nucleic acid sequencing. Wnek et al. [37] proposed the use of fibrinogen nanofibers as a tissue-engineering scaffold, wound dressing, or hemostatic bandage, by electrospinning the solutions of human or bovine fibrinogen fraction I mixed with 1,1,1,3,3,3-hexafluoro-2-propanol (HFIP) and minimal essential medium (Earle's salts). The electrospun fibrinogen fibers possessed an average diameter of 80–700 nm. Carlisle et al. [38] employed an integrated atomic-force-microscopy (AFM) and fluorescence-microscopy technique to examine the mechanical properties of electrospun fibrinogen fibers in aqueous buffer. The spun fibers with viscoelastic characteristics could be extended to 2.3 times their initial length before failure. The maximum stress, tensile modulus, and relaxation elastic modulus were measured to be 22 MPa, 17.5 MPa, and 7.2 MPa, respectively. In addition, when kept at constant strain, the electrospun fibrinogen fibers exhibited a fast and a slow stress relaxation time of 3 s and 55 s, respectively.

9.4.1.4 Dextran

Dextran is a bacterial polysaccharide consisting of α-1,6-linked D-glucopyranose residues with some α-1,2-, α-1,3-, or α-1,4-linked side chains. Owing to their good biocompatibility and biodegradability, dextran-based formulations are promising candidates for the preparation of scaffolds in tissue engineering or as vehicles for drug delivery. Dextran is soluble in both water and organic solvents. It can, therefore, be electrospun into nanofibrous mats through mixing with either water-soluble bioactive agents or hydrophobic biodegradable polymers for biomedical applications.

Jiang et al. [39] investigated the effects of different process parameters on the properties of electrospun dextran membranes. They proposed that uniform nanofibrous dextran membranes can be successfully fabricated through the use of water, dimethyl sulfoxide (DMSO)/water, and DMSO/dimethylformamide (DMF) mixtures as solvents and the optimization of the process conditions. When water was employed as a solvent, up to 10% (w/w) of bovine serum albumin (BSA) or lysozyme could be directly loaded in the dextran membranes without compromising their morphology. No obvious influence of the electrospinning procedure on the activity of lysozyme was noted.

While traditional cross-linking schemes mostly utilize toxic compounds such as glutaraldehyde, Malini et al. [40] showed that the blended aqueous solutions of dextran and boric acid can be electrospun to mats with different degradation rates. Boric acid covalently connects to the hydroxyl moieties existing in the dextran chains, thus forming boronate ester bridges between dextran chains. In addition, the alteration in the cross-linking density could promote the average drug release period up to 500% relative to pristine dextran electrospun scaffolds. Cengiz-Çallıoğlu [41] proposed the preparation of a dextran nanofibrous layer via needleless electrospinning. Nanofibers could not be electrospun directly from a pure dextran/water solution. Rather, solution drops were deposited on the collector; then, the anionic surfactant was supplemented in different concentrations to begin the nanofiber production. They also examined the influence of the surfactant concentration on the solution properties, spinnability, and fiber properties, and proposed that uniform and fine nanofibers can be obtained from the rod electrospinning method. A 2 wt % surfactant concentration was selected as the optimum concentration to manufacture a dextran nanofibrous layer by roller electrospinning.

Ritcharoen et al. [42] fabricated ultrafine fibrous dextran membranes by electrospinning, using water as solvent. The influence of the solution concentration (0.7–1.3 g/mL) and the applied electric field (9–21 kV/15 cm) on the morphology and dimensions of the spun fibers was investigated. Under a fixed electric field of 15 kV/15 cm and a fixed solution flow rate of 0.25 mL/h, beaded fibers were found up to a critical concentration of approximately 0.9 g/mL, beyond which only smooth fibers were observed. The average diameter of the electrospun fibers increased monotonically with the solution concentration (from ~290 nm to ~1950 nm). Additionally, for the dextran solutions investigated, the spun-fiber diameters increased with the electric field, with the average diameter ranging from 520 nm to 1760 nm. Shawki et al. [43] developed antimicrobial dextran nanofibers by electrospinning, using water as solvent. Moxifloxacin-loaded antimicrobial dextran nanofibers that provided sustained release of moxifloxacin were successfully prepared. In addition, the dextran-moxifloxacin nanofibers displayed a wide antimicrobial efficacy against Escherichia coli and Staphylococcus aureus bacteria.

9.4.1.5 Chitin and Chitosan

Chitin and chitosan are polysaccharides that possess good biocompatibility and biodegradability, with adaptable bioactivities including antimicrobial activity, low immunogenicity, and low toxicity. They have been used in cosmetics, food additives, and medical materials for human health. Chitin and chitosan fibers have attracted much attention due to their applications in textile materials, sutures, and as scaffold materials for tissue engineering [44].

Min et al. [45] fabricated chitin nanofibrous mats for wound dressings by electrospinning. Chitin was depolymerized by gamma irradiation to enhance its solubility. HFIP was used as the spinning solvent. The electrospun fibers showed an average diameter of 110 nm (ranging from 40 nm to 640 nm). Barber et al. [46] detailed the electrospinning of high-molecular-weight chitin fibers directly from a 1-ethyl-3-methylimidazolium acetate solution of chitin extracted from dried shrimp shell, thus saving the need for the separation of chitin from crustacean shells.

Regarding the preparation of chitosan nanofibers, Qasim et al. [47] reviewed the electrospinning of chitosan and chitosan composites to prepare fibers combined with other natural polymers, including silk, collagen, cellulose, gelatin, and hyaluronic acid, for tissue engineering. Antaby et al. [48] reviewed the antibacterial efficacy of electrospun chitosan-based nanofibers in the biomedicine, filtration, and food protection industries. Ohkawa et al. [49] investigated the influence of solvents and chitosan concentrations on the morphology of electrospun mats. The solvents that were investigated included dilute hydrochloric acid, acetic acid, neat formic acid, and trifluoroacetic acid. They found that when the chitosan concentration was elevated, the morphology of the deposited mats changed from spherical beads to interlinked fibrous networks. In addition, the incorporation of dichloromethane into the chitosan-trifluoroacetic-acid solution enhanced the uniformity of the spun chitosan fibers, with a mean diameter of 330 nm. Salazar-Brann et al. [50] studied the conductivity of various chitosan-organic-acid solutions and its influence on fiber formation in the electrospinning process. It was found that the conductivity increases linearly as the chitosan concentration increases. However, no direct correlation was found between the electrospinnability of the solutions and their conductivity. Meanwhile, the high viscosity of chitosan may limit its spinnability for electrospinning. Homayoni et al. [51] proposed the use of an alkali treatment that hydrolyzes chitosan chains to decrease its molecular weight. Solutions of the treated chitosan in aqueous 70-90% acetic acid could be electrospun into nanofibers with suitable quality and processing stability.

9.4.1.6 Collagen

Collagen is an essential component of the extracellular matrix (ECM) in organs and tissues, and has been widely employed for applications in tissue engineering. Electrospun collagen nanofibrous mats also provide unique features, including high surface-area-to-volume ratio, adjustable diameter and porosity, and extraordinary biological activity to promote cell function and tissue formation [52]. Matthews et al. [53] explored the fabrication of tissue-engineering scaffolds that consisted of collagen nanofibers. Through the optimization of process parameters, a calfskin type I collagen matrix consisting of 100 nm fibers and exhibiting a 67 nm banding pattern that mimics the characteristics of native collagen could be successfully fabricated. The structural properties of the prepared collagen were found to vary with the tissue of origin (type I from skin vs. type I from placenta), the isotype (type I vs. type III), and the concentration of the collagen solution used to spin the fibers. Electrospun collagen was also shown to enhance cell proliferation and cell penetration into the fabricated matrix. Sizeland et al. [54] analyzed the morphology of the electrospun collagen fibers and compared it with that of native collagen. They found that the electrospinning procedure forms fibers that are short of internal fibril configuration, when compared to the hierarchical composition of native collagen.

As collagen nanofibers are mechanically weak in nature, collagen is commonly cross-linked or blended with synthetic polymers to enhance the mechanical strength without compromising the bioactivity. Liu et al. [55] dissolved PLGA/collagen in HFIP to electrospin biomimetic nanofibrous membranes for wound dressing. The average diameter of the electrospun nanofibers was 250 nm (range of 150-650 nm). They showed that the nanofibrous PLGA/collagen mats are functionally active in human fibroblasts and are efficacious as accelerators in early-stage wound healing.

9.4.1.7 Gelatin

Gelatin is a protein obtained from hydrolyzed and denatured animal skin. It is a condensation polymer consisting of a variety of amino acids; thus, it exhibits good water-solubility. Gelatin is also one of the most important natural biomaterials for the design and fabrication of wound dressings, due to its excellent biocompatibility and biodegradability and great exudate-absorbing capacity. As with collagen, different approaches, including physical, chemical, and biological methods, have been employed to enhance the mechanical stability of gelatin-based mats.

The preparation of gelatin nanofibrous membranes via electrospinning has been reported. Aoki et al. [56] proposed a technique to incorporate protein reagents into the nanofibers using mild solvents. They screened 22 mixtures of 50% organic solvent–50% water, less denaturing neutral dipolar aprotic solvents (specifically, *N,N*-dimethylacetamide, *N,N*-dimethylformamide, and *N*-methyl-2-pyrrolidone), to examine their applicability to electrospin gelatin nanofiber scaffolds through their ability to retain gelatin in a sol state at ambient temperature. By choosing the optimum solvents and concentrations, gelatin nanofibers could be successfully spun with different constructions, from a thick, wide, porous nanofibrous structure to a thin, fine, nanofibrous mesh structure. Topuz and Uyar [57] studied the influence of process parameters on the morphology of electrospun gelatin nanofibers and found that the variations in the applied voltage and the concentration of gelatin result in the transition of the fiber shape from round to flat/ribbonlike. The formation of flat-shaped fibers was caused by the rapid evaporation of the solvent (formic acid) from the fiber matrix when increasing the applied voltage and the gelatin concentration. Meanwhile, round fibers were due to the steady evaporation of formic acid throughout the cross section of the fibers. Recently, Li et al. [58] provided a review on innovative electrospinning approaches, including blend electrospinning, emulsion electrospinning, and coaxial electrospinning, to enhance the mechanical, physicochemical, and biological properties of gelatin-based nanofiber mats. Electrospun gelatin-based nanofiber mats in combination with exceptional functions, used as dressings for burn and diabetic wounds, have been reported.

9.4.1.8 Hyaluronic Acid

Hyaluronic acid (HA) is a naturally occurring biodegradable polymer that has a great potential for biomedical applications [59]. Besides its biodegradability, HA also possesses extraordinary biological properties. In the nanofibrous form, this polymer can mimic biological tissues, especially the layers of the skin, and is an ideal material for the fabrication of wound dressings. However, the preparation of fibers from hyaluronan via electrospinning remains a challenge owing to the polymer's hydrophilic characteristic and the high viscosity of its solutions [60]. Nano-/microfibers from hyaluronan may be obtained by employing special techniques such as binary/ternary solvent or multipolymer systems.

Li et al. [61] explored the electrospinning of HA and HA/gelatin (GE) blends using *N,N*-dimethylformamide (DMF)/water-mixed solvents. They reported that the HA solution can be successfully electrospun into nanofibrous membranes with a volume ratio of DMF to water ranging between 0.5 and 1.5. The average diameter of the spun HA fibers was approximately 200 nm. Blended HA-GE nanofibers with HA/GE weight ratios of 100/20–100/100 were also successfully fabricated, with an average diameter of 190–500 nm. Additionally, it has been reported that nanofibers of HA blended with various polymers, including HA/polyethylene oxide [60], HA/polyvinyl alcohol [60], HA/chitosan [62], and HA/collagen [63], have been successfully electrospun.

9.4.1.9 Cellulose

Cellulose is one of the most abundant natural polymers. This polymer is nontoxic and has good biocompatibility and biodegradability, making it and its derivatives ideal materials for the development of multifunctional systems [64]. Celluloses have applications in biomaterials, water filtration, sensors, electroconductive materials, active packaging, among others.

The integration of the electrospinning technique and cellulose/cellulose derivatives offers the possibility of preparing nanostructured porous materials with good functionalities, flexibility, renewability, and biodegradability. However, cellulose is very difficult to dissolve in common solvents, despite its highly hydrophilic characteristic, owing to an abundance of hydroxyl groups. Therefore, a special approach has been proposed to manufacture cellulose nanofibers from solutions, and the optimal conditions to control the final morphology have been investigated [65]. Zhang et al. [66] employed the periodate-oxidation–adipic-acid-dihydrazide crosslinking technique to electrospin cellulose nanofibers. The periodate oxidation of cellulose was used to create water-soluble aldehyde cellulose, thus allowing the electrospinning in an aqueous solution and avoiding the employment of adverse organic solvents. The cellulose nanofiber mats prepared exhibited a moderate wet mechanical strength of approximately 1 MPa. The spun fibers exhibited good cytocompatibility, and could absorb an amount of water equivalent to 30 times their own weight and degrade gradually by hydrolysis.

9.4.2 Synthetic Polymers

Synthetic biodegradable polymers such as polycaprolactone (PCL), poly(lactic acid) (PLA), and poly(lactic-*co*-glycolic acid) (PLGA) have been widely electrospun into nanofibers for different applications [67]. A few synthetic polymers such as polyglycolic acid (PGA) are more difficult to be electrospun due to their insolubility in most organic solvents [68]. Nanofibers made of synthetic degradable polymers are able to mimic biological environments because they are on the same scale as biological molecules. These synthetic nanofibers have various biomedical applications, including tissue engineering, drug delivery, biosensors, and immunoassay [69].

9.4.2.1 Poly(lactic acid) (PLA)

Poly(lactic acid) or polylactide (PLA) is a synthetic biopolymer that is widely used in the biomedical field. It is a biodegradable and biocompatible thermoplastic aliphatic polyester with good thermal and mechanical performance. These unique features, including nontoxicity for humans, make it an ideal material for biomedical applications. This polymer can be obtained from natural raw substances such as sugarcane, rice, and cornstarch [70]. PLA is produced via condensation polymerization or ring-opening polymerization of the lactic acid. It has gained the approval of the Food and Drug Administration (FDA) for human usage in diverse applications, including implantable medical devices, drug delivery carriers, and tissue regeneration scaffolds. PLA involves three isomeric forms, poly(D-lactide) (PDLA), poly(L-lactide) (PLLA), and racemic poly(DL-lactide) (PDLLA). In some applications, such as medical implants, targeted drug delivery, and tissue engineering scaffolds, stereocomplex PLA (SC-PLA) has gained great attention. SC-PLA can be formed by the strong interaction created by the side-by-side order of the molecular chains of enantiomeric PLA polymers, such as PDLA and PLLA. The creation of the stereocomplex improves the mechanical performance and the thermal and hydrolysis resistance of the PLA-based materials.

The electrospinnability of the PLA-based solutions and the morphological features of the nanofibers fabricated are determined by a set of factors relevant to the solution, such as viscosity, conductivity, polymer concentration, surface tension, and solvent system. Casasola et al. [71] investigated the influence of different solvent systems on fiber morphology and diameter of electrospun PLA fibers. The solvent properties – boiling point, viscosity, conductivity, and surface tension – were found to have a major impact on the process productivity, morphology, and diameter distribution of the PLA nanofibers. Among various solvent systems employed, acetone/dimethylformamide provided the highest fiber production and finest defect-free nanofibers. Beaded nanofibrous structures could be obtained from solutions with concentrations around the critical chain entanglement concentration, while defect-free fibers were fabricated as the concentration was elevated to approximately two times the entanglement concentration. Coles et al. [72] investigated the influence of process parameters – conductivity, concentration (% w/w), electrostatic potential, and collection distance – on electrospun PLA nanofibers using a design of experiments (DoE) approach. They proposed that the interaction between the processing parameters permits the tuned fabrication of the micro-/nanofibers to satisfy the demanded application.

The electrospinning of PLA composites has also been reported. Li et al. [73] prepared composite bilayer nanofibrous mats of PLA/polyurethane (PU) by electrospinning, which can be used as wound dressing scaffolds for controlled drug loading and release. Gajic et al. [74] used electrospun PLA fibers as a carrier of the phytoestrogen biochanin A. The rate and time of the release of biochanin A from PLA fibers were found to be in correlation with the amount of the active ingredient in the matrix of the carrier and follow zero-order kinetics. Sabzi et al. [75] adopted the dual-electrospinning technique to concurrently spin PLA and poly(vinyl acetate) (PVAc), to obtain an interwoven composite with two well-separated thermal transitions. Additional information regarding the electrospinning of PLA nanofibers can be found in the review article by Maleki et al. [70].

9.4.2.2 Poly(lactic-*co*-glycolic acid) (PLGA)

The intrinsic characteristics of PGA such as high hydrophilicity, rapid degradation, insolubility in most organic solvents, and brittleness have hindered the polymer's practical applications. PGA has been mainly used in the form of copolymers, such as poly(lactic-*co*-glycolic acid) (PLGA).

PLGA is a copolymer that has been approved by the Food and Drug Administration (FDA) for making therapeutic devices, mainly due to its extraordinary biodegradability and biocompatibility. It has demonstrated great potential as a drug delivery vehicle and as a scaffold for tissue engineering [76]. Depending on the lactide-to-glycolide ratio, different forms of PLGA can be obtained. PLGAs may change from completely amorphous to completely crystalline based on the block structure and molar ratio. PLGA generally exhibits a glass transition temperature of 40–60 °C and can be dissolved by a wide range of solvents, depending on the composition. This polymer degrades by hydrolysis of its ester linkages in the presence of water. It has been shown that the time needed for PLGA degradation is dependent on the monomers' ratio; the higher the percentage of glycolic acid units, the shorter the time needed for degradation. A special case is the copolymer with a 50:50 monomers' ratio, which shows a faster degradation (about 1–3 months) [77]. This versatility in terms of degradation has made PLGA an ideal material for the preparation of various medical devices, including grafts, sutures, implants, prosthetic devices, surgical sealant films, and micro- and nanoparticles. Additionally, PLGAs have been extensively investigated as delivery vehicles for drugs, proteins, and biomolecules such as DNA, RNA, and peptides [78].

Among all the biomaterials, PLGA is the most researched polymer for electrospun nanofibers. Chen et al. [79] explored the influence of different processing variables on the electrospinning of PLGA nanofibers, based on an orthogonal matrix design approach. The solvents investigated included HFIP, DCM, and TCM, while the processing variables involved PLGA concentration, solution's flow rate, distance between nozzle and grounded collector, applied voltage, and type of solvent. Their results suggested that, among the variables used, the type of solvent and the applied voltage are the principal variables affecting the diameter distribution of the spun PLGA fibers. Liu et al. [80] examined the electrospinnability of PLGA solutions, including the influence of the polymer's molecular weight, the solvent type, and the solvent composition on the physical properties of electrospun nanofibers. Two combinations of solvent mixtures, DMF–tetrahydrofuran (THF) and DMF–chloroform, offered a stable cone-jet during electrospinning. The high-molecular-weight polymer could result in sufficient intermolecular interaction to produce uniform fibers. The solvent was also found to influence the structure and mechanical properties of spun fibers by varying the properties of the PLGA solution and the drying rate of fibers in the electrospinning process. Yu et al. [81] prepared PLGA nanofibrous arrays in random or aligned orientations on glass slips via electrospinning. Their results showed that integrated PMMA microfluidic chips with random PLGA nanofiber arrays can efficiently capture and recover circulating tumor cells (CTCs) from whole blood samples.

Nanofibrous scaffolds have been employed widely in tissue engineering and drug delivery systems. Owing to the deficiency of recognition sites on cells, hydrophobicity, and single function, various efforts have been completed to achieve functionalized PLGA scaffolds, including surface modifications, and composite and drug-loaded scaffolds. These functionalized PLGA nanofibrous scaffolds have significantly promoted cell adhesion, attachment, and proliferation [82]. Additionally, electrospun nanofibers can incorporate antimicrobial agents, such as antibiotics or natural antimicrobials, for infection control during the healing process. Buck et al. [83] investigated the release characteristics of antimicrobial agents from electrospun PLGA nanofibers containing either blended or physisorbed ciprofloxacin. The results demonstrated that physically adsorbed ciprofloxacin offers antibacterial properties superior to those of blended ciprofloxacin, up to 48 h against P. aeruginosa PA14, S. aureus, and S. epidermidis. Chen et al. [84] prepared sandwich-structured nanofibrous drug-eluting membranes with PLGA/collagen as the surface layers and PLGA/drugs as the core. The spun membranes provided sustained release of vancomycin and gentamicin and lidocaine in vivo for more than three weeks. The nanofibrous mats were functionally active in the treatment of infected wounds and were very efficacious as accelerators in early-stage wound healing. Tseng et al. [85] reported the development of bilayered PLGA nanofibrous membranes – by adopting 50:50 and 75:25 copolymers – that enabled the sequential and sustained release of chemotherapeutic and antiangiogenic agents for the treatment of glioblastoma multiforme, by employing an electrospinning technique. Their results showed that the spun nanofibers display a sequential drug-eluting behavior, with the release of high drug levels of chemotherapeutic carmustine, irinotecan, and cisplatin from day 3, followed by the release of high concentrations of the antiangiogenic combretastatin from day 21.

Sheath-core-structured nanofibers prepared by the coaxial electrospinning method provide numerous advantages over those by the traditional electrospinning process, especially in that they can be used to encapsulate drugs that are only water-soluble and will likely lose the bioactivity when they dissolve in organic solvents. Lee et al. [86] developed core-shell insulin-loaded nanofibrous scaffolds for repairing diabetic wounds. The core-shell insulin-loaded nanofibrous scaffolds lessened the quantity of type I collagen in vitro, elevated the transforming growth factor-beta content in vivo, and enhanced diabetic wound repair. Yu et al. [87] coaxially electrospan vancomycin/ceftazidime/bone-morphogenetic-protein (BMP) sheath-core nanofibers, combined with 3D printed scaffolds, for the repair of segmental bone defects during the Masquelet procedure. Their results showed that the drug-eluting composite scaffold allowed the sustained discharge of embedded medications for over 30 days in vitro. Abundant BMP-2 was also induced within the bioactive membrane stimulated by the applied scaffold, and satisfactory bone healing potential was noted on radiological examination and biomechanical evaluation.

9.4.2.3 Polycaprolactone (PCL)

Polycaprolactone (PCL) is a resorbable polyester with a low melting temperature of approximately 60 °C and a glass transition temperature of around −60 °C [88]. The extraordinary rheological and viscoelastic features relative to several of its aliphatic polyester counterparts make PCL suitable for the fabrication of various implants and devices. This material has been used extensively in the biomaterials field and a number of drug-delivery devices. Successful preparations of PCL nanofibers by electrospinning have been reported. Jiang et al. [89] electrospan PCL fibers with a porous surface using a chloroform/dimethyl sulfoxide mixed solution. Gomes et al. [90] explored the effect of up to 15 wt % water in the solvent mixture for electrospinning, and proposed that the desired properties of electrospun PCL can be realized via tuning the acetic-acid-to-water ratio. Pouladchang et al. [91] found that, with only chloroform as the solvent for PCL, thiram-containing microfibers could be electrospun. To electrospin PCL nanofibers containing thiram, methanol was added as a co-solvent. They also found that microfibers show a higher release rate with lower amounts of the loaded thiram, while nanofibrous matrices exhibit an opposite trend. Asadian et al. [92] electrospan PCL nanofibers and found that the electrospinning process parameters modify the physical characteristics of the nanofibers as well as their biological performance regarding, for example, protein adsorption, cell attachment, and cell spreading. Dobrzanski and Hudecki [93] proposed the use of a 5–10% solution of PCL (molecular weight: 70,000–90,000 Da) dissolved in a mixture of formic acid and acetic acid at a ratio of 70:30 (m/m) for the electrospinning of nanofibers. They claimed that the adopted solvents are significantly less toxic than the solvents currently used for PCL, such as chloroform, methanol, dimethylformamide, or tetrahydrofuran. Ma et al. [94] reported the successfully electrospun core-shell PCL/chitosan composite nanofibers by the adoption of a surfactant-free and low-toxic water-in-oil emulsion. Meanwhile, Rawat et al. [95] investigated the electrospinning of graphene nanofiber composites and showed that PCL with exiguous graphene presented an ideal nanotopography with an unprecedented combination of guidance stimuli and substrate cues, aiding the enhanced differentiation of mesenchymal stem cells into dopaminergic neurons.

9.5 Summary

Electrospinning has become one of the leading methods for making biodegradable fibers of micro-/nanosize, and is expected to continue to grow annually for more applications in novel fields such as sensors, engineered tissues, drug delivery systems, and nanoelectronics. By introducing nanoscale components with different dimensions and/or morphologies (e.g., nanoparticles, nanorods, nanowires, nanotubes, and nanosheets) into polymer solutions, the mixtures have also been used for electrospinning. Meanwhile, other process variants such as needleless electrospinning, electrospraying, or spinning with a rotating drum or pin will continue to be explored and developed. The electrospinning technology will play a very important role in contributing to the growth and success of biodegradable polymers and polymer composites.

References for Chapter 9

[1] Xue, J., Wu, T., Dai, Y., and Xia, Y., "Electrospinning and electrospun nanofibers: Methods, materials, and applications", *Chem. Rev.* 2019; 119(8): 5298–5415.

[2] Haider, A., Haider, S., and Kang, I. K., "A comprehensive review summarizing the effect of electrospinning parameters and potential applications of nanofibers in biomedical and biotechnology", *Arab. J. Chem.* 2018; 11(8): 1165–1188.

[3] Fadil, F., Affandi, N. D. N., Misnon, M. I., Bonnia, N. N., Harun, A. M., and Alam, M. K., "Review on electrospun nanofiber-applied products", *Polymers* 2021; 13(13): 2087.

[4] Ibrahim, H. M. and Klingner, A., "A review on electrospun polymeric nanofibers: Production parameters and potential applications", *Polym. Test.* 2020; 90: 106647.

[5] Jiang, S., Chen, Y., Duan, G., Mei, C., Greiner, A., and Agarwal, S., "Electrospun nanofiber reinforced composites: a review", *Polym. Chem.* 2018; 9: 2685–2720.

[6] Campa-Siqueiros, P. I., Madera-Santana, T. J., Castillo-Ortega, M. M., López-Cervantes, J., Ayala-Zavala, J. F., and Ortiz-Vazquez, E. L., "Electrospun and co-electrospun biopolymer nanofibers for skin wounds on diabetic patients: an overview", *RSC Adv.* 2021; 11: 15340–15350.

[7] Lee, D. M., Kao, C. W., Huang, T. W., You, J. H., and Liu, S. J., "Electrospinning of sheath-core structured chitosan/polylactide nanofibers for the removal of metal ions", *Inter. Polym. Process.* 2016; 31: 533–540.

[8] Yu, Y.H., Lin, Y.T., Hsu, Y.H., Chou, Y.C., Ueng, S.W.N., and Liu, S.J., "Biodegradable antimicrobial agent/analgesic/bone morphogenetic protein-loaded nanofibrous fixators for bone fracture repair", *Int. J. Nanomed.* 2021; 16: 5357–5370.

[9] Lee, C.H., Hsieh, M.J., Roth, J.G., Fu, X., Lu, C.H., Hung, K.C., Kuo, C.C., and Liu, S.J., "Hybrid core-shell nanofibrous scaffolds/stent delivery angiotensin II receptor blocker to treat diabetic artery disease", *ACS Appl. Polym. Mater.* 2022; 4: 4199–4207.

[10] Han, D. and Steckl, A.J., "Coaxial electrospinning formation of complex polymer fibers and their applications", *ChemPlusChem* 2019; 84(10): 1453–1497.

[11] Huang, Z.M., Zhang, Y.Z., Kotaki, M., and Ramakrishna, S., "A review on polymer nanofibers by electrospinning and their applications in nanocomposites", *Comp. Sci. Technol.* 2003; 63: 2223–2253.

[12] Sahay, R., Kumar, P.S., Sridhar, R., Sundaramurthy, J., Venugopal, J., Mhaisalkar, S.G., and Ramakrishna, S., "Electrospun composite nanofibers and their multifaceted applications", *J. Mater. Chem.* 2012; 22: 12953–12971.

[13] Zhou, T., Chen, S., Ding, X., Hu, Z., Cen, L., and Zhang, X., "Fabrication and characterization of collagen/PVA dual-layer membranes for periodontal bone regeneration", *Front. Bioeng. Biotechnol.* 2021; 9: 630977.

[14] Lee, C.H., Yu, C.Y., Chang, S.H., Hung, K.C., Liu, S.J., Wang, C.J., Hsu, M.Y., Hsieh, I.C., Chen, W.J., Ko, Y.S., and Wen, M.S., "Promoting endothelial recovery and reducing neointimal hyperplasia using sequential-like release of acetylsalicylic acid and paclitaxel-loaded biodegradable stents", *Int. J. Nanomed.* 2014; 9: 4117–4133.

[15] Teo, W.E. and Ramakrishna, S., "A review on electrospinning design and nanofibre assemblies", *Nanotechnology* 2006; 17: R89–R106.

[16] Wei, L., Sun, R., Liu, C., Xiong, J., and Qin, X., "Mass production of nanofibers from needleless electrospinning by a novel annular spinneret", *Mater. Des.* 2019; 179(5): 107885.

[17] Partheniadis, I., Nikolakakis, I., Laidmae, I., and Heinamaki, J., "A mini-review: Needleless electrospinning of nanofibers for pharmaceutical and biomedical applications", *Processes* 2020; 8: 673.

[18] Yu, M., Dong, R.H., Yan, X., Yu, G.F., You, M.H., Ning, X., and Long, Y.Z., "Recent advances in needleless electrospinning of ultrathin fibers: From academia to industrial production", *Macromol. Mater. Eng.* 2017; 302(7): 1700002.

[19] Gaskell, S.M., "Electrospray: principles and practice", *J. Mass Spectrom.* 1997; 32: 677–688.

[20] Nguyen, D.N., Clasen, C., and Van den Mooter, G., "Pharmaceutical applications of electrospraying", *J. Pharm. Sci.* 2016; 105: 2601–2620.

[21] Hao, S., Wang, Y., Wang, B., Deng, J., Liu, X., and Liu, J., "Rapid preparation of pH-sensitive polymeric nanoparticle with high loading capacity using electrospray for oral drug delivery", *Mater. Sci. Eng. C* 2013; 33: 4562–4567.

[22] Enayati, M., Chang, M.-W., Bragman, F., Edirisinghe, M., and Stride, E., "Electrohydrodynamic preparation of particles, capsules and bubbles for biomedical engineering applications", *Colloids Surf. A Physicochem. Eng. Asp.* 2011; 382: 154–164.

[23] Bock, N., Woodruff, M.A., Hutmacher, D.W., and Dargaville, T.R., "Electrospraying, a reproducible method for production of polymeric microspheres for biomedical applications", *Polymers* 2011; 3: 131–149.

[24] Prabhakaran, M.P., Zamani, M., Felice, B., and Ramakrishna, S., "Electrospraying technique for the fabrication of metronidazole contained PLGA particles and their release profile", *Mater. Sci. Eng. C* 2015; 56: 66–73.

[25] Hsu, Y.H., Chen, D.W., Li, M.J., Yu, Y.H., Chou, Y.C., and Liu, S.J., "Sustained delivery of analgesic and antimicrobial agents to knee joint by direct injections of electrosprayed multipharmaceutical-loaded nano/microparticles", *Polymers* 2018; 10: 890.

[26] Lee, D., Hsu, M.Y., Li, M.J., Liu, Y.W., Liu, S.J., Liu, C.Y., and Ito, H., "Factors affecting the co-axial electrospraying of core-shell-structured poly(D,L-lactide-*co*-glycolide) microparticles", *Japan J. Appl. Phys.* 2018; 57: 116702.

[27] Chen, Y.P., Liu, H.Y., Liu, Y.W., Lee, T.Y., and Liu, S.J., "Determination of electrospinning parameters' strength in poly(D,L)-lactide-co-glycolide micro/nanofiber diameter tailoring", *J. Nanomat.* 2019; 2019: 2626085.

[28] Angel, N., Guo, L., Yan, F., Wang, H., and Kong, L., "Effect of processing parameters on the electrospinning of cellulose acetate studied by response surface methodology", *Journal of Agriculture and Food Research* 2020; 2: 100015.

[29] Dorati, R., Chiesa, E., Pisani, S., Genta, I., Modena, T., Bruni, G., Brambilla, C.R.M., Benazzo, M., and Conti, B., "The effect of process parameters on alignment of tubular electrospun nanofibers for tissue regeneration purposes", *J. Drug Deliv. Sci. Technol.* 2020; 58: 101781.

[30] Panchal, S.S. and Vasava, D.V., "Biodegradable polymeric materials: Synthetic approach", *ACS Omega* 2020; 5(9): 4370–4379.

[31] Bhovi, V.K., Melinmath, S.P., and Gowda, R., "Biodegradable polymers and their applications: A review", *Mini Rev. Med. Chem.* 2022; 22(16): 2081–2101.

[32] Fang, X. and Reneker, D.H., "DNA fibers by electrospinning", *J. Macromol. Sc. Phys.* 1997; 36(2): 169–173.

[33] Bellan, L.M., Strychalski, E.A., and Craighead, H.G., "Electrospun DNA nanofibers", *J. Vac. Sci. Technol. B: Nanotechnol. Microelectron.* 2007; 25: 2255–2257.

[34] Min, B.M., Lee, G., Kim, S.H., Nam, Y.S., Lee, T.S., and Park, W.H., "Electrospinning of silk fibroin nanofibers and its effect on the adhesion and spreading of normal human keratinocytes and fibroblasts in vitro", *Biomaterials* 2004; 25(7–8): 1289–1297.

[35] Liu, Z., Zhang, F., Ming, J., Bie, S., Li, J., and Zuo, B., "Preparation of electrospun silk fibroin nanofibers from solutions containing native silk fibrils", *J. Appl. Polym. Sci.* 2015; 132(1): 41236.

[36] Biagiotti, M., Bassani, G.A., Chiarini, A., Vincoli, V.T., Pra, I.D., Cosentino, C., Alessandrino, A., Taddei, P., and Freddi, G., "Electrospun silk fibroin scaffolds for tissue regeneration: Chemical, structural, and toxicological implications of the formic acid-silk fibroin interaction", *Front. Bioeng. Biotechnol.* 2022; 10: 833157.

[37] Wnek, G.E., Carr, M.E., Simpson, D.G., and Bowlin, G.L., "Electrospinning of nanofiber fibrinogen structures", *Nano Lett.* 2003; 3(2): 213–216.

[38] Carlisle, C.R., Coulais, C., Namboothiry, M., Carroll, D.L., Hantgan, R.R., and Guthold, M., "The mechanical properties of individual, electrospun fibrinogen fibers", *Biomaterials* 2009; 30(6): 1205–1213.

[39] Jiang, H., Fang, D., Hsiao, B.S., Chu, B., and Chen, W., "Optimization and characterization of dextran membranes prepared by electrospinning", *Biomacromolecules* 2004; 5(2): 326–333.

[40] Malini, R.I., Lesage, J., Toncelli, C., Fortunato, G., Rossi, R.M., and Spano, F., "Crosslinking dextran electrospun nanofibers via borate chemistry: Proof of concept for wound patches", *Eur. Polym. J.* 2019; 110: 276–282.

[41] Cengiz-Çallıoğlu, F., "Dextran nanofiber production by needleless electrospinning process", *e-Polymers* 2014; 14(1): 5–13.

[42] Ritcharoen, W., Thaiying, Y., Saejeng, Y., Jangchud, I., Rangkupan, R., Meechaisue, C., and Supaphol, P., "Electrospun dextran fibrous membranes", *Cellulose* 2008; 15: 435–444.

[43] Shawki, M.M., Hereba, A.M., and Ghazal, A., "Formation and characterization of antimicrobial dextran nanofibers", *Romanian J. Biophys.* 2010; 20(4): 335–346.

[44] Pillai, C.K.S. and Sharma, C.P., "Electrospinning of chitin and chitosan nanofibers", *Trends Biomater. Artif. Organs* 2009; 22(3): 179–201.

[45] Min, B. M., Lee, S. W., Lim, J. N., You, Y., Lee, T. S., Kang, P. H., and Park, W. H., "Chitin and chitosan nanofibers: electrospinning of chitin and deacetylation of chitin nanofibers", *Polymer* 2004; 45(21): 7137–7142.

[46] Barber, P. S., Griggs, C. S., Bonner, J. R., and Rogers, R. D., "Electrospinning of chitin nanofibers directly from an ionic liquid extract of shrimp shells", *Green Chem.* 2013; 15: 601–607.

[47] Qasim, S. B., Zafar, M. S., Najeeb, S., Khurshid, Z., Shah, A. H., Husain, S., and Rehman, I. U., "Electrospinning of chitosan-based solutions for tissue engineering and regenerative medicine", *Int. J. Mol. Sci.* 2018; 19(2): 407.

[48] Antaby, E., Klinkhammer, K., and Sabantina, L., "Electrospinning of chitosan for antibacterial applications – Current trends", *Appl. Sci.* 2021; 11(24): 11937.

[49] Ohkawa, K., Cha, D., Kim, H., Nishida, A., and Yamamoto, H., "Electrospinning of chitosan", *Macromol. Rapid Commun.* 2004; 25(18): 1600–1605.

[50] Salazar-Brann, S. A., Patino-Herrera, R., Navarreta-Damian, J., and Louvier-Hernandez, J. F., "Electrospinning of chitosan from different acid solutions", *AIMS Bioeng.* 2021; 8(1): 112–129.

[51] Homayoni, H., Ravandi, S. A. H., and Valizadeh, M., "Electrospinning of chitosan nanofibers: Processing optimization", *Carbohydr. Polym.* 2009; 77(3): 656–661.

[52] Blackstone, B. N., Gallentine, S. C., and Powell, H. M., "Collagen-based electrospun materials for tissue engineering: A systematic review", *Bioengineering* 2021; 8(3): 39.

[53] Matthews, J. A., Wnek, G. E., Simpson, D. G., and Bowlin, G. L., "Electrospinning of collagen nanofibers", *Biomacromolecules* 2002; 3(2): 232–238.

[54] Sizeland, K. H., Hofman, K. A., Hallett, I. C., Martin, D. E., Potgieter, J., Kirby, N. M., Hawley, A., Mudie, S. T., Ryan, T. M., Haverkamp, R. G., and Cumming, M. H., "Nanostructure of electrospun collagen: Do electrospun collagen fibers form native structures?", *Materialia* 2018; 3: 90–96.

[55] Liu, S. J., Kau, Y. C., Chou, C. Y., Chen, J. K., Wu, R. C., and Yeh, W. L., "Electrospun PLGA/collagen nanofibrous membrane as early-stage wound dressing", *J. Membr. Sci.* 2010; 355(1–2): 53–59.

[56] Aoki, H., Miyoshi, H., and Yamagata, Y., "Electrospinning of gelatin nanofiber scaffolds with mild neutral cosolvents for use in tissue engineering", *Polym. J.* 2015; 47: 267–277.

[57] Topuz, F. and Uyar, T., "Electrospinning of gelatin with tunable fiber morphology from round to flat/ribbon", *Mater. Sci. Eng. C* 2017; 80(1): 371–378.

[58] Li, T., Sun, M., and Wu, S., "State-of-the-art review of electrospun gelatin-based nanofiber dressings for wound healing applications", *Nanomaterials* 2022; 12(5): 784.

[59] Castro, K. C., Campos, M. G. N., and Innocentini Mei, L. H., "Hyaluronic acid electrospinning: Challenges, applications in wound dressings and new perspectives", *Int. J. Biol. Macromol.* 2021; 173(15): 251–266.

[60] Snetkov, P., Morozkina, S., Uspenskaya, M., and Olekhnovich, R., "Hyaluronan-based nanofibers: Fabrication, characterization and application", *Polymers* 2019; 11(12): 2036.

[61] Li, J., He, A., Han, C. C., Fang, D., Hsiao, B. S., and Chu, B., "Electrospinning of hyaluronic acid (HA) and HA/gelatin blends", *Macromol. Rapid Commun.* 2006; 27(2): 114–120.

[62] Sun, J., Perry, S. L., and Schiffman, J. D., "Electrospinning nanofibers from chitosan/hyaluronic acid complex coacervates", *Biomacromolecules* 2019; 20(11): 4191–4198.

[63] Fischer, R. L., McCoy, M. G., and Grant, S. A., "Electrospinning collagen and hyaluronic acid nanofiber meshes", *J. Mater. Sci. Mater. Med.* 2012; 23(7): 1645–1654.

[64] Zhang, Y., Zhang, C., and Wang, Y., "Recent progress in cellulose-based electrospun nanofibers as multifunctional materials", *Nanoscale Adv.* 2021; 3: 6040–6047.

[65] Kramar, A. and Gonzalez-Benito, F. J., "Cellulose-based nanofibers processing techniques and methods based on bottom-up approach – A review", *Polymers* 2022; 14(2): 286.

[66] Zhang, H., Liu, Y., Cui, S., Zhou, Y., Hu, J., Ma, J., and Liu, Y., "Cellulose nanofibers electrospun from aqueous conditions", *Cellulose* 2020; 27: 8695–8708.

[67] Panchal, S.S. and Vasava, D.V., "Biodegradable polymeric materials: Synthetic approach", *ACS Omega* 2020; 5(9): 4370–4379.

[68] Samantaray, P.K., Little, A., Haddleton, D.M., McNally, T., Tan, B., Sun, Z., Huang, W., Ji, Y., and Wan, C., "Poly(glycolic acid) (PGA): a versatile building block expanding high performance and sustainable bioplastic applications", *Green Chem.* 2020; 22: 4055–4081.

[69] Kai, D., Liow, S.S., and Loh, X.J., "Biodegradable polymers for electrospinning: Towards biomedical applications", *Mater. Sci. Eng. C* 2014; 45(1): 659–670.

[70] Maleki, H., Azimi, B., Ismaeilimoghadam, S., and Danti, S., "Poly(lactic acid)-based electrospun fibrous structures for biomedical applications", *Appl. Sci.* 2022; 12(6): 3192.

[71] Casasola, R., Thomas, N.L., Trybala, A., and Georgiadou, S., "Electrospun poly lactic acid (PLA) fibres: Effect of different solvent systems on fibre morphology and diameter", *Polymer* 2014; 55(18): 4728–4737.

[72] Coles, S.R., Jacobs, D.K., Meredith, J.O., Barker, G., Clark, A.J., Kirwan, K., Stanger, J., and Tucker, N., "A design of experiments (DoE) approach to material properties optimization of electrospun nanofibers", *J. Appl. Polym. Sci.* 2010; 117(4): 2251–2257.

[73] Li, W., Tan, X., Luo, T., Shi, Y., Yang, Y., and Liu, L., "Preparation and characterization of electrospun PLA/PU bilayer nanofibrous membranes for controlled drug release applications", *Integr. Ferroelectr.* 2017; 179(1): 104–119.

[74] Gajic, I., Stojanovic, S., Ristic, I., Ilic-Stojanovic, S., Pilic, B., Nesic, A., Najman, S., Dinic, A., Stanojevic, L., Urosevic, M., Nikolic, V., and Nikolic, L., "Electrospun poly(lactide) fibers as carriers for controlled release of Biochanin A", *Pharmaceutics* 2022; 14(3): 528.

[75] Sabzi, M., Ranjbar-Mohammadi, M., Zhang, Q., Kargozar, S., Leng, J., Akhtari, T., and Abbasi, R., "Designing triple-shape memory polymers from a miscible polymer pair through dual-electrospinning technique", *J. Appl. Polym. Sci.* 2019; 136(19): 47471.

[76] Makadia, H.K. and Siegel, S.J., "Poly lactic-*co*-glycolic acid (PLGA) as biodegradable controlled drug delivery carrier", *Polymers* 2011; 3(3): 1377–1397.

[77] Hines, D.J. and Kaplan, D.L., "Poly(lactic-co-glycolic) acid-controlled-release systems: experimental and modeling insights", *Crit. Rev. Ther. Drug Carrier Syst.* 2013; 30(3): 257–276.

[78] Kapoor, D.N., Bhatia, A., Kaur, R., Sharma, R., Kaur, G., and Dhawan, S., "PLGA: a unique polymer for drug delivery", *Ther. Deliv.* 2015; 6(1): 41–58.

[79] Chen, Y.P., Liu, H.Y., Liu, Y.W., Lee, T.Y., and Liu, S.J., "Determination of electrospinning parameters' strength in poly(D,L)-lactide-co-glycolide micro/nanofiber diameter tailoring", *J. Nanomater.* 2019; 2019: 2626085.

[80] Liu, X., Baldursdottir, S.G., Aho, J., Qu, H., Christensen, L.P., Rantanen, J., and Yang, M., "Electrospinnability of poly lactic-co-glycolic acid (PLGA): the role of solvent type and solvent composition", *Pharm. Res.* 2017; 34(4): 738–749.

[81] Yu, C.C., Chen, Y.W., Yeh, P.Y., Hsiao, Y.S., Lin, W.T., Kuo, C.W., Chueh, D.Y., You, Y.W., Shyue, J.J., Chang, Y.C., and Chen, P., "Random and aligned electrospun PLGA nanofibers embedded in microfluidic chips for cancer cell isolation and integration with air foam technology for cell release", *J. Nanobiotechnol.* 2019; 17: 31.

[82] Zhao, W., Li, J., Jin, K., Liu, W., Qiu, X., and Li, C., "Fabrication of functional PLGA-based electrospun scaffolds and their applications in biomedical engineering", *Mater. Sci. Eng. C* 2016; 59: 1181–1194.

[83] Buck, E., Maisuria, V., Tufenkji, N., and Cerruti, M., "Antibacterial properties of PLGA electrospun scaffolds containing ciprofloxacin incorporated by blending or physisorption", *ACS Appl. Bio Mater.* 2018; 1(3): 627–635.

[84] Chen, D.W, Hsu, Y.H., Liao, J.Y., Liu, S.J., Chen, J.K., and Ueng, S.W.N., "Sustainable release of vancomycin, gentamicin and lidocaine from novel electrospun sandwich-structured PLGA/collagen nanofibrous membranes", *Int. J. Pharm.* 2012; 430(1-2): 335-341.

[85] Tseng, Y.Y., Yang, T.C., Wang, Y.C., Lee, W.H., Chang, T.M., Kau, Y.C., and Liu, S.J., "Targeted concurrent and sequential delivery of chemotherapeutic and antiangiogenic agents to the brain by using drug-loaded nanofibrous membranes", *Int. J. Nanomed.* 2017; 12: 1265-1276.

[86] Lee, C.H., Hung, K.C., Hsieh, M.J., Chang, S.H., Juang, J.H., Hsieh, I.C., Wen, M.S., and Liu, S.J., "Core-shell insulin-loaded nanofibrous scaffolds for repairing diabetic wounds", *Nanomed.: Nanotechnol. Biol. Med.* 2020; 24: 102123.

[87] Yu, Y.H., Lee, D., Hsu, Y.H., Chou, Y.C., Ueng, S.W.N., Chen, C.K., and Liu, S.J., "A three-dimensional printed polycaprolactone scaffold combined with co-axially electrospun vancomycin/ceftazidime/bone morphological protein-2 sheath-core nanofibers for the repair of segmental bone defects during the Masquelet procedure", *Int. J. Nanomed.* 2020; 15: 913-925.

[88] Woodruff, M.A. and Hutmacher, D.W., "The return of a forgotten polymer - Polycaprolactone in the 21st century", *Prog. Polym. Sci.* 2010; 35(10): 1217-1256.

[89] Jiang, L., Wang, L., Wang, N., Gong, S., Wang, L., Li, Q., Shen, C., and Turng, L.S., "Fabrication of polycaprolactone electrospun fibers with different hierarchical structures mimicking collagen fibrils for tissue engineering scaffolds", *Appl. Surf. Sci.* 2018; 427(A): 311-325.

[90] Gomes, S., Querido, D., Ferreira, J.L., Borges, J.P., Henriques, C., and Silva, J.C., "Using water to control electrospun polycaprolactone fibre morphology for soft tissue engineering", *J. Polym. Res.* 2019; 26(9): 222.

[91] Pouladchang, A., Tavanai, H., Morshed, M., Khajehali, J., and Shamsabadi, A.S., "Controlled release of thiram pesticide from polycaprolactone micro and nanofibrous mat matrix", *J. Appl. Polym. Sci.* 2021; 139(6): 51641.

[92] Asadian, M., Rashidi, A., Majidi, M., Mehrjoo, M., Emami, B.A., Tavassoli, H., Asl, M.P., and Bonakdar, S., "Nanofiber protein adsorption affected by electrospinning physical processing parameters", *J. Iran. Chem. Soc.* 2015; 12: 1089-1097.

[93] Dobrzanski, L.A. and Hudecki, A., "Structure, geometrical characteristics and properties of biodegradable micro- and polycaprolactone nanofibers", *Arch. Mater. Sci. Eng.* 2014; 70(1): 5-13.

[94] Ma, L., Shi, X., Zhang, X., and Li, L., "Electrospinning of polycaprolacton/chitosan core-shell nanofibers by a stable emulsion system", *Colloids Surf. A Physicochem. Eng. Asp.* 2019; 583: 123956.

[95] Rawat, S., Jain, K.G., Gupta, D., Raghav, P.K., Chaudhuri, R., Pinky, Shakeel, A., Arora, V., Sharma, H., Debnath, D., Kalluri, A., Agrawal, A.K., Jassal, M., Dinda, A.K., Patra, P., and Mohanty, S., "Graphene nanofiber composites for enhanced neuronal differentiation of human mesenchymal stem cells", *Nanomedicine* 2021; 16(22): 1963-1982.

10 Heat Sealability of Biodegradable Polymers for Packaging Applications and Their Degradation

Ramin Yousefzadeh Tabasi and Abdellah Ajji

This chapter aims to corroborate the suitability of PLA and its blends with other biodegradable polyesters such as PBAT and PCL as viable seal-layer materials in a biodegradable packaging assembly. A comparison is drawn between the seal performance of polyolefins and biodegradable polyesters, comprising the nature of wetting and the effect of interfacial diffusion on both their heat-seal and hot-tack performances. A foreseeable growth in the U. S. market size of PLA is expected and depends mainly on the packaging industry. The basic know-how of the seal and hot-tack properties of PLA and its blends would enhance its incorporation in different packaging designs and the application of seal and hot tack for a variety of strengths needed, thus dictating the rate of production. As blending would modify the mechanical and heat-seal behavior of PLA, it might as well modify its composting rate and degradation mechanism.

10.1 Introduction

A global move toward circular economy and more sustainable solutions has raised questions about the suitability of biodegradable plastics to replace conventional polymers in commodity applications. Guidelines from the U. S. Environmental Protection Agency (EPA) have put recycling and composting second on their priority list, after "source reduction and reuse", to emphasize their importance in a global approach to a more sustainable future. The EPA's European section published a survey report in 2018 showing that biodegradable resins are mainly used in bags and packaging applications. The worldwide capacity to produce biodegradable plastics has increased significantly and they are currently a substantial portion of the packaging resin market [1].

To support the transformation of packaging systems to compostable ones, PLA (poly(lactic acid)) is becoming increasingly attractive, owing to its high mechanical strength and favorable optical clarity. The drawback of PLA is its brittleness at room temperature, which endangers the integrity of films and flexible packages made from it by increasing the chance of cracks and pinhole formation. Seal layers in packaging systems are also important in ensuring the integrity of the package as they are the last part of the package that hermetically seals the whole assembly. Amorphous PLA grades are introduced to help provide better seal performance when PLA is the material of choice for seal applications [2]. Although they provide better seal performance than semicrystalline PLA, these amorphous grades still exhibit brittleness.

In practice, the successful incorporation of PLA in most applications requires it to be toughened. The toughening of PLA through blending and reactive extrusion has been reported in many studies but, to our knowledge, rarely this attempt has been combined with studies of sealing properties [3–8].

This study is aimed to provide a fundamental understanding of the seal behavior of biodegradable polyesters, that is, PLA and its blends with PCL (polycaprolactone) and PBAT (polybutylene adipate terephthalate). These resins are designed to be used in conventional processes and constructions and run on the same machinery as commodity plastics do. As part of our study, the performances of these blends are compared to that of a nondegradable system based on polyethylene (PE) and ethylene vinyl acetate (EVA). We will also provide a brief discussion of the effect of blending on biodegradation and the selectivity of microorganisms in an active compost to favor one phase through the course of degradation, revealed by periodic sampling and FTIR (Fourier transform infrared) measurements.

The measurement of the performance of a seal layer is done by either heat-seal or hot-tack test protocols, which are defined by the ASTM F02 Committee, 1998. The main difference between the two measurements is in the included cooling interval in heat seal before the exertion of load and the seal strength measurement. The hot tack is performed while the seal is freshly made and still hot. Hence, the effect of elevated temperature on both the mechanical properties of the seal material and its self-wetting, in semi-molten state, is more pronounced.

EVA and LDPE were provided by Dupont and Dow Chemical, respectively. The LDPE had a melt flow index (MFI) of 2.3 g/10 min and the EVA had an MFI of 8 g/10 min and 18% VA (vinyl acetate) content. Polyethylene and its blends with EVA were melt-blended and processed on a blown line using a 45 mm Killion single-screw extruder equipped with a helical blown film die. The PLA/PCL blends were prepared using NatureWorks 4060D PLA and PCL from Union Carbide with a molecular weight of 80,000. The same PLA sealant grade was blended with PBAT (ecoflex® c1200 from BASF) to obtain the second set of biodegradable blends also used for the biodegradation study [9–10]. A rheological characterization and a morphology study have revealed the immiscibility of the blends leading to either matrix disperse or co-continuous morphology, which can be found in detail in our previous publications [9].

All the films were drawn to a thickness of about 40 ± 5 µm. The measurement of the crystalline content of each component in the blend is critical as it will directly affect the chain mobility and sealing properties of samples. Differential scanning calorimetry (DSC) (TA Instruments Q1000) was used to study the crystallinity level of the blends and neat polymers. The ASTM F88 and F2029 experimental methods were followed for our heat-seal and hot-tack measurements. All the sealing experiments were performed in a range of jaw temperatures from 50 °C to 140 °C at a constant pressure of 0.5 N/mm^2 and for a dwell time of 0.5 s. An FTIR spectrometer (Spotlight 400, PerkinElmer) equipped with an attenuated total reflection (ATR) accessory was used in the IR range from 600 cm^{-1} to 4000 cm^{-1} to investigate the film samples. For each spectrum, the data was collected with a resolution of 4 cm^{-1}. The quantitative measurements used are detailed in our previous investigation [11].

10.2 Results and Discussion

The hot-tack behavior of pure and blended polymers is affected by their physicochemical and morphological properties as well as the process conditions (i.e., dwell time, pressure and temperature of sealing). A minimum pressure is required to force the semi-melted interface to become entangled and uniform. An optimum pressure of 35–70 psi has been reported and pressures above those values are shown to distort the surface and push the material out of the sealed area. The temperature and dwell time have to be worked in combination to develop an optimum processing condition. A detailed discussion of how these process conditions would affect the seal strength of each blend was previously published [9–11].

Figure 10.1 shows the maximum achievable hot-tack strength using PE/EVA, PLA/PCL, and PLA/PBAT blends of different compositions. A few studies have reported that the hot-tack strength of polyethylene can be improved by using different EVA grades. The level of VA has been shown to improve seal through contamination [12–13]. The lower melting point of EVA decreases the seal initiation temperature of the PE/EVA blends. Additionally, the higher polarity of EVA provides enhanced wetting, compared with pure PE. This is favorable particularly in hot-tack testing, where the material is in an amorphous semi-molten state.

In theory, the polyethylene surface remains intact while the EVA dispersed phase melts and softens at 30 °C below PE's melting point. This allows the formation of tacky but rigid-enough interfaces to bond and not deform under pressure of the sealing jaws.

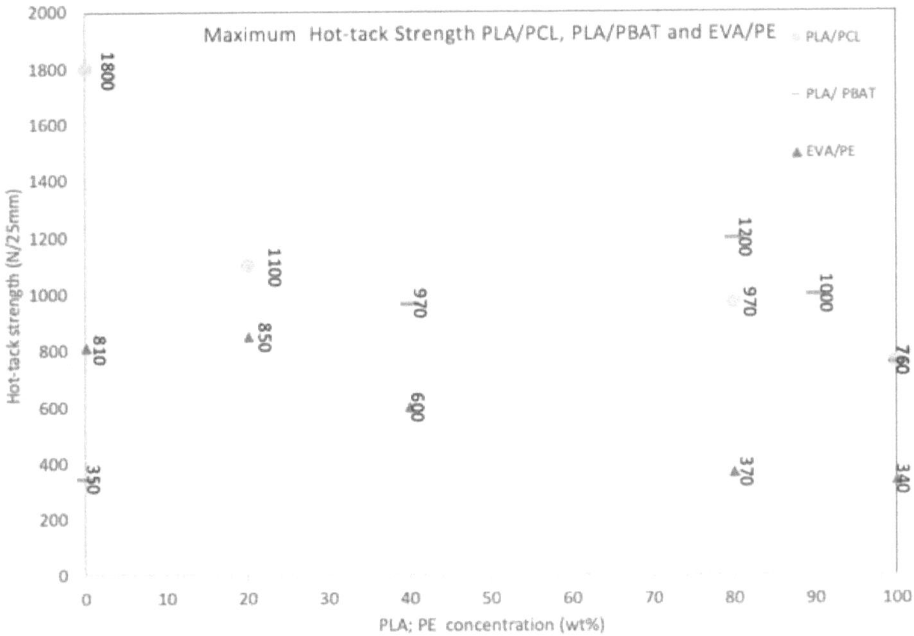

Figure 10.1 Maximum hot-tack strength of PE/EVA, PLA/PCL, and PLA/PBAT blends

The same concept was materialized when PLA/PCL blends were evaluated as a viable route to enhance the hot-tack performance of PLA. Figure 10.1 shows that both blends of PLA/PCL and PLA/PBAT have significantly higher adhesion strength (hot-tack) than PE and its blends with EVA. PCL, being the lower-melting-temperature polyester, is efficient in lowering the initial hot-tack temperature. However, these seals have limited temperature endurance and their performance in sterilization and reheating, if required, is questionable.

Both PBAT and PCL have high surface-tension polar components, which improves the surface wetting and high-temperature adhesion. The PCL grade used in our study has a low melting temperature of 59 °C and a high melt strength that depends on its concentration in the blends and the hot-tack strength. After the sealing area is cooled to room temperature, an ideal welded line develops with properties similar to those of the bulk film, possessing high resistance to crack propagation, manifested in film failure at the edges of the seal. When the failure mode is film breakage, the heat-seal strength is mainly dictated by the film strength at break as reported in our previous work [9]. As PBAT melts at about 120 °C, which is above the glass transition temperature of PLA, it has been shown that a partial melting of the crystals is required to obtain maximum mobility and hot-tack strength [15]. It has been anticipated that a partial melting of PBAT crystals would soften the PLA phase and reduce its resistance to deformation. This is the case when the PBAT content is above 40%, as can be observed in Figure 10.1. A deviation is noticeable in blends with lower PBAT content.

While studying PLA/PBAT blends and their thermomechanical behavior, a significant reduction in the crystalline content of PBAT at low loading in blends with PLA was observed [9]. This could be due to a hindrance effect of the PLA chains as well as the reduced size of the dispersed PBAT phase in these blends. The reduction in crystalline content leads to lower temperatures and heat required to initiate the mobility of the dispersed PBAT phase during the adhesion process. The adhesion takes place while a large portion of surface, made of PLA, is still rigid enough to bear higher peel stresses. To examine this hypothesis, we quenched a neat film of PBAT from the molten state on an aluminum plate cooled in liquid nitrogen. This film was then used for hot-tack measurement and the results are reported in Table 10.1. The quenched film had a lower hot-tack initiation temperature and higher strength, but not high enough as the one it would reach in blends with PLA. The lower final hot-tack plateau temperature is attributed to the fact that the quenched film has a significantly lower crystalline structure and softens at lower temperature.

Table 10.1 Hot-Tack Plateau Initiation and Final Temperatures as a Function of PBAT Content in Blends with PLA

PBAT concentration in blends	Hot-tack plateau initiation temperature (T_{Hi}) [°C]	Hot-tack plateau final temperature (T_{Hf}) [°C]
100%	90	114
100%, quenched	63	108
10%	72	105
20%	65	108
40%	75	111

The main concern in the case of PLA, besides its seal performance, is its brittleness. Blending PLA with rubbery polyesters such as PBAT improves the elongation at break and puncture resistance of films made of it. As mentioned above, the drive to use biodegradable polymers is the environmental concerns and not their cost incentive, as it is with polyolefin-based seal layers. As companies are ready to overcome the cost barrier, they could benefit from the improved resistance to puncture and tensile strength at yield of PLA blends, which are comparable to those of commercial multilayer co-extruded nylon films used in food packaging. This is in addition to the higher level of the oxygen barrier displayed by PLA compared with that of polyolefin-based films, and could facilitate the incorporation of PLA blends in flexible packaging applications.

Blending, although beneficial to the mechanical toughness and hot-tack of PLA, has an adverse effect on the rate and mechanism of biodegradation of blends made with PBAT. Samples of pure PLA and blends with PBAT have been studied for composting performance. Through the course of composting, scanning electron microscopy (SEM) images were taken of film blends of PLA/PBAT, as illustrated in Figure 10.2. As can be noticed, the microorganisms have preference toward one phase. A co-continuous morphology, captured in SEM images of PLA/PBAT blends, was shown by Deng et al. [14] to be the dominant morphology in PLA blends of 20–60% PBAT. FTIR has been shown to be a reliable method to determine concentration of PLA in its blends [15–19].

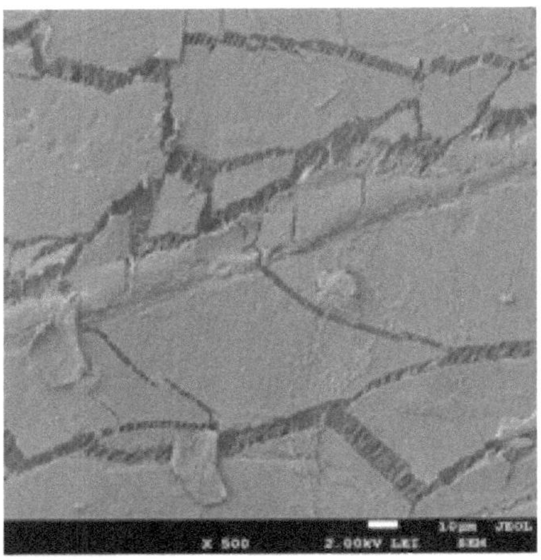

Figure 10.2
SEM micrograph of PLA/PBAT (50/50) after 20 days of composting

We have used a similar methodology to determine the decreasing concentration of PLA as a result of biodegradation in composting and compared its rate to the rate of PBAT degradation [11]. The results, shown in Table 10.2, indicate that the PLA phase undergoes faster degradation in composting. The table presents the change in absorption ratio of PLA's main peak at 1750 cm^{-1} over the corresponding peak of PBAT at 1711 cm^{-1}. The total composting period is dictated by the slower component as shown in our previous work [11].

Table 10.2 Calculated Peak Area Ratio from Deconvoluted FTIR Spectrums for 50/50 PLA/PBAT Blends through Composting Process

Ratio of peak areas	Day 0	Day 3	Day 6	Day 10
(PLA 1750 cm^{-1}/ PBAT 1711 cm^{-1})	1.46	1.30	1.28	1.18

10.3 Conclusions

PLA blends with both PCL and PBAT have a promising hot-tack performance that exceeds that of common olefin-based blends, currently dominating the flexible packaging industry. Films prepared from PLA/PBAT blends present a puncture resistance comparable to that of commercial co-extruded and cast films used in vertical form-fill-seal machines. The combination of improved mechanical and sealing properties while maintaining a high composting rate demonstrates the potential of these biodegradable blends to be utilized in packaging applications.

References for Chapter 10

[1] *https://www.european-bioplastics.org/wp-content/uploads/2020/11/Global_Production_Capacities_2020_by_material_type.jpg* (accessed: 15/06/2023).

[2] M. Jamshidian, E. A. Tehrany, M. Imran, M. Jacquot, and S. Desobry, "Poly-Lactic Acid: Production, Applications, Nanocomposites, and Release Studies", *Comprehensive Reviews in Food Science and Food Safety* 9(5) (2010), pp. 552–571.

[3] J. Odent, Y. Habibi, J.-M. Raquez, and P. Dubois, "Ultra-tough polylactide-based materials synergistically designed in the presence of rubbery ε-caprolactone-based copolyester and silica nanoparticles", *Composites Science and Technology* 84 (2013), pp. 86–91.

[4] K. Hashima, S. Nishitsuji, and T. Inoue, "Structure-properties of super-tough PLA alloy with excellent heat resistance", *Polymer* 51(17) (2010), pp. 3934–3939.

[5] W. Chinsirikul, J. Rojsatean, B. Hararak, N. Kerddonfag, A. Aontee, K. Jaieau, P. Kumsang, and C. Sripethdee, "Flexible and Tough Poly(lactic acid) Films for Packaging Applications: Property and Processability Improvement by Effective Reactive Blending", *Packaging Technology and Science* 28(8) (2015), pp. 741–759.

[6] X. Zhao, H. Hu, X. Wang, X. Yu, W. Zhou, and S. Peng, "Super tough poly(lactic acid) blends: a comprehensive review", *RSC Advances* 10(22) (2020), pp. 13316–13368.

[7] P. Liewchirakorn, A.-O., Duangdao, and C. Wannee, "Practical Approach in Developing Desirable Peel-Seal and Clear Lidding Films Based on Poly(Lactic Acid) and Poly(Butylene Adipate-Co-Terephthalate) Blends", *Packaging Technology and Science* 31(5) (2018), pp. 296–309.

[8] N. Sundar, P. Keerthana, S. Ananda Kumar, G. Ananda Kumar, and S. Ghosh, "Dual purpose, bio-based polylactic acid (PLA)-polycaprolactone (PCL) blends for coated abrasive and packaging industrial coating applications", *Journal of Polymer Research* 27(12) (2020), pp. 1–18.

[9] Z. Najarzadeh, R. Y. Tabasi, and A. Ajji, "Sealability and Seal Characteristics of PE/EVA and PLA/PCL Blends", *International Polymer Processing* 29(1) (2014), pp. 95–102.

[10] R. Y. Tabasi and A. Ajji, "Tailoring Heat-Seal Properties of Biodegradable Polymers through Melt Blending" *International Polymer Processing* 32(5) (2017), pp. 606–613.

[11] R. Y. Tabasi and A. Ajji, "Selective degradation of biodegradable blends in simulated laboratory composting", *Polymer Degradation and Stability* 120 (2015), pp. 435–442.

[12] R. B. Neel and B. G. Stout, "ULDPE blends with LDPE and EVA for film extrusion", *Journal of Plastic Film & Sheeting* 5(3) (1989), pp. 209–221.

[13] B. A. Morris, 4 – Commonly Used Resins and Substrates in Flexible Packaging, in: *The Science and Technology of Flexible Packaging*, William Andrew, Oxford (2017), pp. 69–119.

[14] Y. Deng, C. Yu, P. Wongwiwattana, and N.L. Thomas, "Optimising Ductility of Poly(Lactic Acid)/Poly(Butylene Adipate-co-Terephthalate) Blends Through Co-continuous Phase Morphology", *Journal of Polymers and the Environment* 26(9) (2018), pp. 3802–3816.

[15] N. Noroozi, J.A. Thomson, N. Noroozi, L.L. Schafer, and S.G. Hatzikiriakos, "Viscoelastic behaviour and flow instabilities of biodegradable poly (ε-caprolactone) polyesters", *Rheol. Acta* 51 (2012), pp. 179–192.

[16] Z. Najarzadeh and A. Ajji, "Role of molecular architecture in interfacial self-adhesion of polyethylene films", *Journal of Plastic Film & Sheeting* 33(3) (2017), pp. 235–261.

[17] B.W. Chieng, N.A. Ibrahim, W.M.Z.W. Yunus, and M.Z. Hussein, "Poly(lactic acid)/Poly(ethylene glycol) Polymer Nanocomposites: Effects of Graphene Nanoplatelets", *Polymers* 6(1) (2014), pp. 93–104.

[18] A.M. Torres-Huerta, D. Del Angel-López, M.A. Domínguez-Crespo, D. Palma-Ramírez, M.E. Perales-Castro, and A. Flores-Vela, "Morphological and Mechanical Properties Dependence of PLA Amount in PET Matrix Processed by Single-Screw Extrusion", *Polymer-Plastics Technology and Engineering* 55(7) (2016), pp. 672–683.

[19] S. Shojaei, M. Nikuei, V. Goodarzi, M. Hakani, H.A. Khonakdar, and M.R. Saeb, "Disclosing the role of surface and bulk erosion on the viscoelastic behavior of biodegradable poly(ε-caprolactone)/poly(lactic acid)/hydroxyapatite nanocomposites", *Journal of Applied Polymer Science* 136(10) (2019), 47151.

Index

Symbols

3D printing
- of polysaccharides 182

A

agarose 16
alginate 175
aliphatic polyesters 1, 3
antimicrobial properties 51
applications 1

B

batch mixing
- of polysaccharides 181
batch processing 147
biocomposite food packaging films 51
biodegradable PLA nanocomposites 24
biodegradable polymer blends, copolymers, and additives 20
biodegradable polymers 1
- composition, processing, and properties 3
- heat sealability 281
- managed end-of-life 115
- manufacture by reactive extrusion 113, 116
- packaging applications 281
- summary of properties 18
biodegradable polymers and biodegradability
- definition 2
biodegradable polymers for electrospinning 263
biodegradation of PHA-based materials 58
biodegradation studies
- on PHA-based biocomposites 62
- on PHA-based blends and composites 60
- on PHAs and their blends 61
bioplastics
- average cost compared to petrochemical-compared to petrochemical-based plastics 81
- global production capacities 83
biosynthetic pathway to obtain PHBV and PHB 76

C

cellulose 171
- electrospinning 269
- extrusion and batch processing 149
- extrusion film-blowing 153
- injection molding 157
- thermal processability 145
cellulose ether 15
chemical blowing agents 210
chemical modification by grafting 122
chemical modification of starch 128
chitin 173
- electrospinning 266
chitosan 17, 173
- electrospinning 266

classification of biodegradable and non-biodegradable polymers 35
classification of natural fibers 53
coaxial electrospinning 257
commercial applications of PHAs 63
compatibilization between PHBV, PBAT, and switchgrass 54
composition-properties-processing relationship 1
composting 3
compression molding
- of polysaccharides 181
copolymerization by reactive extrusion 134

D

degradation
- fundamental mechanisms 2
dissolution and plasticization to formulate polysaccharides 177

E

electrospinning 255
- effects of processing parameters 261
- of biodegradable polymers 263
- of cellulose 269
- of chitin and chitosan 266
- of collagen 267
- of dextran 265
- of DNA 263
- of fibrinogens 264
- of gelatin 268
- of hyaluronic acid 268
- of natural polymers 263
- of polycaprolactone (PCL) 273
- of poly(lactic acid) (PLA) 270
- of poly(lactic-co-glycolic acid) (PLGA) 271
- of polysaccharides 184
- of silk fibroin 264
- of synthetic polymers 269
- process 256
electrospinning process variants 257
ex-situ compatibilization 41
extrusion 147
- of natural polymer blends 151
- of polysaccharides 181
extrusion film-blowing 152

F

foaming agents 210
foam injection molding 205
- control of back pressure 212
- machine requirements 212
- process variables 214
- screw modification 213
- shut-off nozzle 213
- special modification of the injection system 213
- surface appearance 206
foam injection molding of PLA-based nanocomposites 230
foam injection molding of poly(lactic acid) 203, 215
- additives and compounds 227
- back pressure 221
- injection flow rate 219
- mold temperature 217
- PLA blends and compounds 231
- with mold opening 225
- with natural fibers 230
- with talc 227

G

grafting 122

H

heat sealability 281
hot-tack performance 281
hot-tack strength 283

I

improving performance of PHAs 94
improving production reproducibility of PHAs 94
injection molding 156, 204
in-situ compatibilization 41

L

lignocellulosic fibers 52

M

maleic anhydride graft co-polymerization 49
mechanical properties of bioplastics 19, 37
microfibrillar composites
- microstructure 241
microfibrillation 239
- effects on crystallization behavior of bioplastics 246
- effects on foaming ability of bioplastics 251
- effects on mechanical properties of bioplastics 249
- effects on rheological properties of bioplastics 243
modified PET 8
- degradation mechanisms 8

N

needleless electrospinning 260

P

packaging applications 281
pectin 176
PHA-based biocomposites and nanocomposites 47
PHA-based blends 35
- morphology 40, 43, 45, 46
PHA-based blends and composites
- biodegradation studies 60
PHA-based materials
- biodegradation 58
PHBH polyester 13
PHBV
- thermal degradation 86
photosynthetic microorganisms 89
physical blowing agents 211
- solubility in PLA 215
PLA blends
- packaging applications 282
poly(3-hydroxybutyrate-co-hydroxyvalerate) (PHBV) 77
poly(3-hydroxybutyrate) (PHB) 9, 78
- properties 11
polyacrylamide 16
polybutylene succinate (PBS) 7
- degradation mechanisms and properties 7
polycaprolactone (PCL)
- electrospinning 273
polyethylene glycol (PEG) 15
polyethylene oxide 17
polyhydroxyalkanoate (PHA) 9, 31
- additives 26
- advancements in blends, biocomposites, and their commercialization 31
- as biodegradable biopolymers for a circular economy 73
- biocomposites and nanocomposites 47
- biodegradability 12
- blends 25, 35
- blends, copolymers, and additives 25
- branched 11
- chemical structure 36
- commercial applications 63
- compatibilized blends 41
- composition and properties 9
- filler-based biocomposites and nanocomposites 47
- lignocellulosic filler- and fiber-reinforced biocomposites 52
- non-compatibilized blends with other biodegradable polymers 38

- processability 12
- stabilization 10
poly(lactic acid)
- functionalization 124
- stereocomplexation 131
- synthesis by ring-opening polymerization 121
poly(lactic acid)-grafted maleic anhydride 123
poly(lactic acid)-grafted vinyltrimethoxysilane 126
poly(lactic acid) (PLA) 4
- additives 22
- biodegradation mechanisms 6
- blends, copolymers, and additives 20
- chemical modification 227
- composition and properties 4
- electrospinning 270
- fillers 23
- foam injection molding 203
- microfibrillation 239
- packaging applications 282
- processability 6
polymer blends and composites
- applications 32
polysaccharide-based blends 177
polysaccharide-based materials
- formulation strategies 177
- processing 180
polysaccharide-based nanocomposites 179
polysaccharides
- chemical structures 166
- fundamentals 168
- processing 165
polyvinyl alcohol (PVOH) 14
poly(ε-caprolactone)-grafted chitosan 123
processing of polysaccharide-based materials 180
processing of polysaccharides 165
processing of structural foams
- gas counterpressure 208
- high-pressure process 209
- low-pressure process 207
- low-pressure with co-injection 208
- mold opening 209
processing techniques of natural polymers 147
proteins
- extrusion and batch processing 149
- extrusion film-blowing 153
- injection molding 158
- thermal processability 146
PTMAT blends 27

R

reactive extrusion 113
- copolymerization 134
- preparation of composites 135
- ring-opening polymerization 119
recombinant bacteria 91
reducing production costs of PHAs 86
- cheaper carbon sources 87
- improved extraction techniques 92
- mixed cultures 90
- photosynthetic microorganisms 89
- recombinant bacteria 91
- valorization of by-products 92
renewable feedstocks 87
ring-opening polymerization
- by reactive extrusion 119
- of lactide 121
- of ε-caprolactone 119

S

sodium polyacrylate 18
spherulites 78
starch 13, 168
- chemical modification 128
- extrusion and batch processing 147
- extrusion film-blowing 152
- injection molding 156
- thermal processability 144
starch-based polymers 3, 13
stereocomplexation 131
structural foams 203, 205
- processing 207

T

thermal processability of natural polymers 144, 159
thermal processing 143
thermomechanical processing
- of polysaccharides 181
twin-screw extruder 117

W

water-soluble, biodegradable polymers 13
wet processing
- of polysaccharides 180

Z

zein 146